Industrial Safety Management

A Practical Approach

Jack E. Daugherty

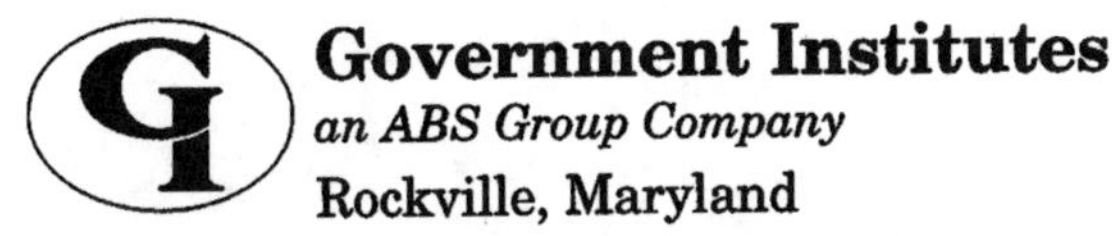

Government Institutes, Inc., 4 Research Place, Rockville, Maryland 20850, USA.
Phone: (301) 921-2300
Fax: (301) 921-0373
Email: giinfo@govinst.com
Internet address: http://www.govinst.com

03 02 01 00 99 5 4 3 2 1

Library of Congress Cataloging-in-Publication Data

Daugherty, Jack E.
Industrial safety management : a practical approach / by Jack E. Daugherty.
p. cm.
Includes bibliographic references and index.
ISBN: 0-86587-636-3
1. Industrial safety–management. I. Title.
T55.D38 1998
658.3'82--dc21

98-43298
CIP

Printed in the United States of America

To those whose unfortunate deaths helped to make the workplace safer for others.

Summary of Contents

TABLE OF CONTENTS

10. Hazard and Operability (HAZOP) Analysis 251

11. The Occupational Safety and Health Act 261

List of Figures and Tables

PREFACE

This book was written for those nonexpert generalists— commonly found in small to medium industries throughout the United States—who are given the task of industrial safety management. These individuals often are not safety professionals by degree, experience, or aspiration, but come to the field by assignment only. Most of them, it seems, would rather be purchasing agents, personnel managers, or plant engineers—really *anything* but safety professionals. While others also may benefit from the information in this book, this "nuts and bolts" book is intended for those nonexperts in safety and health management.

Admittedly, some of the topics I cover may be considered beyond my expertise—such as work psychology and behavioral safety. But I have dared to share my experience and opinions on these matters from the viewpoint of a safety professional. While this may open me up to criticism from true experts, let me say that what I have written has come directly from what I have successfully practiced and has withstood the scrutiny of numerous compliance officers.

Also keep in mind that this book is not a free consulting service. I can only describe safety management problems and solutions in very general terms, which may not be applicable to any given situation. Even the specific examples that I have provided, while they may seem similar to a situation that exists in your plant, can not be used for advice due to unique factors that may not be mentioned in the book. Also, discussions of OSHA standards and regulations are made from my personal viewpoint as an author and safety professional, and OSHA may very well disagree with some of those interpretations and opinions. Finally, please keep in mind that your legal counselor may also disagree with my interpretations of compliance issues in respect to your particular situation, and you should always consider him or her to be the resident experts in your safety compliance situation.

Nonetheless, I think that much of the information in this book is very useful, and I hope that you will enjoy it and use it often. While you may not become a safety expert yourself, you will certainly have a better grasp on a majority of industrial safety and health management issues.

Jack E. Daugherty
Jackson, Mississippi

About the Author

The environmental safety and health engineering career of Jack E. Daugherty began in the U.S. Navy where he studied the human health and safety effects of explosions and nuclear, biological, and chemical agents. Since then he has worked in the mineral processing, chemical processing, and manufacturing industries. Presently, he is owner and principal engineer of AEGIS, a Jackson, Mississippi compliance management and engineering service for industry.

Jack holds bachelor's and master's degrees in Chemical Engineering from Auburn University and the University of Mississippi, respectively. He is a Certified Industrial Hygienist, Certified Hazardous Materials Manager, and a professional engineer licensed in the State of Mississippi. He is a member of the American Industrial Hygiene Association, the American Academy of Industrial Hygiene, and the Academy of Certified Hazardous Materials Managers.

Jack's articles on environmental and occupational safety and health topics have appeared in several trade journals, including *Occupational Hazards* and *Industrial Safety and Hygiene News*. He has also published two books previous to this one, including *Industrial Environmental Management* (Government Institutes, 1996).

ACKNOWLEDGMENTS

I wish to thank the staff of Government Institutes again for their patience and assistance with this project. Wherever he is, I want to express my gratitude to Bill Poole, my safety mentor of three decades ago, who started me on this journey. Thanks to colleagues Mike Guisfredi, CIH, and Dan Markowicz, CIH, for their professional support in recent years.

1

Safety Management Theory

Will Rogers once said that everyone is a philosopher. Like philosophy, everyone seems to have his or her own opinions about safety. If everyone is a philosopher, then everyone is a safety expert, too. For those who earn a living as safety specialists, however, the fact that everybody considers themselves to be safety experts is a source of much frustration. While this universal attitude might seem to be an organizational strength, in actuality, most everyone ends up working at cross-purposes with the real safety professional, because everyone's concept of what is safe and what is hazardous differs.

Industrial safety management has two fundamental management aspects: administration and engineering. A certain percentage of industrial facilities, especially larger ones, maintain a professional safety staff to address each of these aspects. Smaller facilities typically assign the administrative management to the human resources staff and the engineering management to the plant engineering or maintenance staff. Neither staff typically has much preparation for or formation in the administration and engineering principles of safety. These individuals, though, may be well qualified for their principle, non-safety, assignments.

Safety Management Terminology

In order to understand better how to manage the safety and health of human resources in industrial settings, let us briefly review safety management systems that have been in vogue since the industrial revolution. While some of these systems have worked well in specific circumstances, classic safety management systems are still seriously considered useful by only a few safety professionals and others.

More importantly, a great deal of confusion can be prevented by knowing, at the outset, the definitions of the terms hazard, accident, resource, risk, and safety.

A *hazard* is a situation, condition, or sequence of events that, if left uncontrolled, could lead to an accident. How likely an accident may occur is not part of the definition. Roger Brauer speaks of a changing set of circumstances and emphasizes that hazard is the potential to produce adverse or harmful results. Firenze, cited in the National Safety Council's *Accident Prevention Manual,* adds that a hazard does not necessarily need to exist, but merely have the potential to exist.

An *accident* is any unplanned event that results in a resource loss. A *resource* may be any person, property, production, or profit valuable to the company. If a hazard is a trap, then an accident is the springing of that trap. Brauer says that the accident event (or series of events) is unexpected, unforeseen, or unintended. Actually, with the proper training, many accidents can be expected and foreseen, even if unintended. That's why the safety profession exists: to anticipate, prepare for, and prevent accidents. Accident prevention, in other words, is the mission of our profession.

Risk is the combination of probability and consequence of an uncontrolled hazard causing an accident. The consequence of concern to safety professionals is any loss of resource, most particularly the loss of human resources. *Safety,* then, is an acceptable level of risk to all exposed resources. Notice that safety is not the absence of hazard, which is, in fact, a condition that never exists.

Safety in an Industrial Setting

Herein is the basic reason why the best safety programs sometimes do not allow employees to actually "feel" safe. As safety professionals, we assist our companies in defining safety based on risk. Unfortunately, what is safe to you and me is not necessarily safe to any given employee. Safety is always relative. Likewise, risk exists in the eye of the beholder. For example, one employee may find the risk of exposure to five parts-per-million trichloroethylene unacceptable even though she smokes three packs of cigarettes per day and skydives on her weekends off. At the same time, *do not* lightly dismiss her fear of trichloroethylene in light of the other personal risks she may take outside of work.

Safety in an industrial setting is about the relationship of people, machines, and the environment that surrounds those people and machines. When people and machines interface, the potential for traumatic injury and property damage will always exists. When people and the work environment interface, the potential for health problems and environmental damage will always exist. When the environment and machines interface, the potential for functional degradation of either the machine or the environment will always exist. When the internal environment (i.e., that place within the confines of a production operation) is affected, we safety professionals are the ones who must act or get others to act to ensure the safety of individuals within that environment. However, when the external environment (i.e., the ecology) is affected, then our sister profession, the environmental managers, must act or get others to act to protect that environment. At times our professions share some of these responsibilities and at others we are entirely responsible for both at once.

Why Accidents Occur

How do accidents happen? Do they occur merely by chance, or are they preventable?

Before the Industrial Revolution accidents were thought to be "acts of God." In other words, if you had an accident, it was your destiny to do so. In the early nineteenth century, the thinking changed and certain workers were considered "accident prone." Those who had multiple accidents were considered dangerous to all and were therefore fired. These employees were considered to pose a threat to property and the safety of others. If they were killed or maimed themselves, it was believed that they had brought it on themselves.

A subsequent theory held that accident repeaters were responsible for most accidents. In other words, something in the individual's personality and psychology was defective. Since it was believed these employees caused harm to property and jeopardized the safety of others, they, too, were terminated. The nineteenth century manager, therefore, believed that accidents were largely preventable by properly managing his people. He mostly blamed accidents on the carelessness of individual employees and held these employees responsible for the majority of accidents that occurred.

The trouble with this theory is that we all recognize carelessness when we see it, but can we define it? In general, we think we understand what is meant when someone says, "Be careful!", but exactly what do they mean? Unfortunately, some professionals still manage safety and health programs with this outdated philosophy.

The development of more rational administrative and engineering theories on why and how accidents happen brings us to today's better-managed safety processes.

Domino Theory

In the second decade of the twentieth century, H.W. Heinrich proposed the *Domino Theory of Accident Causation* (see Figure 1-1). He saw accidents as a series of five steps that occur one after another.

Figure 1-1
Heinrich's Dominoes

The employee's dismal social environment leads to

Their loathsome mannerisms, resulting in

Their unsafe acts and/or development of unsafe conditions from which

Accidents are bound to ensue, leading to the

Injury or death of these hapless creatures.

According to Heinrich, the employees' social environment (the first domino) leads him or her to develop undesirable traits (second domino) such as recklessness, nervousness, ignorance, a violent temper, and unsafe behavior in general. Such traits cause unsafe acts or lead to unsafe conditions (third domino). Eventually, these acts and conditions lead to an accident (fourth domino) and an injury (fifth domino). By taking a series of counteractions, Heinrich believed, accidents could be prevented.

If you managed safety and health by the Heinrich Domino Theory, you would inspect the workplace until you uncovered an unsafe act or unsafe condition. Next, you would identify the undesirable trait in the employee that led to the observed situation and counsel the employee involved to be more careful. You would also feel called upon to lecture this employee on workplace manners and desirable behaviors or even to relieve his or her ignorance by providing some on-the-job training.

You do not have to be an engineer to manage safety by the Heinrich method, you just need a pencil and notepad, a big stick, and an aversion to writing. Take names and kick some . . . —well, you know the rest. Primitive as it may appear, this was the preferred safety management style for many years and it still has many proponents in American workplaces. Unfortunately, even the most obvious needs for the comfort and safety of workers are ignored by this safety management method.

Naturally, such management techniques do not sit well with collective bargaining philosophy and methods. After all, brutal human resource policies and unsympathetic management methods provided the fertilizer for the seeds of labor unrest early in the Industrial Revolution.

With the advent of the Occupational Safety and Health Administration (OSHA), established under the Occupational Safety and Health Act (OSHAct) of 1970, the methods of safety management began to change. Slowly at first, and overcoming great cultural inertia, the Domino Theory began to give way to other theories. Not to be totally displaced by new theories, however, the Domino Theory was, at first, only ready for a face lift.

The Domino Theory was first modernized in 1976 by Bird and Loftus who presented their own five dominoes (see Figure 1-2).

According to this theory, lack of management control over a situation is the root cause of accidents. Once control is lost, workers fall into substandard practices. The origins of this situation are lack of knowledge or skill to perform a job properly and safely, lack of motivation, physical or mental problems, inadequate work standards, poor design or maintenance, worn or damaged tooling or equipment, or abnormal tool use. The symptoms of these intolerable conditions are either unsafe acts or unsafe conditions, or both.

To progress from potential for loss to loss occurred, contact is made with one of eleven accident types, summarized in Figure 1-3.

Finally, the adverse effects of any of these accidents are the loss of human resources or property. Notice the engineering terms: source of energy, contact, loss. Gone are the pitiable beings of Heinrich who must constantly be corrected and shown the righteous and holy way of management. In this

Figure 1-2
Bird and Loftus' Dominoes

Management lacks control of the situation, leading to

Basic causes such as personal or job factors indicated by

Intermediate causes or symptoms (human errors) waiting for an

Incident or contact with a source of energy, resulting in

Loss of resources: people and/ or property

Figure 1-3
Summary of Accident Types

struck-by
struck-against
contact-by
contact-with
caught-in
caught-on
caught-between
foot-level-fall
fall-to-below
overexertion
exposure

model they are considered to be resources that bring value to the company treasury and are therefore deserving of protection, assuming the economics justify spending the money to do so.

Instead of placing all the blame on ignorant, careless employees, locating the blame for accidents shifted solely to management, in keeping with the national policy pronouncement of the OSHAct. Company management became suspect and was expected to fail to protect its resources due to incompetence, negligence, or economic choice—hence the need for government intervention via standards and regulations with compliance inspections and monetary penalties.

Single and Multiple Factor Theories

The Single Factor Theory is another alternative to the Heinrich theory. It contends that once you discover the true single cause of an accident you need not search further. Sometimes this may be the case, but most accidents are quite complicated in their structure and sequence.

V.L. Gross proposed in 1972 that accidents are caused by many factors working together (see Figure 1-4). He called his factors the Four Ms.

Figure 1-4
Gross' Four Ms of Safety

Man
Machine
Media
Management

Compare Gross' Four Ms with the even more modern three Es: employee, equipment, and environment. Some would add a fourth E: energy (Figure 1-5).

Firenze also advocated multiple factors as the cause of accidents. His four factors are summarized in Figure 1-6.

Energy Release Theory

In 1970, William Haddon broke with domino thinking and proposed that accidents and injuries involved the uncontrolled transfer of energy. The amount of energy transferred and the rate at which it transfers, it was proposed, is

Figure 1-5
The Four Es of Safety

Employee
Equipment
Environment
Energy

Figure 1-6
Firenze's Multiple Safety Factors

Management Factors
Situational Work Factors
Human Factors
Environmental Factors

directly proportional to the kind and severity of injury received. Some call this the Energy Release Theory. Haddon uniquely called for parallel preventive measures rather than the series of actions preferred by Heinrich and other domino theorists.

Now, all these theories may have had their time and place in the history of humankind—and, admittedly, some of these theories may still be made to fit many of our situations—but what the novice safety manager should always keep in mind is that these are *models* of real life situations and, like any model, they have flaws and limitations. These models may still be useful tools for learning about safety management, but they are poor tools of the trade for actually making safety happen. We'll discuss this more in Chapter 2, Safety Management Reality.

The Supervisor and Safety, Quality, Productivity, and Discipline

Why are we interested in safety and health management in the first place? Isn't it every employee's responsibility to care for his or her own hide? After all, we have clean, modern work areas in our new plants, don't we?

True, since the industrial revolution began in the nineteenth century, industrial firms have come a long way toward creating cleaner, safer workplaces. However, it is far from true to say that industry, on the whole, is clean and safe. For instance, from 1985—which was fifteen years after the establishment of the Occupation Safety and Health Administration (OSHA)—until now, four major accidents have occurred in U.S. industrial facilities in which a dozen or more lives were lost at once. The loss of human life in those communities where these accidents occurred is staggering to imagine, not to mention the economic losses to industrial companies, commercial businesses, and families. Consider, for instance, lost productivity, lost tax dollars, and the increase in benefit payments to survivors and widows' pensions. While the economics are certainly not comparable to the loss of life, the price tag is nevertheless staggering when considered in addition to the human loss.

If it was proposed to your company president that the company should write a check to cover the cost of replacing a piece of production equipment due to a mistake, heads would roll. Yet, even more money can be lost due to a human fatality, and little is being done to prevent this beyond coping with OSHA and other authorities and feeling grief for mistakes made. Yet, these managers are not uncaring people, are they? No. You and I work along side them and know that they do care.Then why is safety like a stepchild in your daily business? Why do good men and women care so little about health and safety? It seems they do not see the connection between safety, productivity, and quality. As the safety professional, you need to know the connection and communicate it to others. That is the mission of this book.

No matter where the job of safety manager may be listed on the organizational tree, safety always takes place on the production lines, or on the production end of staff or office positions. Safety is an inherent feature of any operation, whether it be the production line or a secretary's desk. Therefore, safety, quality, productivity, and discipline are all intimately interrelated and none of these basic elements can be separated without sacrificing the others. Rather than dominoes or contributing factors, or any other model, safety is an intricate part of a tapestry that defines the work culture of a plant. You can no more separate safety from this tapestry than you can quality and productivity. No properly informed manager would consider separating safety from quality and productivity.

Identifying the Work Culture

After spending only a few minutes in your plant, and by observing your people and your process, I would be able to tell you what your work culture is and what its weakest threads are—be it safety, quality, productivity, or discipline. This is no idle brag, it's a guarantee. Why? Because if any one of these threads is weak, they all are, and the source of that weakness is easy to identify.

Traditionally, the safety professional has been held solely responsible for safety in the plant. He or she has been considered the safety cop or the safety cheerleader, or sometimes both. But if one person is responsible for safety, then all safety activities are necessarily reactive and defective no matter how capable that one person may be.

Factories with bargaining units will rarely achieve a state where safety is perfectly integrated into the corporate work culture. This is because too often union employees see safety as a tool, or weapon, to be used against management. Safety complaints can be manipulated as evidence that management is inherently evil and needs to be contained by any legal means available.

On the other hand, management invariably chokes on laws and regulations that force safety upon it. "They said I have to do what?—Over my dead body!"

But even the most insensitive and seemingly uncaring management team senses that safety is an inherent part of its corporate culture, even when it cannot verbalize its heartburn over the situation. The corporate culture, in fact, belongs to neither management nor labor but it gains its life from both. It arises from the relationship between management and labor. This can be a good relationship where there is complete cooperation and everything works well. Or it can be conflict-ridden, where both parties blame the other and deny any accountability of their own. Most often it lies somewhere in between.

A culture arising from a poor relationship between management and labor suffers, develops poorly, is maladjusted, or may even be malevolent. OSHA may contribute to this poor relationship by forcing management to be entirely culpable for safety. In plants with exemplary safety programs, safety is a shared responsibility of management and labor.

Only if all employees, from the president right down to the janitorial staff, are held responsible and accountable for safety can an active positive safety process even begin to happen. Everyone owns the safety process—it is not the sole property of management or a tool for labor. Either everyone owns it or no one does. In striking this balance of responsibility, the work culture is always a tricky issue when safety as an issue is involved.

Implementing a Safety Process

If the responsibility for safety is so difficult to balance, then how can we ever implement a safety program? Isn't it always doomed to failure? No. The place to begin is to identify the key players in putting safety to the forefront of the company culture.

The key player in any successful safety process is not the safety manager but the line supervisor. Only the line supervisor can use his or her knowledge of accident causation to maintain an acceptable level of risk to employees and property resources by correcting unsafe conditions.

But the supervisor can only understand what an acceptable level of risk is if he or she communicates freely and openly with employees. In order for everyone to be understood, the safety professional must actively provide information and interpretations about safety to the workers. If supervisors do not look to the safety professional as a mentor regarding the safety thread of the work culture, then something is inherently wrong with the process that will be hard to fix. If employees do not regard their supervisors and safety professionals as trusted friends, then something is also wrong.

Management is not about technology, or the scientific application of theories; it is about human relationships. Science and technology are neat and clean, even antiseptic. Human relationships, on the other hand, can be irrational, disorganized, and messy—very septic at times. You can study management theories—and I recommend that you do—but until you wade into the workplace where empires, cliques, undercurrents of discontent, petty jealousies, and outright hatreds exist, you haven't the vaguest idea what you are up against. Lament that the real world is not as neat as the textbook if that makes you feel better, but roll up your sleeves and become part of the fabric of your work culture. That, my friends, is as good as it gets.

Who Are the Safety Leaders?

Learning about safety through personal experience is more reliable than learning from a textbook. If learning occurs from experience, then it is essentially 100 percent subject to our recall at any time in the future. By comparison, recital, or textbook learning, is subject to about 10 percent recall a mere 24 hours after training. But if personal experience does not produce learning, or if what has been learned is somehow defective, then it can actually be more harmful than learning from a textbook, or from not learning at all. This is not profound stuff, just common sense.

In general, experienced supervisors tend to be better safety supervisors than new supervisors, just as they tend to be better at quality control, production performance, and discipline. This is not true across the board because you cannot always equate experience and wisdom with years on the job, but it is generally the case.

Nevertheless, unless line supervisors are aware of safety issues, knowledgeable about safety rules and policies, cognizant of hazards in their workplace, and can communicate well with their employees about these matters, the safety manager is merely a one-person police force or cheerleading squad at best. Educate the supervisors, demand that they perform their tasks of meeting production and quality goals with safety in mind, and expect them to communicate hazards and provide instructions in hazard control to their employees, and the safety process in your plant will be healthy, even if less than perfect.

Nevertheless, even when line supervisors are held accountable for safety the process will flounder unless teams of employees assume the authority to make their own workplace safe, and are empowered to do so. When the supervisor enables his or her team to identify, assess, and control hazards, or to get help from outside the team if unable to assess, the team is better able to perform safely. The role of the safety professional is to meet with the supervisor and his or her team in order to clarify requirements, and to provide mandatory training.

The work team, with the assistance of its supervisor and the safety professional, discusses the impact of the requirements and develops a list of issues to be resolved to ensure safety is provided. Figure 1-7 lists some of the generic issues to be included.

Reluctant Compliers

Is your company a reluctant complier? According to Veltri, based on a survey he conducted in the late 1980s, 77 percent of industrial companies

Figure 1-7
Generic Issues to be Resolved to Ensure Safety

- Areas that need special precautions
- Tasks that need to be assessed
- Job procedures that need to be reviewed
- Team members who will need special training
- Equipment/engineering systems required to comply

are *reluctant compliers,* meaning they do only what is absolutely necessary to get by and stay out of trouble with OSHA, but no more. Another 16 percent of companies implement creative solutions to safety problems that have been developed by the industrial safety leaders. This second group is called the *followers.* A mere 7 percent of industries are true *leaders* in safety management, meaning they develop and implement their own solutions for safety problems. If all industrial facilities were true leaders we would have no need for OSHA.

Figure 1-8 summarizes the National Safety Council's list of five generic responsibilities of industrial supervisors.

Figure 1-8
National Safety Council Generic List of Supervisory Responsibilities

1. Establish work methods.
2. Provide job instructions.
3. Assign people to jobs.
4. Supervise people at work.
5. Maintain equipment and the workplace.

Figure 1-9 is a list of Peterson's responsibilities of supervisors, quoted by Weber.

Figure 1-9
Peterson's Four Key Safety Tasks of a Supervisor

1. Investigate accidents; determine root causes.
2. Inspect work areas; uncover hazards.
3. Coach employees on how to work safely.
4. Motivate employees to work safely.

Figure 1-10 is a list of Denton's responsibilities of supervisors, quoted by Weber.

Figure 1-10
Denton's Ten Specific Safety Activities of a Supervisor

1. Safeguard employees; protect equipment.
2. Provide necessary safety equipment; enforce its use.
3. Concentrate on good housekeeping.
4. Provide and check for safe tools.
5. Know the medical and physical limits of employees.
6. Provide on-the-job training.
7. Develop continuous participation by each employee in safety process.
8. Help prepare, use, and update job hazard analysis (JHA).
9. Conduct accident investigations.
10. Provide positive rewards for safety performance.

Companies that wish to achieve a top-notch, world-class safety process could learn from Occidental Chemical's company slogan, adopted in 1985:

Safety + Quality + Productivity = Success

This is a very simple slogan, but a proven formula for success.

Top Down Interest, Involvement, and Commitment

One way to demonstrate management's commitment to safety is to implement a Written Safety Program. OSHA regulations do not currently require an employer to have a Written Safety Program, but implementing such a document not only serves as evidence of commitment, it also demonstrates good faith. It provides a framework for organizational and administrative efficiency, and provides an additional planning tool.

Some features of a Written Safety Program include:

- Designation of a safety coordinator
- Delineation of safety committee responsibilities and authorities
- Expression of management's safety policy
- Provision for general safety rules
- Procedures necessary to minimize risk of injury and illness in the workplace

Traditional safety programs have four key requirements:

- Top management must support the program.
- The safety professional must be committed to the program.
- Management must be convinced that production of goods and services and protection of employee health are not mutually exclusive goals.
- Qualified safety and health employees must be developed and retained.

A management team that is truly committed to safety is not motivated by laws and regulations, rather it is driven by its own vision for the safety process. One way to spot the level of commitment a management team has for safety and health is to pinpoint where in the organization tree the safety staff is located.

Typically, uncommitted managers provide minimal staffing to address major legal requirements only. Hansen refers to this as *reluctant compliance.* In this scenario, a single safety professional is assigned to a lower level staff position with more responsibility than he or she can handle and with no authority whatsoever.

Management teams that are more committed to safety provide staffing and resources beyond the minimum in order to reduce actual and perceived risks. Safety professionals are found at an appropriate structural level, up to and including the executive level. The safety staff has a reasonable budget and, more importantly, the authority to get things done.

Not only does committed management provide resources for organization staffing, but it also serves as the motivating force behind the success of the safety process. A half-hearted management team cannot hide its failure to motivate behind efforts to intimidate employees into the appearance of compliance when the chips are down. Committed management regards employee safety and health as a fundamental value of the organization and applies its commitment to safety and health protection with as much vigor as it pushes other organizational purposes.

Management Commitment to Safety

Show me the ongoing safety budget and, without walking into the plant, I'll tell you how it stacks up. Uncommitted managers do not assign a budget to the safety staff. Perhaps they divvy out some money to solve a specific problem from time-to-time, but the staff has no ongoing budget. These plants experience cyclical safety—that is, they have periods of accidents and periods of no accidents. If the incident rate is plotted over a period of time, however, the curve bounces back and forth between the two extremes with no real improvement in the overall safety environment.

Management must clearly state a policy on safe and healthful work conditions so that all personnel with responsibility understand the priority of safety and health protection relative to other organizational values such as productivity, quality, and customer service. Where safety and health programs are effective, the senior managers are visibly involved in the process. They make safety and health inspections themselves. They sit in on safety committee meetings. They know the site's incident rate and lost work day rate. Everyone knows they know, and everyone knows they care. That is the foundation of an effective safety and health program. Even the most enthusiastic, energetic safety manager cannot simulate management commitment on his or her own. You know management is committed to safety from the moment you walk into the plant. Its walls exude a "safety culture."

The driving mechanism for continuous safety improvement, according to Krause, is the appropriate use of statistics and behavioral science coupled with employee involvement. He sees safety happening much as in Figure 1-11.

Culture, from Krause's example, is nothing more than the attitude of management in practice. Not what they say, in other words, but what they

Figure 1-11
Krause's Safety Process

Culture
Vision
Values
Common Goals
Assumptions

Management System
Training
Safety Measurement
Facility Design
Discipline - Reward
Feedback
Accountability
Priorities
Resources
Attitudes

Exposure
Safety-related Behavior
Conditions
Equipment
Facilities

End-Point
Incidents and Accidents

do. It colors the attitudes of everyone involved. True, each individual has his or her own attitude, but the work culture provides an influence over that attitude that can be pervasively positive or negative, depending on the individual's personal reaction to management.

However, the management team cannot merely sit in an ivory tower and dictate a culture of its own vision and attitudes. Management must instead be actively involved and physically present in the safety process. Is your management involved? Does the plant manager routinely sit in on safety committee meetings? Does he or she attend training sessions? Does he or she include safety observations in notes when walking through the plant on informal inspection tours? No? Then I find it hard to visualize just how involved your management is in your safety process.

Safety Teams

In creating an ideal safety organization, Trautlein and Milner would like to see safety made part of the responsibility of every self-directed work team, which would elect a safety leader, who is accountable for accident investigation, reporting, and feedback.

Killimett, Spigener, and Hodson see a management team that knows how to involve all of their employees in the process. They also get beyond the either/or limitations of *hard solutions,* such as engineering changes, and *soft solutions,* such as cultural changes. Engineering changes are primarily equipment changes; cultural changes involve raising awareness. Such a management team would provide the means to reorganize all organizational levels to initiate and nourish a *total* or *integrated safety effort.*

Managers of plants that have effective safety and health programs encourage employee involvement in the structure and operation of the program. Employees are thereby involved in the decisions that affect their safety and health. Their insight and energy are tapped to achieve the goals and objectives of the safety and health program.

Peterson lists the safety obligations of management, as summarized in Figure 1-12.

A trend that has continued from the early 1990s is corporate downsizing. The board of directors want profit, they want it now, and do not tell them to wait until next quarter. Consequently, safety managers have seen reduced budgets and staffing. More safety must happen with less staffing and with less budget at the disposal of the remaining staff.

No matter what the size of the safety staff, or even if none exists as such, the chief role of management is to create a culture that encourages safe behavior and rewards prompt corrective action to improve safety. Culture lets workers know whether safety is really the priority that management says it is. Written safety programs and policies are not the least part of culture. How management acts—that's culture, too. More than that, culture is how everyone acts collectively; but, keep in mind, most employees will key their behavior off management.

Figure 1-12
Peterson's Safety Obligations of Management

1. Concentrate on the long-range goal of being a safety model and industry leader.
2. Accidents are not acceptable.
3. Use statistics to identify sources of accidents.
4. Institute thorough job skills training.
5. Eliminate dependence on accident investigation.
6. Provide employees with, and expect them to use, knowledge of statistics to identify areas needing study.
7. Expect employees to report system defects and help find solution?
8. Design safety into the process.
9. Eliminate slogans, posters, incentives, and gimmicks.
10. Examine work standards to remove accident traps.

Safety Programs

Why does management persistently cling to failed methods and processes for safety? Hansen suggests four reasons: lack of preparation; microwave mentality; the Lemming effect; and the Fourth Symptom.

American management, by and large, receives an education based on past practices that worked at various times and places, and can still work in various times and places, but do not work in all times and places. That is not adequate preparation for the safety-integrated management of factories. Innovative safety management techniques and procedures are left to a few cutting edge lecturers on the continuing education seminar circuit. Paraphrasing Deming, Hansen claims that many managers get high marks in all their safety management courses, yet do not have the vaguest idea about how to improve the safety process when they come face-to-face with one.

The microwave mentality, according to Hansen, is the tendency of modern managers to go for quick fixes instead of determining and correcting root causes. If it looks good on paper and seems to satisfy the appropriate OSHA standard, microwave managers go with it.

What Hansen calls the Lemming Effect, others call the myth of paradigm: "we've always done it that way and it always has worked well, or at least as good as any alternative—which, by the way, no one has ever tested." A paradigm is a model, or set method, established as a traditional way for performing a task, and, as such, can be a good thing. The danger is that a paradigm typically becomes outdated and extended beyond the original set of circumstances for which it was designed.

Regarding the Fourth Symptom, Hansen refers to John Graham, President of Graham Communications. The Fourth Symptom is a tendency to think in the past. This, in my mind, is paradigm paralysis. We have a tendency to think about the things that worked for us in the "good old days." We cling to old ways and means—never mind the many circumstances that have since come together to make them ineffective. Things change, and the alert safety manager recognizes this and adapts to keep up with change. This is not to suggest that all paradigms should be destroyed or replaced or that all things from the past are worthless. Not so. "Continuous improvement" means that we have to examine, evaluate, and modify the safety process constantly.

The fact is, we never actually ever throw away a paradigm. A paradigm, as Winn points out, is partly culture, partly rules and beliefs, partly science, and partly art. We cannot really speak of modifying or abandoning the paradigm because it is too complex to define and its threads run too deep. But let me say that I agree with Hansen that we can see the future of safety management from where we are. We want and need a paradigm shift to get to that place we see. A the same time, I have to agree with Winn that the paradigm shift Hansen wants has not occurred. A lot of people are writing like hell to force a change that, in the end, will probably evolve all by itself, in a way that we never dreamed possible.

According to Winn, comprehensive safety ultimately means including the last guy. Peterson emphasizes this, too, and Winn agrees that it is the key to recognizing a new paradigm when it comes. That is the future. Safety in the future will include each and every employee in the process. Sounds rather messianic, doesn't it?

To summarize, visible management leadership provides the motivating force for effective safety and health programs but has often been immobilized by failed methods and processes. Figure 1-13 summarizes tools provided by a committed management staff for effective implementation of policy.

Figure 1-13
Implementation Tools Provided by Management

- budget
- information
- personnel
- assigned responsibility
- adequate expertise and authority
- line accountability
- program review procedures

Bottom Up Practice

Safety is everyone's job. The personal dedication of each employee to safety is absolutely essential for safety to be successfully implemented. All the management interest and commitment and the best supervisory skills in the world will get you nowhere unless the employees are 1) well motivated to perform their tasks safely, 2) aware of the existence of hazards in the workplace, 3) suitably prepared with knowledge, skill, and equipment to make rational decisions about hazards on the job, and 4) otherwise behave safely.

Safety must be practiced from the lowest ranks of the company to the top manager. The greatest personal risks are taken by those with the least say in the organization, so that is where the concentration of effort must be made with regard to practicing safety.

Preventive Actions

Comprehensive baseline surveys must be conducted for the purpose of identifying safety and health hazards. These surveys must be updated periodically. Once hazards are known, assign and communicate responsibility for hazard control so that management, supervisors, and employees in all parts of the organization know what performance is expected of them.

Analysis of worksite hazards must extend to planned and new facilities, processes, materials, and equipment. Routine job hazard analyses (see Chapter 9) must be performed.

Too often, a job-related injury is caused by a condition that either the injured worker, or some coworker, saw just before the accident, yet no action was taken. Why? Because those who saw the unsafe condition did not consider it an unreasonable risk to take, so they took it. According to Ulmer, this is the most common reason given for why no action was taken to prevent an accident.

There is a second common reason given for not taking preventive action: management, for whatever reason, had not responded adequately to reports of unsafe conditions. Workers see this as evidence of a breakdown of the management system, and, typically, give up on the notion that management cares about protecting them. Supervisors either do not correct unsafe acts or they are overwhelmingly harsh, giving both the workers concerned and observers the opinion that the supervisors either do not care or that they are meting out gratuitous punishment. Therefore, employees eventually stop reporting all but the most imminently serious hazards, and, in the worst situations, not even those unsafe conditions get reported.

People will only be rejected or punished, or be made to feel rejected or punished, once or twice. After that, they go their own way and do their own thing. The long and the short of it is that if you are receiving no reports of unsafe conditions, you may not have a safe plant after all.

Without the benefit of measurements or audits, how do you know that your workplace is safe? When you enter a work area, how do you know it is safe? If you are honest with yourself, you will admit that you either feel safe or unsafe. The safety professional views safety as the opposite of risk—as an engineering or management science. To everyone else, safety is a matter of "feeling" safe or not. To derive the most benefit from bottom-up practice, you must make those who spend their days in the workplace—supervisors and their workers—feel safe, yet be constantly vigilant for clues and evidence of changing conditions and new hazards.

This means that sometimes you must accommodate workers and fix things that may not be required to be fixed by OSHA. If all you do is comply with standards and regulations, you will be like a puppy chasing its tail—you will have a grand time, but you won't get anywhere. I am not suggesting, however, that you merely make some cosmetic changes in the workplace to make employees feel safe and neglect more significant corrections and modifications. You must first do the basic things that really make the workplace safe; then sometimes you can reap benefits of making some cosmetic changes that just make the employees feel better. Do not carry this to the point of patronizing them, however. If you do so you will shoot yourself in the foot.

Employee Participation in Safety

Management safety objectives must be met from the bottom up through employee participation, just as quality, productivity, and customer service are. Tyler suggests implementing employee participation on four levels.

At the first level, employee audit teams are formed to assess the safety process. For instance, one such team audits unsafe acts, while another might be formed to audit key responsibilities within operations and maintenance. The next level, Education Involvement Teams, present safety topics of interest to the Involvement Task Teams, the third level. In each work area the task teams attempt to solve ongoing safety problems and address areas needing improvement. Finally, at the fourth level, the involvement process attempts to make each employee a safety leader, and to involve everyone in the safety process. Each employee must be held accountable and responsible for safety, not only for themselves but for others.

Participation, recognition, and accountability go hand-in-hand. Employees, in general, will not participate wholeheartedly in any program unless they perceive incentives for good safety performance. They need to see instances of management rewarding safe work behavior, such as by recognition in front of other employees. They also need to see management holding persons accountable for safety performance by making promotion to higher-level jobs dependent on good safety performance. This can be done simply by mentioning good safety performance along with other accolades when a promotion is announced.

The Role of Job Pride

An important motivation factor that is often overlooked is job pride. People who take pride in what they do give extra effort in productivity, quality, customer service, and safety. This is why line supervisors make or break the safety program. It is the supervisor who infuses job pride into his or her employees. Some people have more natural job pride than others, but a supervisor worth his or her salt can instill pride in others.

For example, a new supervisor took over a demoralized buildings and grounds department in a factory. The department, which had the worst safety record in the plant, was responsible for the appearance and maintenance of the building and grounds. The previous supervisor was constantly in trouble with the top managers for poor housekeeping, inside and out. The first thing the new supervisor did was to conduct a series of training sessions in which he taught the basics of painting, crafting canvas awning, and several other similar topics, including safety. When a spotless new awning was installed on the front windows and doors, the new supervisor passed on the compliments he received to his crew. Soon they were working with a new energy and enthusiasm. In no time the housekeeping was spotless everywhere, and the department had the lowest accident rate of any department in the plant.

The turnaround that occurred in this example was due to much more than sending employees to a couple of training sessions. It is also not merely due to improved housekeeping—though that gives us some clues. Training and awareness and housekeeping were very important to the turnaround, but they were not the entire picture. Equally important was the attitude and concern of the new supervisor. Also, the compliments received were powerful positive behavior reinforcement tools in the skillful hands of the supervisor.

Wilkinson has a model for management that he calls the PRIDE management principles.

Professionalism
Respect
Interesting
Direction
Ethics

Don't coerce. Don't argue. Reinforce these positive principles, give your employees the facts, tell them their options, explain the consequences, and let them make the choice.

Behavior Reinforcement

According to industrial psychologists, employees who entered the workforce before 1965 are motivated differently than employees who began working in 1965 or afterwards. Generally, post-1965 employees are suspicious of being observed while working, and also are suspicious of management's motivation in auditing the workplace. Those who entered the workplace before 1965 are more likely to rely on management to inform them that the workplace is safe or hazardous and trust management to make it so. The newer worker relies primarily on his or her own feelings and experiences to decide whether or not he or she feels safe. Any unsolicited input by management is held suspect.

The problem with this newer attitude is that human feelings do not necessarily or accurately reflect reality. Secondly, the experiences that we base these feelings on may be limited or misinterpreted, especially if we are uninformed about principles and theories of safety. Perhaps management cannot always be trusted, certainly, but relying on feelings and experience alone can be just as deadly as even the most Machiavellian management. In my opinion, we Americans have become so savagely individualistic that we now mistrust anything that does not come from ourselves. We rarely trust or rely on cooperative effort anymore, or work for the good of a community.

Besides individualism, American employees post-1965 are products of the slogan imprinting that occurred during our youth, much of which compels us to become risk takers. Figure 1-14 lists some of these slogans and catch phrases.

While risk taking is encouraged and rewarded in general society, in the workplace risk taking is typically harmful or punishable. Therefore, the post-

Figure 1-14
Slogans and Catch Phrases that Encourage Risk Taking

Do or die!
Go for broke!
Go for it!
Just do it!
Grab the gusto!
No guts - no glory!
No pain - no gain!
Push the envelope!
Rules are made to be broken!

1965 worker is more likely to have a built-in inner conflict—the tendency to take risks even in an environment where it is discouraged rather than encouraged.

Couple this inner conflict with a general distrust of civil leadership—caused by being exposed to unpopular wars, assassinations of public figures, and dishonest leadership at all levels of government and management—and you have a workforce that lacks respect for authority and distrusts rules and policy statements.

New Compliance Management Styles

The traditional approach to compliance was to conduct compliance training, audits, and inspections. The control system was a hierarchical organization tree. Management reacted to problems and solved them mainly by punishing offenders. The new approach is more positive and systematic. Management shares the decision-making process and encourages workers to own the process. Leadership and accountability are established at all levels of the organization. In the past, management assumed that people were guided by self-interest and treated safety as another duty to be done. New management teams understand that people are guided by higher standards, values, and beliefs, and treats them accordingly.

Safety Incentives

Many traditional safety programs utilize safety incentives. If a department has no accidents for a specified period of time, or achieves a specified number of payroll hours worked without a lost time accident, for example, everyone gets a T-shirt, coffee mug, baseball cap, or team jacket as a reward for good behavior. If all that interests you is numbers, put an incentive system into place. They almost always lower accident rates and usually lower lost workdays, too. But this form of good news is like the emperor's new clothing—it does not exist except in the mind of the wearer. The reduction in numbers is typically due to less accident reporting, not to actual improved safety performance. Sure, the lost workdays are probably accurate. After all, if someone has severe enough an accident to miss work, it is going to be reported, regardless. But minor accidents and near misses will magically disappear from the records.

From a safety improvement standpoint, what does that mean? Too often, in incentive program plants, it means that a large number of hazards and accidents go unabated because they go unreported. No matter the incentive program, every near miss or minor accident is a symptom of a larger problem that cannot be ignored if safety is to be ensured.

Incentives work best in traditional management systems that are organized along military fashion, including a strict chain-of-command where employees are dependent on management. Where the workforce tends to be individualistic, however—"I'll do-things-my-way-to-hell-with-the boss"—incentives are not realistic. In fact, in these workplaces, incentive programs are generally received antagonistically and the workers suspect management of coercion or of covering something up. In any environment where the workforce was mostly born after the mid-1960s (is that not most workplaces?), I would be extremely careful about implementing new incentive programs.

For existing incentive programs, incentives are only effective for the duration. Once an incentive program falls into apathy and indifference, my bet is that you will not be able to restart it with the same results. The reason is that incentives do not really change the temperaments and emotions that underlie employees' reactions to supervision. As if that were not enough, incentive rewards are often viewed as punishment or neglect by those workers who do not receive them. Instead of inducing them to try harder, these employees feel left out and rebel against the program by not trying at all. They may even develop negative behaviors in order to punish management. Thus a conflict arises that simmers, festers, and debilitates the overall safety effort.

Safety incentive programs also distract management from focusing on the true causes of accidents and illnesses in the workplace. Employees are not ignorant of this dissipation of resources and resent it. Incentives discourage workers from changing the safety process for the better, opting for the status quo instead. An incentive is an overjustified means of inducing safe behavior that can be counterproductive because it destroys the natural motivation to protect oneself. It is based on the theory that money motivates more than anything else and it incorporates competition and fear in the workplace.

If incentives are used, they should balance the measurement and control of the safety process with the performance of individual employees. Changes should be attributed to the appropriate level of control: system or individual. System level management and control requires the commitment of groups and the individuals within them. Individuals must be empowered to contribute to the organizational performance. They must also be optimistic that their individual effort will be useful.

ABCs of Safety Evaluation: Attitude, Behaviors, Conditions

Geller's ABCs of Safety Evaluation

Attitudes
Behaviors
Conditions

Attitudes mirror the reality of the workplace. To achieve safety improvements, then, get the supervisor to make the workplace safer. Next, get the employees involved in making it safer still. Attitudes will begin to change in a positive direction if these safety steps are put in place. Workplace behaviors will become safer as a result. With a positive attitude and safe behavior, conditions are corrected in the workplace one step at a time until a degree of safety is achieved that is satisfactory to the team.

The problem with behavior-based safety programs is that they are nearsighted and let management off the hook. Deming's 85-15 Rule, cited by Walton, states that 85 percent of problems within any operation are systemwide and the responsibility of management, while the worker is accountable for only 15 percent. In his own book, Deming states that 94 percent of troubles and possibilities for improvement belong to the system—the responsibility of management. The other 6 percent are special, not necessarily the responsibility of the worker. Where then is the payback for behavior-based programs?

Everyone tends to get fired up when a behavior-based safety program is first introduced. Eventually, though, employees see these programs as manipulative. This usually happens when a few new employees come along who were not privy to the initial hoopla and rah-rah sessions, and want to know what all the fuss is about, and where is the benefit? Also, these programs stifle initiative by making employees focus only on themselves.

The chief weakness of behavior-based safety programs is the failure to deal with real root causes, opting instead to get employees to "behave" their way out of hazardous situations. Managers eventually tire of behavior-based safety because "all the rah-rah and hoopla does not save the moolah." Employees tire of it because the real root causes are not corrected and the exposed workers will still be harmed no matter how careful they are and how well they behave. Finally, behavior-based safety lacks the power to be self-sustaining. The cheerleaders go home when the team has lost.

That is not to say that positive behavior should never be reinforced. In fact, behavior reinforcement is a very necessary part of the overall safety process. But only a *part,* not the main part. No matter what, good behavior rewarded and undesirable behavior must be corrected.

In Figure 1-15, Geller tells us how to reward good behavior. I made one change in his procedure. Good behavior ought to be rewarded in the presence of the worker's peers. Being ex-military, I believe very strongly in the practice and procedure of giving rewards in public and punishments in private.

Figure 1-16 is Geller's method of giving correcting feedback to decrease at-risk behavior.

Figure 1-15
Geller's Rewarding Feedback Procedure (modified)

1. Give the feedback face-to-face and in public.
2. Give the feedback as soon as possible after observing the desired behavior.
3. Identify the safe behavior(s) observed.
4. Be sincere and genuine.
5. Express personal appreciation for setting the right example for others.

Figure 1-16
Geller's Correcting Feedback Procedure

1. Give the feedback one-on-one and privately.
2. Give the feedback as soon as possible after observing the undesirable behavior.
3. Begin with an acknowledgment of safe behavior(s) observed.
4. Identify the at-risk behavior(s) observed.
5. Specify the safe alternative(s) to the at-risk behavior(s).
6. Indicate concern for the person's welfare.
7. Request commitment to avoid the at-risk behavior(s).
8. Thank the person for his/her commitment to continuous improvement.

SAFETY LEADERSHIP

Leaders are not born, they are developed. Although some people seem to have a natural affinity for leadership, or have charismatic personalities that make following them attractive for others, the military, which depends heavily on good leaders, has long been a proponent of developing leaders. Industry should take a cue from the military and spend more time developing their supervisors to be leaders, rather than throwing untrained leaders into the fray.

Sarkus proposes that leadership and culture are inextricably linked. Certainly, leaders affect change, and without leadership change is virtually impossible among any group. To the extent that safety culture is established by executives, managers, and supervisors, how that culture develops among the rank and file will evolve from the beliefs, values, behaviors, and assumptions of those men and women in the leadership positions.

Competent leadership, along with training, hazard communication, motivation, behavior modification, and discipline, is a critical element of safety management, without which superior results cannot be achieved.

Safety Leadership Traits

Safety leaders, like other leaders, share some common traits. These traits are more often learned or acquired by leaders rather than innate. The five traits shared by all leaders are:

1. The spirit of initiative
2. The ability to risk
3. A sense of responsibility
4. Authenticity
5. Generosity

First, *leaders have a spirit of initiative.* They are able to reach their own decisions, prepared by training, experience, and innate initiative. They also have the courage to act upon their decisions. They are not people who, when the chips are down, do as an old Navy saying: "When in worry, when in doubt, run in circles, scream and shout!" Rather, they are the ones who make, initiate, and stick by sometimes difficult decisions and plans. Leaders do not seek the spotlight at all cost. They are people who know how to cooperate with others. Safety leaders, too, are doers, not spectators.

Second, *leaders are willing and able to take risks.* That does not mean they perform wild, irresponsible actions. They are instead able to consider all the facts available and to infer assumptions about missing facts. A leader does not fail to act just because of the possibility of being wrong. A leader observes, judges what seems right, and then acts. No hand wringing. No "Woe is me." A leader is willing to stand alone, if need be, to take an unpopular stand. Therefore, a leader upholds the ethics of his or her profession, undaunted by fear or the majority opinion. The leader is willing to risk himself or herself where justice is needed. When necessary, the leader is willing to work quietly to prepare for action.

Third, *a leader has a developed sense of responsibility.* The leader feels responsible for the mission of his or her team. If no one else does, the leader owns the goals and objectives of the team. The leader is unwilling to point fingers or look to others for blame when mistakes are made or accidents occur.

Authenticity is the fourth true indicator of leadership. A leader is not merely truthful, but especially true to himself or herself. A leader is what he or she is suited for and does not pretend to be anything else. A leader is what he or she says he or she is. What you see is what you get, and you can count on that!

Finally, *a leader has a certain generosity of spirit* that is compelling to the rest of us. "He gives what he has to give." "She gives everything she can." These are the sort of things that you hear about quality leaders. A good leader gives himself or herself totally for the benefit of all.

Communicating Safety

Management, as the organizational leadership, must set and clearly communicate safety and health policy and goals to all employees. Then management must set the example by following its own safety and health rules. Visible support must be given to all safety and health efforts. Annual safety and health goals must be incorporated into all annual planning activities.

A management team that is actively and positively leading participates in committees, teams, inspections, investigations, and other safety activities. Its written safety action plan is updated annually and includes estimated completion dates, assigned responsibilities, and methods for tracking activities.

Management leaders insist on meaningful safety and health meetings on a regular basis. Work teams are assigned safety responsibilities, are given specific safety and health activities as part of the team design, and are held accountable for both

Safety and health performance is written into the job descriptions of managers and supervisors and is reviewed annually. Safety and health topics are routinely discussed at department, shift, production, and quality meetings.

Interaction Dynamics

Supervisors must use their knowledge of the causation of dangerous situations in order to maintain the level of risk at an acceptable level within the boundaries of his or her area of responsibility—the workplace. Knowledge of what causes injuries and illnesses is learned more from experience than from training, as discussed earlier. On-the-job training is very effective, but only if properly and routinely conducted. People typically retain less than 10 percent of what they were taught about a particular subject, but almost 100 percent of what they learned by hands-on activity. In a dangerous work area, which of these employees would you prefer to work with?

Sarkus lists Seven Cs of safety leadership, which are summarized in Figure 1-17.

The key elements of good management will always be planning, organization, and leadership.

Figure 1-17
Sarkus' Seven Cs of Safety Leadership

Caring—genuine concern and support for others.

Coaching—demonstration and repetition.

Correcting—stopping at-risk behavior.

Confirming—praise for working safely.

Collaborating—transfer knowledge that empowers the group.

Clarifying—communication of goals and objectives.

Conciliating—reconciliation of differences.

Motivation

A *motive* stimulates a person to seek to satisfy a perceived need. *Motivation* is a quality of the person motivated, referring to the level or energy or determination with which that person acts because of a motive.

Abraham H. Maslow supposed that all people have five categories of needs as listed in Figure 1-18.

The Maslow theory is that these needs are hierarchical. Every human is motivated first to fulfill his or her physiological needs. When these needs are obtained, he or she can address safety needs. When safety needs are met, then social needs can be addressed. A person is motivated to satisfy self-fulfillment needs only after the lower four categories are met satisfactorily, though not necessarily completely.

Under the Maslow theory, as applied to a workplace, an employee can only be motivated by management in one of two ways: 1.) management can appeal to the particular set of needs which the employee is presently trying to satisfy, or 2.) management can place the employee under a threat that jeopardizes a previously sated need. Loss of pay, for instance, particularly threatens safety and physiological needs, but can be said to threaten higher needs as well.

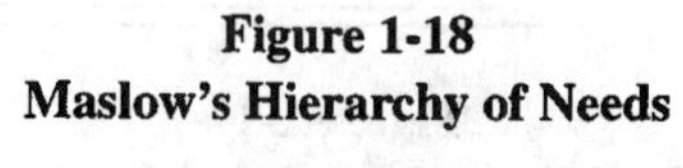

Figure 1-18
Maslow's Hierarchy of Needs

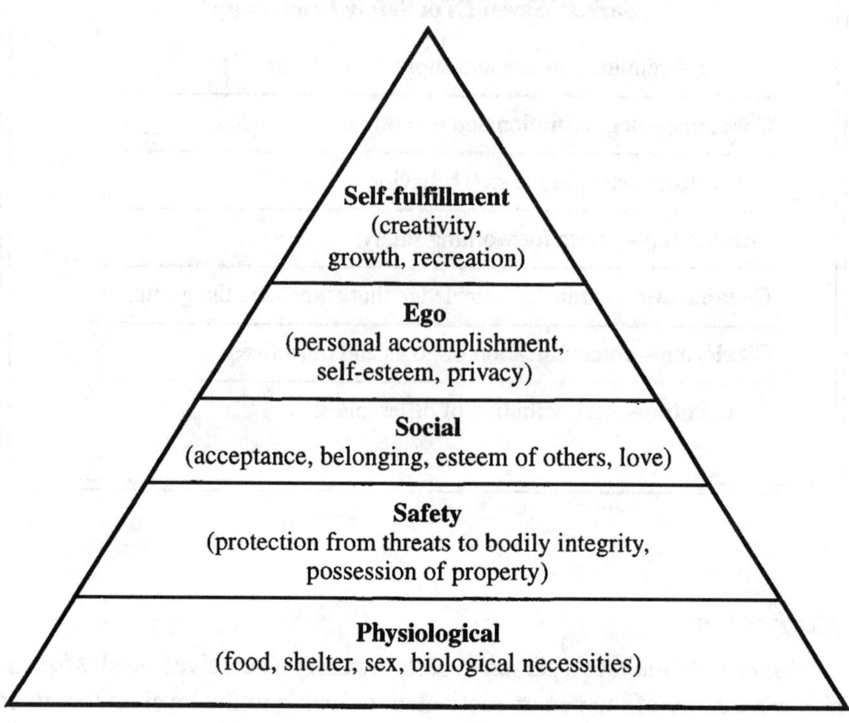

Douglas McGregor developed the Theory X and Theory Y methods of motivating employees. *Theory X* managers assume that employees are motivated primarily by their physiological and safety needs, as defined by Maslow. *Theory Y* managers assume their subordinates strive mainly to meet their social, ego, and self-fulfillment needs. Theory X managers, then, think that people are mainly driven by baser motives, while the Theory Y manager believes the drive comes from higher motives.

Frederick Herzberg theorized that people are motivated to perform a task in order to achieve satisfaction for a job well done. They will not work well, Herzberg continues, if doing so involves too much pain. What causes pain? Bad relationships with other employees or supervisors; inadequate pay; lack of job security; poor or inconsistent supervision; uncomfortable working conditions. Herzberg sees two factors of work motivation: job content and job context. The relationship with your boss is part of "job content," while environmental conditions in the workplace are part of "job context." Figure 1-19 shows a basic motivation model.

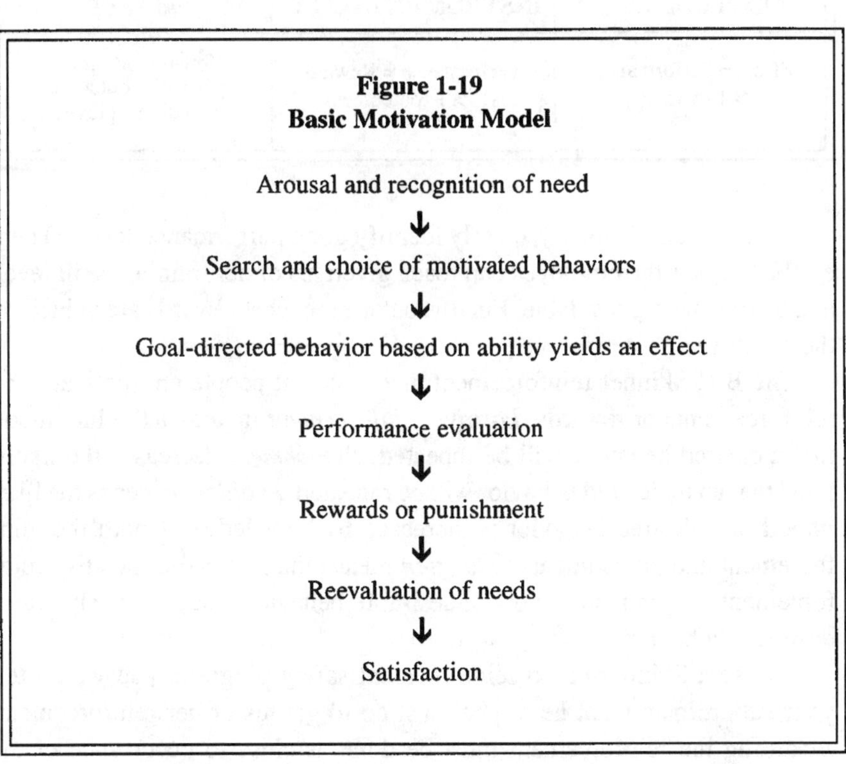

Arousal and recognition of need
Search and choice of motivated behaviors
Goal-directed behavior based on ability yields an effect
Performance evaluation
Rewards or punishment
Reevaluation of needs
Satisfaction

Figure 1-19
Basic Motivation Model

Positive Reinforcement

The expectancy theory of motivation was developed by Victor H. Vroom. According to Vroom, people basically expect rewards and progression from their work. Three main variables make the expectancy theory work: expectancy, instrumentality, and valence. Expectancy is the perceived relationship between effort and performance. Instrumentality is the perceived relationship between performance and rewards. Valence is the strength of preference for a particular type of outcome of expectations. Figure 1-20 shows the model for expectancy theory.

Figure 1-20
Expectancy Theory Model

EXPECTANCY	INSTRUMENTALITY	VALENCE
Effort = Performance 0 -> 1 (certainty)	Performance = Reward -1 -> 1 (outcomes)	How important a particular outcome is to the employee

To use Vroom's theory, clearly identify good performance levels. Then, establish goals the employee can reach given his or her training, skill level, and technical help available. Finally, determine what rewards are valued by the employee.

The B.F. Skinner reinforcement theory is that people are motivated by reinforcements or rewards. *Positive reinforcement* increases the likelihood that a desired behavior will be repeated. *Punishment* decreases the likelihood that an undesired behavior will be repeated. *Avoidance* means the likelihood of a desired behavior is increased by knowledge of both the reinforcement and punishment. *Extinction* means the removal of positive reinforcement to eliminate a now undesirable behavior that previously was a desirable behavior.

To use a Skinner-based reinforcement safety program, you would tell your subordinate what he or she must do to get his or her reinforcement. Awarding the reinforcement must be done as close to occurrence of the behavior as possible. Do not reward all employees in the same way. Make the actual reward based on measured performance. Failure to reward can be a *negative reinforcement* (decreases the likelihood that a desired behavior will be repeated). Therefore, it is important that you keep employees informed of why you are rewarding or not. Use punishment wisely and sparingly. Render punishment as soon after poor performance as possible. Reprimand only in private, not in front of other employees.

Competent discipline is a positive component in accident prevention. Good discipline leads to acceptable conduct. Decisions about conduct that might be difficult without sound discipline become simple with it.

Safety Attitudes

The culture of the old-style safety management is one of compliance with regulations. New safety cultures stress core values and go beyond compliance. These management teams impose higher standards on themselves and measure their performance accordingly. While the objective of the old-style management was avoidance of penalties, the new management objective is to achieve safety excellence based on values. Employees read culture and behave accordingly, though perhaps not consciously.

No safety process will be effective unless employees implicitly approve of the safety culture and of the methodologies of its processes.

Geller's ABCs of Safety Motivation

Activator
Behavior
Consequence

Geller's ABCs of Safety Attitudes

Affect
Behavior
Cognition

Motivate people but hold them accountable for achieving superior results.

In order to motivate, you need to do two things: 1) begin to change the culture if it impedes the safety process and 2) service the needs of your customers. Practically speaking, you cannot change the work culture by yourself. The work culture, as discussed above, is the practiced attitude of management and labor. All you can do is frankly but tactfully point out how management policies and the behaviors of all employees hinder the safety process. Behaviors are definable and changeable on their own without being forced or influenced. The first step is dispensing the knowledge of how attitudes affect behavior.

Next comes showing how this behavior affects productivity and safety.

Service the needs of your customers: who in this case are management and employees. These needs are fear, pride, recognition, involvement, participation, and competition. Posters and newsletters describing the more gruesome details of accidents will instill a healthy fear of unsafe behavior. But remember that fear only becomes a valuable teaching opportunity if the underlying causes of the accident are illuminated and a discussion of prevention options is invited. Pride in safe behavior is promoted by recognition. Employees have a need to be recognized, so photographs and stories in local newspapers and other forms of publicity will meet this need. Despite the grumbling and complaining, employees need to feel involved. They want to feel as if they participate in management of the facility. With respect to safety and health, get them involved with the safety, inspection, and investigation committees. Undertake campaigns to eliminate specific hazards and get as many employees involved as practical in these campaigns.

Do not forget that you, as well as your safety customers, have needs. Eckenfelder, in a recent article on culture, waxed philosophical, but hit the nail on the head in my opinion. Quoting from the *Bridadaranyaka Upanishad,* he wrote, "You are what your deep, driving desire is. As your desire is, so is your will. As your will is, so is your deed. As your deed is, so is your destiny." Quoting from the Christian scriptures, he wrote, "For where your treasure is, there will your heart be also."

What do you treasure? Is it safety? If not, why are you in the safety profession? What does your management treasure? If not safety, then show them how risky hazards and accident potential can adversely affect whatever their treasure may be. Since managers are your customers too, respect their treasure, even if that treasure is not compatible with the safety process. Remember it's easier to get people to change their behavior than to change their treasures, unless you are a pretty good sales person, or a damn good shrink. Even then, I am betting, the results will be mixed.

Conflict Management

Why do people do the risky or harmful things that they do? Why do they break safety rules? Why do they disobey directions to work safely? Why do they ignore warning signs? According to behavior scientist, Don Groover, we must apply the ABC model to change undesirable behavior. ABC, to Groover, stands for Antecedent, Behavior, Consequences, analogous to Geller's Activator, Behavior, and Consequence. The antecedent triggers behavior, which is an observable act, resulting in consequences. The consequences, not the antecedents, control behavior. Antecedents only predict behavior as well as they predict consequences. Figure 1-21 is Groover's Five Step Process for controlling behavior.

Figure 1-21
Groover's Process for Behavior Control

1. Define the crucial behavior.
2. Measure the frequency of the behavior.
3. Provide feedback to the employee.
4. Determine the underlying cause of the behavior.
5. Determine a solution.

Conflict arises from the apparent incompatibility between the supervisor and the safety function. The goal of achieving full production is apparently incompatible with the need to have employees wear protective gear. For instance, a common complaint of employees who fail to wear protective gloves for honing parts under an oil spray is that with the gloves on they cannot feel the part. Obviously feeling the part by the sense of touch is important to them in their perception of honing the part properly. This type of conflict is all too common in the safety field.

What can you do? First, present the employee and his or her supervisor with factual data. What will happen to his or her skin if gloves are not worn? Does another employee successfully hone parts with gloves on? What types of alternative gloves might be tried? Make reasonable recommendations. If these are rejected, do not give up. Put the facts and your recommendations in writing to your superior. Perhaps the superior can get the employee and his or her supervisor motivated.

Conflict-caused frustration comes in three types. *Approach-approach* frustration is created by two desirable conflicts, or conflicts that are equally attractive. *Avoidance-avoidance* frustration is caused by two equally disagreeable conflicts. The responses to avoidance-avoidance are vacillation, flight, daydreaming, or fantasizing. *Approach-avoidance* frustrations are caused by both attractive and repulsive conflicts with the same goal.

In order to cope with these kinds of conflicts, we must understand their four sources. The first source of conflict and probably the most frequent is goal incompatibility. Returning to the example, your goal is to prevent dermatitis. The employee's goal is to hone parts. This conflict presents a basic lack of agreement on the direction of an activity or criteria for evaluating a task. Either way, two goals are involved and the individuals having the separate goals become barriers to each other.

Another source of conflict is the availability of resources. We are either competing for scarce resources or jostling in line waiting for resources that are not available yet—one of these is the case most of the time.

Performance expectation is a third source of conflict. Are you waiting on someone else to proceed or complete a task? Is someone waiting on you? The organizational structure itself can be a source of conflict. Line and staff management generally have two different viewpoints that can lead to conflict.

Techniques for resolving conflict include avoidance, defusion, and confrontation. *Avoidance* is accomplished by either non-attention (ignore it until it goes away), physical separation (break it up), or limited interaction of the parties to the conflict (regulated, structured interaction). *Defusion* attempts to buy time. Smoothing defusion plays down the conflict. Compromise sacrifices some attractive parts of the conflict in order to end up with an overall agreeable situation. *Confrontation* is a mutual personnel exchange, such as temporarily swapping jobs. In a confrontation, subordinate goals are used to draw both sides into the picture. Confrontation ought to be a problem-solving session.

Leadership Styles

Sarkus reviewed literature on management styles that he finds lacking because moral safeguards, previously found in society, are missing today. In doing so, he noticed three types of leaders who can effect profound organizational change. He refers to these leaders as charismatic leaders, transformational leaders, and servant-leaders.

Charismatic leaders have long been shunned by the military, which lives or dies by teamwork and community spirit. Therefore, the military desires leaders who have all the attractive characteristics of a charismatic leader except for the magnetic personality. That is because the military knows that such a personality can disappear in an instant, by being killed or removed somehow. Besides, within such a regimented community no one personality should stand out that much more than any other, except by virtue of rank with its degree of responsibility and authority.

Transformational leaders are similar to charismatic leaders in effecting change, except they do not rely so much on personality alone, but on persuasion and position. Again, the military does not wish to have too many leaders who effect too much change at once. By and large, neither does industry, even though the current industrial buzz-words are "change," "innovation," "culture," and "paradigm paralysis."

The truest, most effective leaders are the *servant-leaders.* According to Greenleaf, who developed the term as an industrial concept, the servant-leader humbly serves without expecting to be served by those who follow. The servant-leader realizes that all individuals and groups, with all good intentions, sometimes make mistakes. This leader accepts rather than inflicts pain by working to correct mistakes in a learning rather than a punishing process, seeking to heal relationships, and reconciling differences. That is not to say that he or she is a cosmetic mender, but seeks solutions and resolution. The servant-leader is not one who avoids conflict and tension at all costs. Rather, he or she is nonjudgmental. This person would be a "let's pick up and move on from here, but let's learn from our mistakes" kind of person.

Professionalism and Ethics

Every profession has an ethical code by which the practitioners measure themselves. To begin this discussion, let us examine the Ten Safety Principles of DuPont, as summarized in Figure 1-22.

Any code of ethics will include responsibilities to the profession, to employers, to employees, and to clients. In a way, all these groups are customers of the professional.

Figure 1-22
DuPont's Ten Safety Principles

1. All injuries and occupational illnesses are preventable.
2. Management through all levels is responsible for preventing injuries and illnesses.
3. Safety is a condition of employment.
4. Employees have to be trained to work safely.
5. Safety and occupational health audits must be conducted.
6. All deficiencies must be corrected.
7. React to incidents, not just injuries.
8. Off-the-job safety is as important as on-the-job.
9. Preventing injuries and illnesses is good business.
10. People are the most important element of the safety and occupational health process.

The safety professional who is new on the job, or who comes to work one day asking why a quantum improvement cannot be achieved, needs to determine the root causes of the existing situation. Weber says he or she needs to:

- Secure evidence concerning the existing situation or condition.
- Identify standards or norms with which to compare existing situations or conditions.
- Recommend a plan of action to improve safety performance.
- Determine how to take the next step.

Most safety professionals do not have any problems accomplishing tasks 1 through 3. The glitch in the process, the one that causes so much frustration, is determining how to take the next step. Very often, the safety professional is a staff person with dotted-line reporting to one of the middle managers. He or she has no budgetary discretion of his or her own, and, though weighed with great responsibility, has no real authority.

This leaves the safety professional with one of several responses to take, assuming he or she does not go screaming out the door. These choices are to choose to get by, do as well as you can under the circumstances, or take an active planning stance.

If you choose to just get by, then you will only do what you have to do and then do that just good enough to pass muster. Quite honestly, this is the situation in many of our plants today. For whatever reason, but mostly just being tired of butting his or her head against brick walls, the safety professional at these plants has basically given up and no changes or improvements occur in the safety program.

Doing the best that you can under the circumstances is favored by technicians, those skilled in engineering or computer science who have safety responsibilities. They are going to organize their records and files to the nth degree, use computer programs to sift and sort and manipulate data, and dive enthusiastically into the big projects that are worthy of their time and effort, while ignoring all sorts of small, low priority bush fires. Two problems should be evident here. First, the employees and data about them can be sifted and sorted and filed all kinds of ways, but that does not make them any safer. Secondly, brush fires frequently become forest fires if not attended to.

The last general response by safety professionals—and perhaps the rarist—is to become a long-range planner. This individual realizes that not everything can be accomplished all at once, so he or she develops a plan to take care of all the problems in a reasonable and timely fashion, while using his or her own skills, if appropriate, or the help of others, if needed, to document and administer the process.

Risk Factors and Safety Awareness

The number one cause of occupational injuries and illnesses in the U.S. is the workers' inability to detect rising risk factors in their work environment, or, if recognized, the fact that they became concerned too late. A risk factor is neither a specific hazard nor an unsafe act. People are hazard hunters by nature, causing them to overlook risk factors, or clues, that are pointing to the ultimate appearance of some hazard or other.

Americans, as mentioned earlier, are inherent risk takers. We drive faster than the posted speed limit without compunction. We ignore other traffic rules routinely, while expecting everyone else to obey the same laws strictly.

A comedian once announced that those who drive slower than he does are idiots and those who drive faster are maniacs. It's all relative, isn't it?

Risk taking is encouraged from the time we are toddlers. Baby-Boomers, who entered the workplace after 1965, are imprinted with slogans that motivate them to take risks (See Figure 1-14.)

While it is not merely a matter of training, safety awareness must be instilled in modern employees, more so than in pre-Baby-Boom employees. The jury is still out, but from first encounters, the post-Baby-Boom employee seems pretty much like the Baby-Boomer with respect to risk taking and lack of respect and trust for authority. They also want to be safe, but they do not want to have to think about safety constantly. Whereas older workers could be counted on to look after themselves, given a few pieces of protective equipment, and engineering controls, younger workers do not want to be bothered or have a supervisor "looking over their backs." They don't care as much about the details as they do about the Big Picture. They want to participate in decision making, be consulted by management, and be asked for their opinions, which they tend to hold in high esteem. Heaven help the "intrusive" manager who asks them to constantly be on the vigil for some hazard or another.

Younger workers also want their managers to set the example. Do not ask the younger employee to do something you will not or cannot do yourself. Double standards were once accepted as the privilege of class and rank. Today, the death knell of a safety program will sound the instant a double standard is uncovered.

If the company president will not wear protective gloves, which are hot and make grasping an object difficult, to examine a few parts coming from a dip tank, do not expect the young employee to do it. Never mind logic—or the lack of it. If you want all the people assigned to a specific work area to wear respirators, you had better put one on the supervisor. And the boss had better put one on when he or she enters the departmental area, too.

Effectiveness of Safety Training

Training is important, but it is not everything. A panel of experts, assembled by OSHA to study four major industrial disasters, found that each of the four companies involved, highly recognized for excellent safety records, had invested a significant amount of training per employee per year, amounting to four percent of payroll hours. Unless employees are also motivated to learn and to apply their knowledge, unless they are motivated daily to be alert on the job, and unless they are motivated to think, even the best training program will have dubious value.

In one plant, three groups of employees were given the same training on a new cleaning material that was replacing an ozone depleter. EPA had required the phaseout of the old cleaning material and the plant complied quickly, settling on a new, more immediately hazardous material to employees. All three groups were given the same training on how to safely work with the new material. The first two groups made the change to the new material immediately, and with no trouble. The third group, however, did not experience the change right away, rather the new material was phased in during a four-month period. Everything went smoothly for the first two groups. No safety and health complaints were reported. The third group began filing complaints in droves. They were sullen and resisted the change. They refused to use new operating procedures for the cleaning machinery, which were designed to protect them from exposure. They steadfastly attempted to operate the degreaser with the old procedures. Yet, they complained persistently about their exposures and symptoms of exposure. Their work practices increased the dragout of the dangerous material and exacerbated their health and safety problems. When supervisors and managers used disciplinary measures to force them to comply with the new safety procedures, they filed a series of grievances, claiming harassment. They even conducted several brief work stoppages, though their union asked them not to. In general, they refused to cooperate, whereas the first two groups, in the same plant, used the new material safely and contentedly. The only difference, besides specific personalities involved, was the timing of the changeover.

Traditional safety training is mostly accomplished by recital. That is, compliance training is given in a classroom with a lecture, and perhaps a demonstration. Training aids may or may not be used. Occasionally, a performance test may be administered. Usually, however, recital training means watching a video and signing in on a sheet of paper.

In the new culture, safety training is values based, and leaders are trained to accept safety values as an integral part of their duties and responsibilities. The safety staff facilitates change, and shop floor learning matches desired norms. Values and activities related to safety performance are assessed.

Lischeid and others urge us to use Total Quality Management (TQM) principles and tools to identify safety problems and develop solutions to control hazards. TQM training, they point out, teaches employees how to use problem-identification and problem-solving tools such as affinity diagrams, arrow diagrams, bar graphs, cause-and-effect diagrams, statistical control charts, decision matrices, flow charts, pareto charts, process maps, force-field analysis, scatter diagrams, and fault and event tree diagrams. The safety professional's job is to teach people how to use these tools for identifying and solving safety problems.

Geller's ABCs of Safety Education

Attention
Behavior
Commitment

Training Program Elements

A core safety and health training program should provide employees with the background they need to understand more specialized training. A five session core program is recommended to get your safety program off the ground and flying:

Introduction
Accident causes and control
The role of behavior in safety
Accident prevention
Personal protective equipment

The introductory session covers the functions of the safety organization and demonstrates how communications flow up and down the chain of command. Employees should understand how they fit into the overall safety process. They should understand their basic rights and responsibilities under the OSHAct. Be patient, but defer complaints to the Safety Committee. Try instead to generate enthusiasm for the process of safety and to elevate safety consciousness among employees. Encourage them to obey safety rules and follow safety procedures and to report all accidents or near misses to their supervisors.

The second basic training session covers accident causes and control. Familiarize employees with the types of conditions and events that lead up to an accident and show them how the safety process intervenes to control the hazards.

The basic training continues with a primer on behavior and safety. What is the role of people in accidents and accident prevention? How does the company plan to handle employees who work unsafely?

The fourth session of the basic training program examines the company accident prevention systems.

The fifth and final session covers the reasons for using personal protective equipment (PPE). Various types of equipment should be shown to the trainees, explained, and proper use demonstrated.

Continued training on specialized subjects helps prevent accidents and OSHA recognizes this as a fact. Many OSHA standards contain explicit requirements for training. Training is implied in other standards as well. Figure 1-23 lists some training topics. These differ from the basic safety and health training in that these topics are specialized and may not apply to all hands.

Figure 1-23
Special Safety and Health Training Topics

Access to records
Bloodborne pathogens
Confined space entry
Cranes and hoists
Driver training
Electrical safety-related work practices
Emergency procedures
Ergonomics
Evacuation procedures
Fire brigade duties and operations
Fire prevention

Figure 1-23 *(continued)*

Fire protection equipment
Fire safety
First aid
Forklifts
Hazard communication
Hazardous materials handling
Hearing protection
Housekeeping
Ionizing radiation
Lifting and material handling methods
Lockout/tagout
New employee indoctrination
Personal protective equipment
Power equipment operation
Powered platform operation
Process safety management
Respiratory protection
Storage of hazardous materials

Ergonomics training is outlined in Table 1-1 as suggested by NIOSH.

Training may not eliminate suspicion, resentment of authority, distraction, anger, impatience, indifference, boredom, or behavior-based foolish thinking, but it helps. Safety depends a great deal on management's good faith effort to eliminate or moderate workplace hazards, but it also depends greatly on the desire of employees to work safely. The overall goal of safety training is to stimulate every employee to think safety beyond his or her usual work environment. The expectation is that well-trained employees will not commit unsafe acts when questionable situations arise. Therefore, the Safety Coordinator as a representative of management must continually sell safety to all hands, wage earners and management alike.

Table 1-1
NIOSH Ergonomics Training Guidelines

Training	All hands	Employees at risk	Supervisors of people at risk	Lead employees/ supervisors	Ergonomic task team
Informal awareness	X	X	X	X	X
Formal awareness & job specific		X	X	X	X
Job analysis & risk factor control				X	X
Problem-solving & team approach					X

Source: *Elements of Ergonomics Programs,* NIOSH, January 1997 Draft

Performance Objectives

A good training program will be designed around performance objectives that are expected of the trainees upon completion of the training. Examples of performance objectives are:

Upon completion of this course, the trainee will be able to list five key characteristics each of Class A and B fires.

An effective performance objective has a performance, a condition, and a criterion. The performance is what the trainee is expected to accomplish after the training. In the above objective, the trainee is required to be able to make a list to demonstrate retained learning about the two types of fires. The condition is the situation under which the performance is expected to occur. The condition above is that the trainee will have had the benefit of the training about types of fires in order to make the list. Some objectives will not have a condition or criterion element, but all will have a performance element, if written properly. For an in-depth understanding of training objectives, the following references of the Gebrewold and Sigwart article are provided:

H.H. Jacobs, *Interdisciplinary Curriculum: Design and Implementation,* Washington, D.C.: Association of Supervision and Curriculum Development, 1989.

Robert F. Mager, *Preparing Instructional Objectives,* Belmont, CA: Davis Lake Publishers, 1984.

John Miller and Wayne Seller, *Curriculum Perspective and Practice,* New York: Longman, 1985.

G. Posner and A.L. Rudnitsky, *Course Design: A Guide to Curriculum Development for Teachers,* New York: Longman, 1986.

Table 1-2 summarizes some excellent advice by Topf for dealing with distractions that prevent people from learning and applying safety training in their work. According to Topf, it is essential that the safety trainer be aware of these potential distractions and address them before presenting the subject material. Figure 1-24 below is an OSHA model for training.

Figure 1-24
OSHA Training Model

1. Conduct job analysis to determine if training will solve a specific problem.
2. Identify training.
3. Identify goals and objectives.
4. Develop task-oriented learning activities.
5. Conduct training.
6. Evaluate workers' opinions, observe workers' behavior, and measure improvement.
7. Improve program as necessary.

Table 1-2
Topf's Solutions to Training Distractions

Distractions	Solutions
Past training produced little or no change in safety culture.	Emphasize current genuine commitment to safety and personal well-being.
Productivity slips during training.	Reinforce management's commitment to making safety a top priority; productivity gains should follow.
Employees have heard it all before.	Listen. Emphasize benefits of the skills/information today.
Training is on employees' personal time.	Listen. Encourage employees to make time worthwhile.
Morale is low; confusion reigns due to downsizing/ restructuring.	A safe workplace and open dialogue improve morale. Safety training adds value to a smaller workforce.
Emphasis is on injury rates and totals, not on personal well-being.	The health and well-being of employees is a high priority for their employer.
Tension or poor communication	between labor and management. Involve employees and management in training development.

CONCLUSION

Whatever your management philosophy may be, your role as safety manager is to protect the resources of your company, particularly the human resources. One goal is common in the profession: zero fatalities and zero lost work days.

References

Bailey, Chuck. "Managerial Factors Related to Safety Program Effectiveness: An Update on the Minnesota Perception Survey." *Professional Safety.* August 1997, pp. 33-38.

"Behavior Analysis Model Provides Insights for Industrial Hygiene, AIHCE Told." *The Synergist.* June/July 1996, p. 21.

Brauer, Roger L. *Safety and Health for Engineers.* New York: Van Nostrand Reinhold, 1990.

Daugherty, Jack E. *Industrial Environmental Management: A Practical Handbook.* Rockville, MD: Government Institutes, Inc., 1996.

Deming, W. Edwards. *Out of the Crisis.* Cambridge, MA: Center for Advanced Engineering Study, 1986.

Eckenfelder, Donald J. "It's the Culture, Stupid." *Occupational Hazards.* June 1997, pp. 41-44.

"Elements for an Effective Safety and Health Program." Appendix D.

Elements of Ergonomics Programs. NIOSH. January 1997 Draft.

Gebrewold, Fetene, and Dennis F. Sigwart. "Performance Objectives: Key to Better Safety Instruction." *Professional Safety.* August 1997, pp. 25-27.

Geller, E. Scott. "The ABCs of Safety Management." *Industrial Safety & Hygiene News.* September 1992, p. 15.

Geller, E. Scott. "Safety Coaching." *Professional Safety.* July 1995, pp. 16-22.

Geller, E. Scott. "Who's Responsible for Safety?" *Industrial Safety & Hygiene News.* October 1992, pp. 12-13.

Haimann, Theo and Raymond L. Hilgert. *Supervision: Concepts and Practices of Management.* 2d ed. Cincinnati: South-West Publishing Company, 1977.

Hansen, Larry. "Management Issues: Rate Your B.O.S.S. (Benchmarking Organizational Safety Strategy)." *Professional Safety.* June 1994, pp. 37-43.

Hansen, Larry. "Safety Management: A Call for (R)evolution." *Professional Safety.* March 1993, pp. 16-21.

Heath, Earl D. "The Evolution of Safety Management in the United States." *Professional Safety.* October 1986, pp. 15-21.

Hughes, Richard G. "The Safety Management Maturity Grid." *Professional Safety.* June 1985, pp. 15-18.

"Jumping to Management not Mission Impossible." *The Synergist.* June/July 1996, p. 12.

Killimett, Pat, Jim Spigener, and Stan Hodson. "The Role of Senior Managers in a Safety Change-Effort." *Occupational Hazards.* August 1997, pp. 47-50.

Kohn, James P., Mark A. Friend, and Celeste A. Winterberger. *Fundamentals of Occupational Safety and Health.* Rockville, MD: Government Institutes, Inc., 1996.

Krause, Thomas R. "Driving Continuous Improvement in Safety." *Occupational Hazards.* February 1995, pp. 47-50.

Kuhn, Todd A. "Paradigm for Safety." *Professional Safety.* September 1992, pp. 28-32.

LaBar, Gregg. "Safety Management in Tight Times." *Occupational Hazards.* June 1993, pp. 27-30.

Lischeid, William E. "TQM & Safety: New Buzz Words or Real Understanding?" *Professional Safety.* June 1994, pp. 31-36.

Manuele, Fred A. "Principles for the Practice of Safety." *Professional Safety.* July 1997, pp. 27-31.

McElroy, Frank E., Ed. in Chief. *Accident Prevention Manual for Industrial Operations: Administration and Programs.* 8th ed. Chicago: National Safety Council, 1981.

Minter, Stephen G. "Creating the Safety Culture." *Occupational Hazards.* August 1991, pp. 17-21.

National Safety Council. *Accident Prevention Manual: Volume 1, Administration and Procedures.* 8th. ed. Chicago: National Safety Council, 1986.

OSHA Manual. The Merritt Company, 1988.

Peterson, Dan. "Establishing Good Safety Culture Helps Mitigate Workplace Dangers." *Occupational Health & Safety.* July 1993, pp. 20-24.

Peterson, Dan. "Integrating Safety into Total Quality Management: Changing Safety." *Professional Safety.* June 1994, pp. 28-30.

Quazi, Moumin. "Body and Spirit: Improving Safety Performance by Changing Behavior and Changing Attitudes." *Occupational Health & Safety.* July 1992, pp. 34-39, 52.

Ragan, Patrick T. and Brooks Carder. "Systems Theory and Safety: Safety as a System." *Professional Safety.* June 1994, pp. 22-27.

Sarkus, David J. "Servant-Leadership in Safety: Advancing the Cause and Practice." *Professional Safety.* June 1996, pp. 26-32.

Scotti, Marie. "How to Supervise a Positive Discipline Program for Safety." *Professional Safety.* April 1986, pp. 25-27.

Smith, S.L. "Occidental Chemical: Making Changes for the Better." *Occupational Hazards.* May 1992, pp. 65-68.

Smith, Thomas A. "What's Wrong with Safety Incentives?" *Professional Safety.* May 1997, p. 44.

Topf, Michael D. "10 Lessons for Safety Training." *Occupational Hazards.* August 1997, pp. 37-40.

Traulein, Barbara A. and David W. Milner. "Work System Design: Creating a New Safety Paradigm." *Professional Safety.* December 1994, pp. 27-31.

Tyler, W. Walter. "Total Involvement Safety." *Professional Safety.* March 1992, pp. 26-29.

Ulmer, Phillip E. *The Basic Seminar in Resource Risk Recognition.* Eagle River, AK: Northwest Safety Management, Inc., 1995.

Veltri, Anthony. "Transforming Safety Strategy & Structure." *Occupational Hazards.* September 1991, pp. 149-152.

Walton, Mary. *The Deming Management Method.* New York: Putnam Publishing Company, 1986.

Weber, J. Owen. "Developing a Comprehensive Safety Program." *Professional Safety.* March 1992, pp. 33-38.

Weber, J. Owen. "The Front-line Supervisor's Role in Safety." *Professional Safety.* May 1992, pp. 34-39.

Wilkinson, Bruce S. "The PRIDE Management Principles." Mississippi Manufacturers' Association 3rd Annual Safety Conference and Exposition, May 7, 1997.

Wilkinson, Bruce S. "Proactive Safety & Health Strategies for the 21st Century." Keynote Address: Mississippi Manufacturers' Association 3rd Annual Safety Conference and Exposition, May 6, 1997.

Winn, Gary L. "In the Crucible: Testing for a Real Paradigm Shift." *Professional Safety.* December 1992, pp. 30-33.

2

Safety Management Reality

Understanding the theories of safety management is necessary for ultimately understanding the processes operating to counter the efforts of the safety professional and others attempting to make the workplace safer. These theories are vital and, as safety professionals, we must strive to utilize them, apply them in new ways to improve on safety performance, and communicate our successes and failures to others who might benefit from the knowledge of our experience. Make no mistake, I subscribe to these theories so long as they are dynamic and useful paradigms, and I recommend that you do too. However, the application of these theories can leave us short of our goals and objectives—if they are all we know, and we have no idea how to implement goals and objectives. The "how to" is the meat of this chapter.

How Safety Really Happens

Safety, in practice, is one hundred percent conformance to an employee's real personal security needs. You may expand this to meet the security needs of all your company's resources. Safety has nothing to do with OSHA's needs. Nor does it have to do with management's needs as senior managers sometimes perceive the needs of management. It does, however, have everything to do with the need to protect the resource investments of your company, whether senior managers perceive that or not.

An accident is an undesired event that results in harm to people, damage to property, or loss to process. An accident results when a resource contacts a harmful substance, agent, or source of energy: mechanical, electrical, chemical, radiant, thermal, acoustical, or biological.

Harmful Energy

Harmful energy in the form of ionizing or non-ionizing radiation is present in many workplaces. X-rays, alpha particles, beta particles, and gamma rays are emitted from radioactive isotopes. Also, x-rays may be emitted from by the safety manager. This is far from ideal, but that is the way it is.

Safety as a Team Effort

Safety as a team effort is the ideal situation. Ideally, too, the line supervisor is the glue that keeps the team together and its chief monitor.In a manufacturing plant, the team consists of such diverse elements as engineering, manufacturing, quality control, purchasing, maintenance, and industrial relations. This list is not meant to exclude anyone. Engineers evaluate and design safe processes, equipment, and machinery. Manufacturing implements safety changes in process, tool design, and coordination and scheduling of production. Quality tests and inspects materials and finished products, and conducts studies to determine whether alternative designs will improve safety, quality, and productivity. Manufacturing engineers participate in this study for safety and productivity. Purchasing ensures that materials and equipment meet standards and obtain Material Safety Data Sheets (MSDSs). Maintenance performs preventive maintenance to prevent equipment and machines from becoming unsafe due to mechanical or electrical malfunctioning. Maintenance also follows safety rules and practices in performing its work. The industrial relations staff influences workers to obey safety rules, follow safe work practices, and supports line supervisors in discipline matters, especially as related to safety.

The worker himself or herself is a key player on the safety team. He or she directly monitors and gathers information on malfunctioning safety devices or procedures. As an information processor, he or she must decide on a course of action. Preferably, any accident or near miss, no matter how insignificant it may seem, must be reported to the supervisor. The worker performs a control function in that he or she takes action to correct safety malfunctions within his or her ability, such as housekeeping. Equipment and machines, therefore, must be properly designed and built to support the role of the worker. Also, equipment and machinery need to be designed to be easily maintained and operated. Finally, the workplace environment—that place a worker lives for nearly one-third of his or her life—must be worker friendly. The environment must not be a distraction to the worker, for purposes of productivity, quality, and safety. The environment must also be friendly to the worker. The way the workplace is laid out, housekeeping, illumination, temperature, humidity, noise, chemical exposure, and more must be taken into consideration.

Controlling Harmful Energy

All dangerous exposures may be considered as some form of uncontrolled energy. The successful team devises strategies for energy control. One way to do this is to prevent the accumulation of harmful energy:

- Set limits on noise
- Set limits on temperature
- Set limits on voltage
- Store combustibles properly
- Limit heights of ladders

Another strategy for energy control is to prevent the release of harmful energy:

- Better design engineering
- Containment of vessels
- Toe boards/guard rails
- Seat belts
- Lock outs

Yet another strategy is to modify the release of the energy:

- Shock absorbers
- Safety valves
- Blow-out panels
- Less incline on ramps

You can also separate the humans from the energy:

- Move employees further from the energy source
- Limit exposure time
- Use long-handled tools
- Substitute safe materials

Providing barriers is another way to control energy:

- Sound attenuating materials
- Fire doors
- Machine guards
- Personal protective equipment (PPE)

Raising the injury or damage threshold is yet another strategy for energy control:

- Improve equipment design
- Acclimate to heat or cold

- Immunize against diseases
- Wellness programs
- Facility exercise and warm-up programs

You can also establish contingency resources:

- Early detection of energy resource
- First aid treatment
- Hazard communication training
- Emergency showers and eye wash stations
- Decontamination plans

As an example, consider a power mower. It has a one cylinder gasoline engine and is started by a manually-wound spring starter. It is a push-type rotary mower. A single rotating blade cuts the grass and expels it through an opening in the side of the mower body. Blade movement creates a vacuum to lift and expel the cut grass. List the sources of energy you can identify in this system.

Mechanical:
Electrical:
Thermal:
Chemical:
Radiant:
Acoustical:

To summarize the strategies available to control harmful energy, Figure 2-1 lists them.

Figure 2-1
Strategies to Control Energy

Prevent the accumulation.
Prevent the release.
Modify the release.
Separate the resources from the energy source.
Provide barriers to block or attenuate.
Raise the injury or damage threshold.
Establish contingency response.

In Figure 2-2, match the energy source with the situation:

Figure 2-2
Energy Sources

A. Mechanical	___ Bright light from welder
B. Thermal	___ A moving front-end loader
C. Chemical	___ Rust on a supporting wire
D. Electrical	___ Machine noise
E. Radiant	___ Static spark
F. Acoustical	___ Blistered paint behind stove
	___ Dermatitis from cutting oils
	___ Fall on ice in parking lot

In Figure 2-3, match the corrective action with one of the strategies summarized in Figure 2-1.

Figure 2-3
Corrective Actions

1. Put safety valve on a pressure cooker.
2. Develop a plant emergency plan.
3. Place screens around an electric welder.
4. Exercise the back muscles.
5. Wear a hard hat.
6. Don't buy any gasoline for the mower.
7. Secure the compressed gas bottle with a chain.
8. Inspect fire extinguisher monthly.
9. Conduct first aid training.
10. Store flammable liquid in a remote shed.

Hazard Control

Effective safety and health programs systematically identify, evaluate, and prevent or control general workplace hazards, specific job hazards, and potential hazards that may arise from foreseeable conditions. Compliance with the law, as well as specific OSHA standards, is an important and necessary objective, but effective programs look beyond rules and regulations to address all hazards. Effective programs seek to prevent injuries and illnesses whether or not compliance is at issue. The extent to which the programs are written is less important than how effectively they are practiced. At large work sites and complex hazardous operations, written guidance enhances clear communication of policies and procedures. Written programs also better define priorities. Finally, written documents help ensure consistent and fair application of rules.

Teamwork just does not happen overnight. In reality, a progression takes place unless the management team has stagnated in the mire of status quo and paradigm paralysis. The phases of safety management growth are the same as growth in productivity and quality. Hughes, who reviewed the works of the quality guru Crosby, lists the phases of growth as uncertainty, awakening, enlightenment, wisdom, and certainty.

An uncertain management is confused and uncommitted. It lacks knowledge of safety as a tool it can use. In this phase of safety development, the safety professional is seen as a policeman. The safety function is located deep in an operating department: manufacturing, operations, industrial engineering, maintenance, or human resources. Accident investigators and inspectors are always looking for *who* caused the problem instead of *what* caused the problem. An uncertain management and labor are locked in a struggle and labor uses safety issues as a hammer on management. OSHA compliance inspectors, corporate auditors, and even insurance loss control representatives are considered the bad guys who are out to get management.

Safety Management

The awakening management team is starting to understand that good safety management is beneficial, but is unwilling to devote time and money to let it happen. At this point, a stronger safety leader is recruited or the old one is given more authority, but emphasis on production still overshadows safety, despite the high potential cost of noncompliance. The safety committee suddenly is seen as the answer to all the plant's safety problems. Unfortunately, safety issues are still seen as redheaded step-children and take a back seat to productivity and quality. The awakening management typically delves into motivational promotions and incentive programs, which ultimately loose steam and fail because they are mainly gimmicks, hoopla, and rah-rah. The employees quickly see through them and understand that the more things change the more they remain the same. At this point in time, management is either completely demoralized and falls back to an uncertain stance or it digs deep into its spirit and emerges in the next phase.

Enlightenment happens when management realizes, as the employees already have, that all its efforts thus far have been merely to appease its own ego. A decision is made to bite the bullet and establish a world-class safety management process. A genuine safety department is established as a balanced, well-organized function. It is empowered to lead the new safety crusade so management may atone for its sins.

The safety department is even given resources. Hallelujah! The wonders of divine intervention!

Problems are faced openly for the first time. Line supervisors are finally held accountable for resolving current problems and preventing future ones. The potential cost of noncompliance is getting noticed.

Management teams in the wisdom phase wonder why they used to have safety problems. Problems are routinely handled at the lowest empowered level as they appear. The safety manager is now the safety director, one among the senior managers of the plant. The management call to constantly improve shifts to other weak areas—quality or productivity. This is a critical time, because if the safety process falls into complete inattention, and a temptation to do so will exist at this point, your safety effort may be set back considerably.

The final phase is safety certainty. This company knows exactly why they no longer have festering safety problems. Safety management is an absolute vital part of the management team. The safety manager is equal to production manager and quality manager. The prevention system is so advanced and effective that few, if any, accidents occur beyond first aid cases. The potential cost of noncompliance is almost zero. Safety is the corporate religion and the safety manager is its high priest.

Would you like to be here? Start assembling a team to help you get the growth process off top dead center. Meet requirements. Identify and eliminate noncompliance. Improve continuously. Do not rest. Meeting requirements today has no bearing on being in compliance fifteen years from now.

The Safety Team

The Cheerleader—Provides the Energy

The cheerleader is an executive or senior manager with clout who continually promotes and sells safety in the organization. That may be you, if your position and personality are up to the job. The cheerleader is too important not to recruit one, if your position or personality is not up to it. The cheerleader constantly mentions safety benefits to other managers on the leadership team. This person drives home that productivity, quality, and safety are fully integrated. Everyone is aware of the benefits of a good safety effort, because this person brings it up all the time in many appropriate and effective ways.

The Champion—Provides the Vision

At industrial plants with successful safety processes only the best people are placed in safety positions. Their leader is an influential executive who "goes to bat for safety," as Eckenfelder says. The champion is your best batter. Sometimes the cheerleader and the champion are one and the same. If one must be sacrificed, let it be the cheerleader. The champion is too vital to your success. This senior manager is your ramrod. Your main battle tank. Your big guns. When you are losing the battle, appeal to this person for assistance because he or she is the one who can get things done.

The Mentor—Provides the Knowledge and Wisdom

You need a mentor. You need an internal mentor and probably an external mentor. The external mentor is your close friend and advisor who can guide you in your career. You need this person for your own sanity and mental health. This mentor and friend should be a senior safety professional who speaks your language and has been where you are.

You need the internal mentor for an entirely different reason. This mentor need not be a safety professional, but needs to be knowledgeable about the inner workings of your company. He or she must be an experienced veteran of some of the larger battles that have occurred within the halls and shops of your company. He or she must be intimately familiar with the personalities who are barriers to an effective safety program. The mentor must also be familiar with company policies and procedures.

The Safety Professional: The Right Stuff

Where do you fit in with all these team roles? You provide the work. As the safety professional, you coordinate all the activities concerned with safety. Certain things are essential for your success, but more importantly, for the success of the safety program for which you are responsible. I call this the "right stuff" and I wish it were as simple as having some stuff in a little box that is always available in tough times. Not so, but where successful safety processes are you will find that management and workers are focused on shared beliefs, values, and cultures. This focus, which must be your focus too, drives all legitimate efforts and is the precursor of productivity, quality, and safety.

Common Awareness of Culture

The culture of an organization gets its life from the top, as we discussed in the previous chapter. Work culture influences the degree of safety or absence of risk achievable in a given plant. The amount of risk acceptable to the leadership team reflects the corporate culture and represents an asymptotic goal. You may get near it, but you'll never achieve it. The problem is that this risk goal is far too high in some industrial facilities. Many managers are willing to accept far too much risk.

Leaders form and sustain cultures. But when the work force is only subconsciously aware of their culture, and it seems to be anti-safety, the whole plant will be at cross-purposes with the safety professional. This is too often the case. Employees pick up on the signs and behaviors of top managers, and quickly pin-point their true stand on safety. Written safety policies and programs mean nothing when these negative signs abound in your culture. Erickson summarizes how values reflected in management opinions, actions, and behaviors affect culture and I have put these into Figure 2-4.

Therefore, as the safety professional, look for every opportunity to trot out the top one or two managers. Get them to accompany you or the safety committee on a walk-around safety inspection. Get them to make indepen-

Figure 2-4
How Management Values Are Transmitted to Employees

1. Through the things management does
2. Through the things management pays attention to
3. Through the things management ignores
4. Through measures and controls management uses
5. Through the manner in which management responds to crisis

dent inspections from time-to-time, and to communicate the results to you. If the top two are interested in safety, asking employees and their supervisors about correcting this or that item, and praising people where appropriate, you will get the attention of all but the recalcitrant few, who are not to be confused with the same few the U.S. Marines have been looking for all this time.

This applies to the safety professional as well as to everyone else. In order to communicate with management, the safety professional must speak the right language. He or she should not be ignorant of the bottom line and return on investment, or whatever term his or her company uses for that financial measure.

Schaechtel says to start with a philosophy and I have to agree. Figure 2-5 summarizes the elements of a safety philosophy according to Schaechtel.

The safety professional should understand how the company manufactures its products. This may take time. Begin learning the first day on the job by touring the operation from a product flow viewpoint. What are the raw materials that go into the product? What other materials are used in the manufacturing process, but do not end up in the product? Where does waste go? How is your product priced? What are the critical pricing issues? What is your company's customer base? How does manufacturing the product affect them? How does the product or its use affect them? You do not need to be an environmental manager or product safety manager, but you should know and understand the interrelationships between you and these two professions.

What are your company's goals and objectives? Can you rattle them off? Are the goals and objectives of the safety department, and your personal

Figure 2-5
Elements of a Safety Philosophy

1. Safety is a line management responsibility.
2. Safety standards are written documents that define safe procedures and management practices.
3. Training ensures that everyone understands and practices safety standards.
4. Audits evaluate the implementation effectiveness of standards.
5. Investigations detect problems in the implementation of procedural responsibilities, management practices, training, and auditing.
6. Involvement builds ownership of safety practices.

goals and objectives, in alignment with the company's goals and objectives? Understand the purpose and mission of the safety professional. Know how what you do contributes to the goals and objectives of your company.

As the safety professional, assuming you do not play some other role in the organization, you can exert the most influence on the content of the safety program, which probably has the least influence on the safety culture. Employees just do not pay much attention to large binders full of policies, procedures, and written programs.

You can have a moderate degree of influence on the selection or modification of production equipment and processes, which will have more impact on the overall safety culture than the safety program does. The OSHA inspector likes to see those written programs, though. So, your interaction with maintenance, process engineering, design engineering, and plant engineering will pay dividends in effectiveness.

You can have some influence over individual employees, but, unless you are a charismatic personality, don't expect to influence many. Unfortunately, some may never give you their trust. Many will have their own notion of what being safe means. No amount of monitoring or measurements of stressors will convince them they are safe. Let this latter group win some small battles from time-to-time, and, I have found, that you will get along tolerably well with them. Be careful about patronizing them, though, or you will alienate them permanently. I have done that, too, though with, what I thought to be, good intentions. The point is that I had to get other people to have meaningful feedback sessions with them, due to the fact that I had alienated them. Regardless, your influence on individual employees has limited influence on the overall safety culture.

Where you could get the most for your effort, you will have the least influence: the management style. More than anything else, the management style influences the safety culture. And while this statement is certainly not cast in stone, you as the safety professional probably do not have the wherewithal to affect the management style. Unless, you just happen to be the CEO in addition to being the safety professional.

Common Awareness of Values

The safety program will never get off the ground unless management is committed to the core values of safety. If management listens to employees and responds to their legitimate needs, this commitment is demonstrated. This management team will provide safety awareness to some degree in every activity. Committed management invests in technology to enhance safety. Committed managers take personal initiative for constructive change in the safety posture of the overall organization. Managers who are committed to safety behave ethically and respect the dignity of individual employees. They are responsible corporate citizens. They build teamwork. In promoting safety, they reach across functions, departments, and work cultures.

Determine how your company's core values apply to safety and communicate this to your employees. If you do not know what the core values of your company are, ask yourself, what values do we have as a company that distinguish us from others?

Are you customer oriented? Do you listen to your customers and respond to their needs? Then be safety customer oriented in the plant and teach your employees to be safety customer oriented. If the life they save is their own, then they are their own safety customer. If that life saved belongs to someone else, then that person is the safety customer. Also, the spouse and children, extended family, and friends of each employee are safety customers. The management and stockholders are safety customers, too. They stand to lose a lot of money over a fatality or serious maiming. That may sound cold, but it is a fact of life, and one that any boardroom can understand, no matter what their politics, religion, core values, code of ethics, or morals. Another safety customer is OSHA, which will require that your services meet specific standards, by the way.

Do not forget the essential safety customer: the worker. As industrial safety professionals, the safety and well-being of the worker is what we are all about. Therefore, get out and listen to your workers. What are their safety concerns? What are they afraid of? What are they not afraid of that they should be? Why don't they wear personal safety equipment? Why does so-and-so behave the way he or she does? Finding out these things will not be without some pain, grief, frustration, and other unpleasant emotional reactions on your part. I'm sorry—did management promise you a rose garden if you took this job? Some workers will argue with you, some will yell at you, some will not trust you, and many will treat you as the enemy, but getting out and listening is part and parcel of your job. It is what you do as a safety professional, and you cannot do the rest of your job without face-to-face time with the workers. They are your customers. Treat them as you would want to be treated, for instance, when you next deal with the Customer Service Department of a major department store about defective goods you have purchased.

Is your company quality oriented? Do you have an ISO 9000 series certification? As discussed in the previous chapter, quality, safety, and productivity are all parts of a whole. If any one of these parts is missing, then the whole suffers. Every slogan, motto, buzz word, measurement process, improvement process, and participative management system that is applicable to quality is just as applicable to safety. Substitute the word "safety" for "quality" in all the literature and you have the safety version of a quality policy. Substitute "quality-safety-productivity" for "quality" in the same literature and you have a complete policy. Go among the task teams and participative management teams and let them know how safety impacts their activities. Show them where they can improve safety in their workplace. As the safety professional, deservedly or not, you are the guru of safety. Maybe you do not receive much in the way of respect right now, but if you start teaching and showing, doing the things a guru does, then you will win the respect of others.

Is your company technology oriented? Is technology one of your values? Does your company invest in technology to enhance productivity and effectiveness? Visit with the designers and planners of manufacturing processes, of buildings and structures, and of products, and discuss ways to improve safety before people are exposed to any hazards.

Is innovation your ideal? Do you encourage employees to take personal initiative for constructive change? Then encourage them to take personal initiatives to effect safety improvements. Reward them for implementing safer procedures, for making modifications to the workplace that improve safety, or for recommending a needed change to management, even if it did not make this year's budget.

Integrity? Is that your value? Conduct yourself ethically in all situations and require everyone else to do so, too. It does not take capital punishment to correct an unethical idea. Many times a person is merely exploring options out loud when he or she mentions some unethical alternative. Simply tell him or her that the action just mentioned is not an option because it is not ethical, but do so without making accusations and rejoinders. Respect each person's dignity and treat him or her as if you were in his or her shoes at the moment. Keep your promises and do not promise what is not in your authority and power to keep. Tell the truth even when it is painful to you or the other person. Diplomacy has no place in the safety process, but truth is supreme. The best way to develop integrity is to cultivate leadership traits in yourself and others. That is not to say that there are no corrupt or wicked leaders, but good leaders, in the finest sense of the word, are men and women of integrity.

Whatever your company values are, start thinking beyond compliance. As long as everything you do is aimed merely at complying with the OSHA standards, you will be like a puppy chasing its tail. You will have a grand chase, but you won't get anywhere. But, I've said that before, haven't I?

Team Authority

Employees, through involvement teams, must be given the authority to take care of their own areas with respect to productivity, quality, safety, and customer service. Unless you, as the safety manager, want to be the safety policeman forever, empower your employees to act when they are able to.

Self-Responsibility

Assuming responsibility for safety is a fundamental requirement of a successful safety process. Employees must accept responsibility for getting done what needs to be done, otherwise the safety process is merely a vehicle by which labor can bash management for not caring and not being effective. Since the safety manager is probably overworked already, delegate some of the management responsibility to others. Let a volunteer fireman lead the emergency response team. A maintenance supervisor can manage lockout/tagout. Likewise, a specialist can be put in charge of respirators.

CONCLUSION

Neither OSHA compliance, nor documentation, nor faddish gyrations cause safety. Safety happens because people have a passion for it and a commitment to achieving it. When you go into a plant where safety is practiced well, you will be able to feel it. A safety presence will come at you from every angle.

REFERENCES

Eckenfelder, Donald J. "It's the Culture, Stupid." *Occupational Hazards.* June 1997, pp. 41-44.

Eckhardt, Robert. "Practitioner's Influence on Safety Culture." *Professional Safety.* July 1996, pp. 23-25.

Erickson, Judith A. "The Relationship between Corporate Safety Culture & Performance." *Professional Safety.* May 1997, pp. 29-33.

"Jumping to Management; not Mission Impossible." *The Synergist.* June/July 1996, p. 12.

Manuele, Fred A. "Principles for the Practice of Safety." *Professional Safety.* July 1997, pp. 27-31.

Schaechtel, Don. "How to Build a Safety Management System." *Professional Safety.* August 1997, pp. 22-24.

3

FIXING THE MANAGEMENT SYSTEM

The problem with the traditional safety program is that it is based on a model that is more than fifty years old. As Eckenfelder points out, the traditional safety program is fear driven. Safety audits are used as its principle tool to imitate the regulatory compliance inspection. Unfortunately, traditional safety programs are culture blind. Traditional safety programs are insensitive to both management and labor. Measurement, the lifeblood of management, is after-the-fact. Creativity, the lifeblood of labor, goes unrewarded. Worse yet, traditional safety programs institutionalize bad ideas.

Can this awful situation be fixed? Yes, it can. How can it be fixed? I do not presume to have all the answers, or even the best answers, but for the sake of discussion, let's examine some ideas.

SYSTEMS APPROACH

Traditionally, a general safety program consists of a systematic process that promotes communication of hazard information up and down the line. Each person is responsible for safety, but management will assign a person as the overseer of the process. A complete safety process then consists of a Safety Coordinator, one or more safety committees, regularly scheduled safety meetings, a written safety policy, and general safety rules, as well as specific procedures for safety and health matters. We will discuss these later, but for now let us examine the systems approach.

The systems approach is neither behavioral nor management nor engineering. In the systems approach to safety, safety is seen as a process that takes place in the workplace. It may be a defective process. It may be an incomplete process. The traditional approaches to safety talk about programs or an agenda and timetable of things that have to be accomplished. Traditional safety has a docket that must be adhered. It the safety docket is adhered to, somehow safety happens. At least, that is the theory.

In the systems approach, safety is a dynamic process that each of us is a participant in. Safety is going to happen with or without us. We need no docket, no agenda, no timetable because safety will happen. Unless we cooperate with the process and learn our roles within and help it along, we may not like how the process unfolds. So despite the fact that safety happens, we safety professionals have a distinct place in the process. So, too, does everyone else. For the safety professional, though, it is, as the Chinese say, *ch'uan hsin ch'uan i*—with whole heart and mind.

Comprehensive Process

If safety is a comprehensive process, it will consist, among other things, of the elements listed in the last section: safety coordinator, safety committee, safety meetings, a written policy, rules, and procedures. The process does not lack the components of a traditional program, but it is so much more. The purpose of the safety process is to move information about hazards from the point of detection, say a person in a certain department, to the lowest authorized point of decision making relative to abating or correcting the hazardous condition.

In traditional safety programs all decision making is made at the highest level. The safety coordinator makes recommendations and urges action, but the plant manager or some other high level manager makes the decisions. If authority is not delegated to the lowest practical point in the organization, the workers on the floor get the distinct idea that management does not care. Not only that but they tend to judge management as being inept, unknowledgeable, and unreasonable in its response to safety problems. Does the image of bumbling fools present itself?

Now, you and I both know that this is not the case. In fact, management often *does* care. It cares about of lot of things, including safety. We also know that safety is too often the red-headed step-child. The job of the safety professional is to keep safety in the forefront of managers' minds without bringing the organization to either gridlock or panic. The best way to do this, the very best way, is to allow and encourage, and indeed urge, the lowest level of authority possible to resolve the problem at hand. Let the top managers read about the corrective action—if worthy of notice—in a monthly safety status report. If it can be fixed on the shop floor, let it be. If it takes the plant manager to get action, let it be. Just let it happen at the lowest practical level.

Critical Implementation Issues

The danger of discussing pie-in-the-sky is that when it comes to implementation the ideal can be elusive to the inexperienced. The critical issues of implementing a safety process as a system are communication, empowerment, management availability, and staff authority-line accountability. Let's examine these one-by-one.

Communication

Communication may very well be the most essential element in the safety process. OSHA recognized this, I think, when it promulgated the Hazard Communication standard in 1985. Unfortunately, the standard is so defective that OSHA shot itself in the foot, but it was a wonderful idea. A brief study of the shortcomings of this standard relative to communication is in order.

Modern employees are willing, and want to, take responsibility for themselves. The notion that communication leads to empowerment is not that far-fetched, in my opinion. Where OSHA went wrong was to give a typewriter to a monkey. To make matters worse, OSHA expected Shakespeare plays from the monkey and typewriter. Communication is defective unless both parties to the communiqué are about equally capable of understanding it. This is far from the case in chemical safety, hence the Hazard Communication standard is grossly inadequate. More on that later.

To make communication work for you instead of against you, educate your employees in the basic issues. Don't bring in a rocket scientist to give them an education. Make it interesting and to the point.

You also need to understand that we do not communicate solely on an intellectual level. Anytime you and I have a conversation, other agendas are on the table. Unspoken agendas to be sure, but they are there nonetheless. Our seemingly simple communication with one another is always filled with psychological and sociological undercurrents.

While I am admittedly no expert, let me give you some examples as I understand this communication process. One hidden agenda is sex. As we talk with one another a repulsion/attraction thing is going on. This is not limited to male-female conversation but extends to male-male and female-female. If, for instance, two heterosexual males, who are having a conversation, do not fit each others' mental and emotional conception of masculinity, they experience discomfort that may be so subtle as to go unnoticed, but it affects their conversation. If nothing else, it affects how they listen to one another.

What else affects the conversation of any two individuals? That's not hard. Race, religion, political beliefs, strong opinions, unusual opinions, eccentricity, color of hair, color of eyes, size of nose, facial blemishes, lisping, stuttering, country of origin, region of the country, accent, and on and on and on. Yes, it is mostly prejudice in one form or another. Not to excuse prejudice, but we cannot help it subconsciously. Consciously we can surely make an effort to overcome our prejudices, but subconsciously we never know exactly how they affect us from moment to moment.

So what am I telling you? As safety professionals we need to know that our communications are probably always defective in some way. We need to do what we can to clarify, simplify, and broadcast so that everyone out on the shop floor gets the message we intend them to get. Then we need to repeat, repeat, and repeat until they *do* get it.

Empowerment

The surest way to demonstrate management commitment (where have you heard that before?) is through empowerment. To me, empowerment means nothing more than allowing the employee at the lowest practical level in the organization to fix a problem. This means that management must trust every employee in the organization. A management team that does not have confidence in itself, with respect to safety, is unlikely to let the employees in a particular work station have the authority to identify and abate hazards. Identify? Yes. Abate? Submit a request, we'll consider it. A confident management team is more likely to empower employees. It makes good business sense. Do managers need to micromanage? Your managers need to answer that for your plant. If they do feel the need, they will not be empowering employees to make safety happen in their workplaces.

Empowerment works best where management gives guidance to its empowered employees. Priority setting, for example, should be the job of management. Otherwise each employee runs off in a different direction. Some clear guidance on priorities is the rudder that steers empowered employees.

Management also needs to clearly assert its expectations of accomplishments. Nothing is more frustrating and demoralizing for empowered employees than to get down the road and hear, "this is not what we wanted." The management team should be "up front," as is often said today. In addition, the management team needs to clearly specify limitations so the employees know the constraints they are working under. I am reminded of a company that announced a new monetary incentive program for its employees. Payouts were to be on a quarterly basis. Each quarter sterling performance was reported concerning sales and marketing, billables and receivables, and operating costs. Each quarter for two years the employees were told they would receive no payout due to some limitation or other placed on the incentive system. None of these limitations were announced before the quarter in which they nullified the payout. Do you wonder why they were so demoralized in the third year that they stopped participating? Management must develop and communicate performance expectation for empowered employees.

Management Availability

The next key implementation issue is management availability. The senior management in your plant must be visible and present in the safety process. One of the senior managers should attend every safety committee meeting. This assignment can be rotated as often as every meeting. Someone ought to be in attendance with the authority to say, "Do it!"

That's not all! A senior management representative should be seen regularly making safety inspections on the plant floor. Senior managers, regardless of gender, are to companies what fathers are to traditional nuclear families. In healthy families the father is a presence that represents stability, security, unwavering love, genuine concern, and authority. In healthy factories the senior managers collectively are a presence representing stability, security, unwavering commitment, genuine concern, authority.

Besides being merely present, senior managers must also be available to employees. They should be easily accessible, at least while they are in the plant proper. Yes, some employees will have nagging personalities and some will be incessant complainers. Nevertheless, all employees need access to management. This interaction, however painful it may be at first, is what can set your plant afire (not literally) with zeal for safety. As I have said elsewhere, trot your managers out regularly.

Staff Authority—Line Accountability

The last implementation issue we need to discuss is the staff authority-line accountability dichotomy. Many industrial safety professionals have a staff position in the organization, but are held accountable as if they were line supervisors. This leads to frustration and eventual burn-out. You should have a staff position that reports to the top manager in the plant on the operational level. Line supervisors should be held accountable for safety, not you. Yes, *you* are accountable for administration and the overall safety process, but day-to-day safety that happens out on the floor is the responsibility of the line supervisors. Insist on it.

Successful Safety Management Processes

The characteristics of successful safety management processes are:

- Management commitment
- Employee involvement
- Continuous improvement
- Work site analysis
- Hazard prevention and control
- Health and safety training

We've discussed this in the last two chapters. OSHA propounds it. All sorts of safety experts preach it. But what do you do to fix your safety system if it does not work?

How do you know when your management is committed and your employees are involved? Having a clearly stated safety policy is the first tier of evidence of management commitment. The management team must then establish clear goals and objectives for meeting safety policy requirements. Obviously, in order to do this, someone has to be held accountable for safety (line supervision) and someone has to communicate to management some performance measurements (safety professional). Top managers who are out on the floor identifying hazards themselves or talking to employees about their safety are the surest evidence of commitment. Manager attendance at employee safety committee meetings is another sure indicator of interest and commitment. Successful safety programs are found where management's sincere concern for employees drives the safety process. In these factories, the safety process is in clear harmony with productivity and quality.

Management commitment leads to employee involvement. Convening the safety committee is no hard task if the employees have the idea that the top management is interested in its activities. Committed managers delegate adequate authority and resources to the safety committee and other parties responsible for safety. Committed managers attend safety committee meetings and soon the employees start getting the idea that management does care.

Committed managers review the safety process and its programs at least on an annual basis and they involve employees in the review. Part of this annual review is a measurement of the attainment of goals and objectives. We will discuss measurements in Chapter 5.

Successful safety programs have already conducted a comprehensive baseline health and safety survey. Something akin to the OSHA VPP checklist may be used, although other checklists are available from commercial sources as well. Someone familiar with successful safety systems analyzes the hazards of the planned and new processes, equipment, and materials that enter the plant. Because these successful safety processes have regular, periodic self-evaluating safety inspections, they can promptly investigate and analyze accidents and near-misses to identify root causes and means of prevention.

Two types of inspections are helpful: informal and planned. Informal inspections consist of a supervisor's walk-through of his or her area of responsibility. Some companies have employees fill out reports of unsafe conditions, which allow a paper trail to track the problem reported until it is corrected. Planned inspections include preventive maintenance, housekeeping, and general health and safety inspections.

Successful programs routinely analyze injury and illness trends to identify common causes and means of prevention. Trends are tracked by using safety metrics as discussed in Chapter 5.

The managers of successful safety programs establish procedures to provide control of identified hazards in a timely manner. Routine facility and equipment maintenance is provided to prevent hazards associated with breakdowns. Plans and practices are also developed for emergency situations. A medical program is established that includes availability of first aid.

Training is conducted in order to ensure that all employees understand the hazards to which they may be exposed. Supervisors and departmental safety coordinators understand the safety activities required in their job descriptions. Training allows management to understand and follow their own policies, procedures, and guidelines, and the opportunity to endorse the safety and health program.

Teamwork

The safety professional is the facilitator of a team consisting of all employees—both management and wage earners. Employee participation in problem solving is important in getting ahead. Although each person is responsible for safety, the process needs to be overseen to ensure that it flows smoothly. The person who oversees the process is the safety professional.

A good safety professional does certain things to enable the process to flow smoothly. Overall, he or she coordinates all safety activities throughout the facility. The safety professional typically maintains and analyzes all accident reports for the facility. Even if someone else has that responsibility, the safety professional is the one who receives the reports that analyze accidents. The investigation of accidents and near misses is either conducted by the safety professional or takes place under his or her supervision. He or she will issue reports regularly to inform key people of accident trends and the safety performance of the facility.

The safety professional will conduct or arrange educational and training activities for supervisors and employees. He or she is also a permanent member of the plant safety committee or a regular visiting member if the plant has more than one. The safety professional should not be the chair of the committee, in my opinion, as the committee is his or her customer.

One way to improve the flow of hazard information is to get employees interested in the process. In that effort, many successful safety professionals use safety bulletins and posters to keep up the interest.

Another task for the safety professional is to keep abreast of federal, state, and local safety regulations. This is not as difficult as it sounds, but the company needs to invest some money to achieve this. This means the safety director needs to attend as many local workshops and seminars on the subject as feasible. These one or two day events are relative cheap and a company cannot afford to miss too many of them. The safety director also ought to annually attend at least one nation-wide convention that focuses on safety.

The National Safety Council, the American Society of Safety Engineers, the American Industrial Hygiene Association, and the American Conference of Governmental Industrial Hygienist are among those that present such conventions. These events are invaluable for the education and training of safety professionals and for networking with other professionals. Trade journals and professional magazines are also good sources of information. Other sources of information are discussed at the end of this chapter.

Unless the number of employees is few, they are represented by the Safety Committee. The purpose of the Safety Committee is to continually improve the plant's safety performance and thus enhance the quality of work life. Everyone wants a safe and healthful workplace and, as will be discussed in Chapter 11, this is his or her right. The committee recommends safety policy for management approval, performs inspections, follows up on corrective action, and monitors safety training.

The Safety Committee is a two-way radio between management and labor. Employees and supervisors appeal to the Safety Committee for abatement of hazards and improved quality of work life and the committee reports to employees the decisions made and actions taken. An effective Safety Committee contributes not only to the well-being and protection of employees but also to their morale. A good committee is also able to cut down on the number of safety and health complaints filed with OSHA.

The safety professional must attempt to prevent the Safety Committee from becoming a mouthpiece for conniving managers. If an OSHA compliance inspector gets the idea that the committee is merely a front for management's efforts to avoid spending money on hazard abatement. . . . Well, it would be better for that plant not to have a Safety Committee at all.

The committee can work together as a whole or as subcommittees to develop safety rules and procedures for all employees. At many plants this task is assigned to one person—usually the safety professional—who then develops procedures in a vacuum. Safety procedures too often read like a safety textbook. A committee representing all parties can produce a concise, step-by-step procedure that reads like a recipe.

Safety committees should conduct inspections and review accident reports. A subcommittee can perform an in-depth investigation of major accidents. Any of its duties may be delegated to a designated employee who then makes a report to the committee.

Regular general safety meetings provide training and awareness for all employees. The chairperson of the meeting is appointed by the Safety Committee. This person arranges the meetings and sends out the notification of time, place, and agenda to all hands. He or she also ensures that whoever is presenting the program is ready and that inspection committee reports are ready.

At least two employees should be selected at general safety meetings to be the inspection subcommittee, the membership of which should be rotated on a regular basis. This gets more people involved in the safety and health process. Inspectorship is a training method in itself. Before each general safety meeting the Inspection Subcommittee should inspect its assigned area to identify unsafe conditions such as physical, chemical, and fire hazards as well as unsafe behaviors. A written report should be forwarded to the Safety Committee for follow-up and an oral report should be made at the next general safety meeting.

An Investigative Subcommittee should also have a rotating membership for the same reasons: involvement and training. This committee should provide quality assurance for a supervisor's accident reports. A report of findings is made to the Safety Committee, which, then provides the follow-up. Allow the Investigative Subcommittee to make an oral presentation of follow-up actions to the general safety meeting.

Integration of Metrics and Accident Investigation

A successful safety process will be found where safety measurement is prospective, positive, and credible. Safety metrics is an effective management tool that can be used to correlate safety performance with all that matters in the organization. Primarily, in successful safety organizations, safety is viewed as a profit center, not as overhead and the safety professional is considered by senior managers to be a valuable asset to the organization.

One thing that safety metrics do for you is to assign a value to your efforts. Senior managers tend to think in terms of dollars added to or subtracted from the bottom line. If safety is seen as a perpetual drain on the bottom line, managers naturally will not be enthusiastic when you show up at a meeting. However, you can use safety metrics to show how improved safety performance is, at least, avoiding loss of more dollars from the bottom line, if not adding to it. Safety projects do have a return on investment, or ROI. Safety metrics should support the plant's current business plan. Therefore, safety metrics must be derived from the plan.

Metrics can also be used to describe safety performance in terms of activities or efficiency—for instance, how many hazard communications classes were conducted last year as opposed to the number needed.

Surveys can be utilized to define employee perceptions of safety. The results of such surveys can be described using metrics.

Metrics can be used to define risk. Such a metric particularly brings to the managers an appreciation for the value of safety and industrial hygiene efforts. Chapter 5 contains more on safety metrics.

Controlling the Process

You cannot control a process you do not understand. That's where your personal professional training comes in. Also, you must make the management team aware of issues in a way that allows them to understand the forces at work behind a hazardous situation. A management that understands the issues and is truly committed to safety is guided by logic and does not react emotionally to surprise hazards and break downs in the system. Such a leadership team has complete confidence in its safety process and sticks to it tenaciously. Dependency on the safety profession is shed for routine hazard abatement matters and that function is used instead to optimize performance. As a safety professional yourself, you should like it that way.

Statistical process control (SPC) can be used to improve the performance of any system. A *system* in SPC is anything that changes over time. The safety process is therefore a system that can be controlled by SPC.

Pareto charts are used in SPC to determine the priority of problems. A Pareto chart is a bar chart arranged in descending order. A common safety metric is to plot the frequency of different types of accidents on a Pareto chart to determine which type requires the most immediate attention. Where accident numbers are small, of course, this is obvious without a Pareto chart.

A *control chart,* or *C-chart,* is used to analyze and control a process by charting defects or lack of performance. An accident is a system defect. Plot the OSHA recordables month by month. Compute the mean and plot it on the same chart. The *upper control limit, UCL,* is three standard deviations above the mean. Anytime the number of recordables exceeds the UCL you have serious problems somewhere in your safety system.

Integration of Safety and Quality

Productivity, quality, and safety are the three facets of the business pyramid. At plants with successful safety programs, the safety process is totally integrated with productivity and quality and is accepted by management and employees alike as an essential ingredient for business success. In reality, a formal safety inspection program is a valuable tool for managers. Inspections can detect and correct potential problems before they become accidents, breakdowns, or waste. Inspections can pick out day-to-day wear and tear. Finally, inspections give managers full control of their area.

A comprehensive safety inspection program has many of the same benefits as a good quality inspection program:

- More reliability and efficiency as well as safety in the workplace,
- The manager becomes more visible through inspection,
- Problems get corrected, and
- Fixes are accomplished or at least attempted (in which case they had better be communicated to someone who can make the fix).

Planned inspections identify potential problems, equipment deficiencies, improper employee actions, effects of change, inadequate remedial actions, good practices, positive performance, and demonstrate management commitment or lack of it. Additionally, planned inspections measure the adequacy of preventive maintenance. The efficiency of the work layout is evident by inspection. The orderliness of the workplace and control of damage and waste is also evident.

Where a successful safety program exists, everyone recognizes that safety excellence is never ending. Like productivity and quality, continuous improvement of the safety process is the goal. Fred Manuele, citing a book by Graham M. Brown, Darcy E. Hitchcock, and Marsha L. Willard (*Why TQM Fails and What to Do about It,* Burr Ridge, IL: Irwin Professional Publishing, 1994), paraphrases a statement:

> when safety is seamlessly integrated into the way an organization operates on a daily basis, safety becomes not a separate activity for committees and teams, but the way every employee performs job responsibilities.

Manzella makes a neat comparison between the safety process and Total Quality Management. Figure 3-1 is adapted from Manzella.

Figure 3-1
Relationship of Safety to Quality

Safety	Quality
Goal: zero accidents	Goal: zero defects
Incident analysis	Event analysis
Written policies, procedures, and guidelines	Documented policies, procedures, and work instructions
Safety committees	Quality circles, employee involvement teams
Employee participation Statistical analysis	Empowerment
All accidents are preventable	Control charts, statistical process control
	All non-conformances are preventable

INTEGRATION OF ENVIRONMENTAL, HEALTH AND SAFETY

More and more safety professionals are being assigned responsibility for environmental compliance at the plant level. Conversely, many environmental professionals are being assigned safety management duties. Some safety professionals are vehemently against this situation. Mostly, these opponents are those who advocate behavioral safety.

The physics and chemistry of environmental hazards, whether the environment is the workplace or the great outdoors, is the same. An engineer or other scientist trained in Newtonian mechanics can deal with hazards no matter where they occur. Therefore, engineers and scientists make good candidates for environmental, health, and safety professionals as comprehensive practitioners. Environmental, health, and safety engineering are complementary disciples that interrelate with each other.

The administration and management of neither environment nor safety and health matters requires a degree in rocket science. Any well educated person can manage and administer these matters. Nor does reading and understanding laws and regulations require a law degree. Presenting what you think a law or regulation means before a judge or administrative hearing does emphatically require a law degree. So get yourself a good attorney versed in compliance law if you find yourself in that predicament. However, day-to-day plant level administration and management of compliance should not require such expertise. You can do it yourself.

If you agree with some professionals that knowledge of Freud, Jung, B.F. Skinner, and other behavior specialists is more important that Newtonian mechanics in our profession then perhaps you also agree with them that the mixture of environmental with safety and health duties is not good. I am not one of those. I think that industrial psychology has its place. I also think that a comprehensive practitioner can handle a certain amount of behavior-based safety. But I also strongly believe that if a substantial amount of behavior-based safety is required for your particular plant, then it is time you call in an expert. Behavior-based safety, in my opinion is not a day-to-day part of the general practice of our profession. The howls in the background are those who disagree with me.

Complementary regulations issued by EPA and OSHA in the last few years have made it more sensible to integrate all of these functions at the plant level. For instance, both OSHA and EPA regulate the handling of hazardous waste. A better match of regulations is found in OSHA's Process Safety Management and EPA's Chemical Accidental Release Plan.

Luce gives a good summary of an integrated program. An integrated program of environmental, health, and safety addresses the interrelationship of all these functions as well as their differences. While complying with the law, an integrated program exemplifies a moral commitment to employees and the public. The superior program involves government agencies in planning and problem solving. The integrated program allows the staff to provide detailed training for all employees on the total program as well as specific employees for specific tasks and responsibilities. Willful violations of company policy, rules, and procedures as well as public law and regulations can more easily and consistently be disciplined.

A superior integrated program keeps employees informed and requests their assistance in resolving problems. Everything is documented. Records are open for review by employees and the public agencies. The program head is at the director level and reports directly to the CEO. Environmental, health, and safety issues are line budget items.

Company programs are the institutionalized procedures whereby compliance processes are managed. The processes that may be managed in a single integrated program are:

- Environmental protection
- Pollution prevention
- Occupational safety
- Occupational health
- Industrial Hygiene
- Wellness, (general health)
- Workers compensation
- Loss control
- Fire protection
- Fire fighting
- Disaster preparedness and control
- Emergency medical response
- Toxic substance control
- Fungicide, insecticide, and rodenticide management

Assigning a single generalist to manage all these functions is not likely to produce exemplary results, especially if that person has other duties as his or her primary job assignment. One nonexpert person can do a credible job of overseeing all these functions in small plants if the primary chores can be delegated to others. A handful of comprehensive practitioners are available who are capable of managing an integrated program alone with delegation of some of the work. Larger plants need a staff of specialists, however. Their supervisor could be a comprehensive practitioner.

Using Outside Consultants

Beside industrial psychology, specialists are needed for other occasions. Even as a comprehensive practitioner, I call in specialists from time-to-time. Consultants can do several things for us.

One thing the consultant can do is to relieve some of the work load—at least theoretically. Sometimes you have to spend so much time showing and explaining what you need that you may as well do it yourself. The less time your consultant needs on a learning curve, the more value you receive from his or her services. Hire a comprehensive practitioner for this type of project.

A consultant can also be used to provide specialist expertise that you do not have. Sometimes a comprehensive practitioner can do this type of project for you, but more often you will need to hire a specialist. For instance, you may have a comprehensive practitioner conduct noise monitoring. The consultant may also propose some noise abatement schemes. But suppose you have a situation that has alluded successful abatement? That is when you need the help of an experienced acoustical engineer.

You can also use a consultant to oversee a large project. For instance, say you are having asbestos removed from one wing of your building. Put a state certified asbestos management planner in charge of the project. Or suppose you have agreed with the state to clean up a corner of your property where some past practices have contaminated the soil. An experienced remediation engineer can manage this project for you.

Expect your consultant to be forthright and communicative. You should never be in the dark about the status of your project. Keep in mind that it is your project, not the consultant's. She or he is providing a service based on knowledge and skill. Expect your consultant to give you value for your bucks.

Selection: Getting the Most for Your Dollar

The way to get the most for your dollar when buying services from consultants is not the same way you buy widgets. Many people tend to get proposals from two or three consulting firms and choose the lowest proposed cost as if the product were widgets. Consultants develop proposed costs based on how they see the project unfolding. It may be the lowest proposed cost is the best value. The higher proposed costs may be throwing billable hours into the project that need not be there. On the other hand, the lowest proposed cost may not visualize the project properly and you may get very little value for your dollars.

Obviously, you must read the proposed scope of work carefully in each proposal. It is far better to give each firm a written scope of work as you see it before they estimate their cost proposal. Be suspicious of a cost proposal made from a telephone conversation. While not always true, most of the time someone needs to visit your plant to see for himself or herself what work effort is needed. The larger the project, the more true this is.

Once you weed out any proposal that does not have the scope of work right, look at the qualifications of the people who will work on the project. Are their skills and experience pertinent to the project? Is the firm loading up your project with billable hours for persons of debatable qualifications? You do not need a geologist on a noise abatement project. Beware of proposals that list every person or discipline in the company. Are they trying to impress you with their depth of experience and qualifications? Or have they set up an opportunity to increase the billable hours of some of their lesser utilized personnel?

Use: Controlling the Project

Have your consultant make a weekly telephone report of the project status to you. A written report, unless it was specified in the cost proposal, takes too much of the consultant's time, but he or she should not begrudge you a telephone call. If the consultant is working on site, then a brief face-to-face meeting once per week will suffice. Hold the consultant to the time schedule and agreed cost. Also, hold him or her to the agreed scope of work. Changes to the scope of work will require changes in cost, so get the cost change quoted in writing.

Sources of Information: Documents and Databases

Many companies assign the safety management responsibilities to a person who is not prepared and has too little time because he or she specializes and has time commitments in another field. If that is your situation, helpful resources exist.

OSHA

Single copies of OSHA standards may be obtained for free from regional or area OSHA offices. (See Appendix for regional offices.) Standards are available for purchase from the Superintendent of Documents, U.S. Government Printing Office (GPO), Washington, D.C. (telephone number: 202-512-1800). The standards are also available in hard copy or on a CD-ROM at nominal cost.

The *Federal Register* is one of the best sources of information on OSHA standards, since they are published in the *Federal Register* when adopted, as well as when amended or corrected. The *Federal Register* is available in many libraries but annual subscriptions are available from the GPO.

Each year all the regulations and standards that were published in the *Federal Register* are organized (codified) in the *Code of Federal Regulations* (CFR). Copies of the CFR are available at many libraries and from GPO. OSHA's regulations are collected in Title 29 of the CFR, Parts 1900-1999, usually written as 29 CFR 1900-1999.

Some OSHA documents may be obtained directly from the Department of Labor. Contact the Department of Labor, OSHA Publications, P.O. Box 37535, Washington, D.C. 20013-7535, (202) 219-4667, or (202) 219-9266 (Fax).

One publication you should have is the *Blue Book* (OMB 1220-0029), which gives guidance on recordkeeping.

Some additional OSHA Publications you may need, besides regulations and standards, are:

- 501, *OSHA Training Institute Manual, A Guide to Voluntary Compliance in Safety and Health*
- 2254, *Training Requirements in OSHA Standards and Training Guidelines*
- 3067, *Concepts and Techniques in Machine Safeguarding*
- 3071, *Job Hazard Analysis*
- 3122, *Principal Emergency Response and Preparedness Requirements in OSHA Standards and Guidelines*
- 3138, *Permit-Required Confined Spaces*

NIOSH

Publications of the National Institute of Occupational Safety and Health (NIOSH) are available from the GPO. A useful series of publications from NIOSH is called the *Current Intelligence Bulletin.* These bulletins discuss the toxicity of various chemicals or their metabolites. Publications are also available on noise, lighting, and other physical hazards. Some NIOSH documents may be obtained directly from the U.S. Department of Health and Human Services (HHS). Within the HHS, NIOSH is an agency under the Public Health Service's Center for Disease Control.

EPA

EPA regulations are printed in the *Federal Register.* The GPO provides copies of these regulations for a nominal fee. Private subscription services of varying costs also provide copies of regulations. EPA has produced hundreds of background documents, guidance documents, and model documents. These are available through the National Technical Information Service (NTIS). EPA regulations are found in Title 40 of the CFR.

DOT

DOT regulations on hazardous materials are found in Title 49 of the CFR, Parts 171-180, available from GPO. These comprehensive regulations cover transportation of hazardous materials, specifying basic conditions for filling packages, marking and labeling, packaging, shipping papers, handling, loading, segregating packages, and securing packages. Training requirements and hazard communication procedures are also specified.

NFPA

The National Fire Protection Association (NFPA) is a professional organization that issues consensus standards on fire prevention and protection measures. NFPA issues, in particular, the *National Fire Code* (NFC), which contains the *National Electric Code* (NEC) and the *Life Code,* among other documents. When OSHA came into being in 1970 it adapted the NFC, NEC, and Life Code as its fire protection standard.

ANSI

The American National Standards Institute is another consensus standard body. Many ANSI standards were adopted in whole by OSHA.

ASME

The American Society of Mechanical Engineers (ASME) issues consensus standards on pressurized vessels and tanks, among other things. When a process vessel is said to meet code, it meets one or more ASME standards.

AIChE

The American Institute of Chemical Engineers (AIChE) has established centers for research and consensus standard development on chemical process safety. Their Center for Chemical Process Safety (CCPS) has issued several documents on conducting hazardous operability (HAZOP) studies as well as other process safety management guidelines.

CGA

The Compressed Gas Association (CGA) has many guidelines and consensus standards on compressed gas cylinders, air separation processes, cryogenic liquid cylinders, compressed gas and cryogenic liquid handling, and more.

Books and CD-ROMs

Several publishers specialize in safety and health books or else have departments that specialize in such books. During a typical week at work you probably receive several fliers and catalogs and stacks of post cards marketing these books. Many publishers now list their books online.

Even OSHA standards, interpretations, directives, and other information may be obtained online on the World Wide Web at: http://www.osha.gov and http://www.osha-slc.gov/.

A wide variety of OSHA materials including standards, interpretations, directives, and more can be purchased on CD-ROM from the U.S. Government Printing Office for a nominal price that includes quarterly updates. To order, write to Superintendent of Documents, P.O. Box 371954, Pittsburgh, PA 15250-7954. Specify OSHA Regulations, Documents & Technical Information on CD-ROM (ORDT), S/N 729-01300000-5. The price is $88 per year ($110 to foreign addresses) for a one year subscription with quarterly updates. A single copy is $30 ($37.50 foreign).

OSHA new releases, fact sheets, and other short documents are available by telefacsimile for a nominal charge of $1.50 per minute. Call (900) 555-3400 to access this service.

States

Since some states adopt and enforce their own standards under state laws, you will have to contact those states directly for copies of state laws and standards. See Appendix for states with their own OSHA-approved state plans.

REFERENCES

All about OSHA. U.S. Department of Labor. OSHA Publication 2056. U.S. Government Printing Office, 1995.

Birkner, Lawrence R. and Ruth K. Birkner. "Performance Metrics: Linking Them to Business." *Occupational Hazards.* February 1998, pp. 21-22.

Eckenfelder, Donald J. "It's the Culture, Stupid." *Occupational Hazards.* June 1997, pp. 41-44.

Luce, Zoyd R. "IMP: The Integrated Management Approach to Environmental Protection, Health, and Safety." *Professional Safety.* January 1990, pp. 30-33.

Manuele, Fred A. "Principles for the Practice of Safety." *Professional Safety.* July 1997, pp. 27-31.

Manzanella, James C. "Achieving Safety Performance Excellence through Total Quality Management." *Professional Safety.* May 1997, pp. 26-28.

OSHA Compliance Manual. The Merritt Company, 1988.

Sorrell, Larry W. "Safety and Statistical Process Control: One Practitioner's Perspective." *Professional Safety.* January 1998, pp. 37-38.

Thurber, Sarah. "Safety Directors: Too Many Tasks, Too Little Time." *Safety & Health.* July 1993, pp. 42-45.

4

Cutting Your Losses

Loss control, like quality, is a constant element of business. Continuous improvement, therefore, is the only goal for loss control that makes sense. You can no more start out with a perfect loss control program than you can with any other program. The diligent safety professional works with a constancy of purpose toward improving all aspects of safety and loss control.

Define Loss Targets

You cannot define targets unless you find problems. You cannot find problems sitting in your office or looking the other way. For the most part the squeaky wheel gets the grease. The most nagging problems always get the most attention. Also the problems that cost the most money, or at least have the potential for costing the most money, are the ones that will attract the most attention.

Therefore when you set targets or goals, set as the highest priority those that can save your company the most money. The danger is in thinking only of the most direct cash losses. It is obvious you can save money by eliminating a large potential loss when you install fire protection equipment in a vital space. It is not always so obvious that you are avoiding large losses by eliminating a chemical exposure that is making employees the walking-ill. (Ill enough, that is, to make them less productive and increasingly likely to be absentees, but not ill enough to be a clear money drain.) Sometimes the frustrating part of the safety professional's duties is to sell upper management on these less obvious needs.

Define Scope of the Loss Process

When reporting on a newly identified hazard, be sure to define the scope of the potential losses. You may have to do a little digging, but how much is it costing, or may it cost, directly and indirectly? How is productivity affected? How is quality affected? If the senior managers who hold the purse strings do not understand what is at stake, do not wonder why they are reluctant to spend money on what is, in their minds at least, a dubious problem.

Management Controls

Engineering, work practice, and administrative controls are the primary means of reducing employee exposure to occupational hazards. Engineering controls minimize employee exposure by either reducing or removing the hazard at the source or isolating the worker from the hazard. Engineering controls include eliminating toxic chemicals and substituting non-toxic chemicals, enclosing work processes or confining work operations, and the installation of general and local ventilation systems.

Work practices alter the manner in which a task is performed. Some fundamental and easily implemented work practice controls include 1) changing existing work practices to minimize exposures while operating production and control equipment; 2) inspecting and maintaining process and control equipment on a regular basis; 3) implementing good housekeeping procedures; 4) providing good supervision; and 5) mandating that eating, drinking, smoking, chewing tobacco or gum, and applying cosmetics in regulated areas be prohibited.

Administrative controls include controlling employees' exposure by scheduling production and tasks, or both, in ways that minimize exposure levels. For example, the employer might schedule operations with the highest exposure potential during periods when the fewest employees are present.

When effective work practices or engineering controls are not feasible or while such controls are being instituted, appropriate personal protective equipment (PPE) must be used. Examples of PPE are gloves, safety goggles, helmets, safety shoes, protective clothing, and respirators. To be effective, PPE must be individually selected, properly fitted, and periodically refitted. PPE must be conscientiously and properly worn by all those employees who require protection. Another requirement is that the PPE be regularly maintained and replaced as necessary.

Figure 4-1
Strategies for Engineering Loss Control

1. Prevent the marshaling of the energy form in the first place.
2. Reduce the amount of energy marshaled.
3. Prevent the release of energy.
4. Modify the rate or spatial distribution of release of the energy from its source.
5. Separate in time and space the energy being released from the susceptible structure, whether living or inanimate.
6. Separate by imposition of a material barrier.
7. Modify the contact surface, subsurface, or basic structure, as in eliminating, rounding, and softening corners, edges, and points with which people can come in contact.
8. Strengthen the structure, living or nonliving, that might otherwise be damaged by the energy transfer.
9. Move rapidly to detect and evaluate damage that has occurred or is occurring in order to counter its continuation or extension.
10. Take measures between the emergency period following the damaging energy exchange and the final stabilization of the process after appropriate intermediate and long-term reparative and rehabilitative measures.

Figure 4-1 summarizes ten engineering strategies for controlling losses by containing unwanted energy, according to a safety management theory to that effect.

Identify Loss Exposures

Figures 4-2 and 4-3 give direct and indirect costs, respectively, associated with accidents.

Figure 4-2
Direct Costs of Accidents

- Lost time of injured employee(s)
- Cost of medical services
- Lost time of nearby employees due to response
- Lost time of nearby employees due to curiosity
- Lost time of supervisors/managers due to response
- Lost time of supervisors/managers due to curiosity
- Lost time of accident investigator
- Lost time of emergency response team
- Cost of outside emergency services
- Damaged equipment
- Damaged property
- Damaged product
- Damaged supplies
- Lost profit on sales
- Fines, assessments, and other penalties

After Coletta

Evaluate the Risk

Some degree of financial risk is acceptable or you and your stockholders would have your money in a safer investment. Some degree of risk to life and health is also acceptable or people would not smoke, drive cars, and sky dive, among many other dangerous activities. While some risk is acceptable, the certainty of accidents is not. Yet if we were so conservative that we accepted no risk whatsoever, we would not even get out of bed in the morning, would we?

Risk, remember, is the measure of both the probability of and the consequences of any undesirable event. Risk acceptance is not as easy as coming up with a number. For instance, would you agree that one in million is an acceptable risk? Sounds safe enough, doesn't it? Well, one in a million may seem an acceptable risk, but that risk applied means that nineteen thousand people would get the wrong prescription medicine at the pharmacy this year.

Figure 4-3
Indirect Costs of Accidents

- Training of replacement employees
- Inefficiency of replacement employees
- Time lost by supervisors/managers in follow-up modification of work procedures
- Time lost by supervisors/managers in follow-up response to internal and external inquiries
- Overtime to make up lost production
- Payroll overhead charges applied to all lost time and overtime (runs to 40% base pay)
- Plant overhead charges applied to all lost time and overtime (runs to 350% base pay)
- Cost of salary continuation for injured exempt employees
- Increased insurance premiums
- Lost profit on unrealized sales caused by long-term erosion of market shares (not a factor in every accident)
- Charges for uninsured rehabilitation services
- Diminished good will

After Coletta

A one in a million chance for an electrocution means that three people could die in any given year in a plant of fifteen hundred employees—and that is assuming that the hazardous contact is no more often than once per hour! More frequent contact means more people could die! The point is that when we hear numbers like "one in a million" we safety professionals tend to think one in a million persons. That sounds safe to us. But that is wrong thinking. One in a million *exposures* or *potential contact events* is the risk. How safe does that sound to you?

Risk Taking

A little fear is good, my grandfather used to tell me. Employees who feel safe tend to take more risks than employees who feel unsafe.

According to Lark, employees who feel safe may take risks because they are seeking higher levels of sensation and sensory stimulation. I don't know how much of that really goes on in the workplace. I do not think it prevalent, but I agree it does happen. I have known three cases where employees bent over vapor degreasers in order to get high on the solvent vapors. I have also known a case where an employee using methyl ethyl ketone in a wipe cleaning operation would sniff her wipe tissue to get high. So I agree with Lark that it can happen but I have only seen these four cases in thirty years time! Of course that is probably the tip of the iceberg, not only in numbers but in degree of sensation seeking. Still, it is pretty rare.

Lark quotes research by Professor Marvin Zuckerman of the University of Delaware who studied this sensation-seeking form of risk taking. Zuckerman found that extraordinary sensation seekers, such as the ones I observed, probably have lowered levels of the enzymes monoamine oxidase and dopamine beta hydroxylase as well as increased levels of the gonadal hormones. I am no expert in this, but I would guess that the four persons I encountered had increased levels of gonadal hormones. That's a lay opinion based on the impression their personalities made on me.

Lark also reviewed Professor Frank Farley of the University of Wisconsin at Madison who found evidence of a sensation-seeking, risk-taking personality, which he called a *Type T personality.* T stands for "thrill-seeker." According to Farley, Type T personalities are aroused by uncertainty, unpredictability, novelty, much variety, complexity, ambiguity, flexibility, low structure, high intensity, and high conflict.

The personal acceptance of risk—Philley calls it *risk tolerance*—is based on some underlying factors. For instance, people will accept fairly high *voluntary* risks such as sky diving, while a much lower *involuntary* risk such as exposure to a carcinogen is unacceptable. The *familiar* risk, such as driving to work every day, is acceptable, although high, while a new, unusual, or strange risk, even if low, such as the introduction of a new chemical into the workplace, may not be. *Detectable* risks are also more generally acceptable than those which cannot be detected by the normal body senses: sight, hearing, taste, smell, or touch. *Natural* risks are generally more acceptable than *anthropogenic* risks. If Mother Nature did it, somehow that is all right regardless of the grief caused. If it is high tech risk, it's definitely out. Lightning bolts are acceptable; high voltage power lines are not. Risk that appears to be controllable, such as driving at high speeds, is generally more acceptable than uncontrollable risks, such as the level of lighting in a workplace.

These are typical individual reactions to risk. When you get a group involved, the situation may change entirely. A risk may be acceptable to a group of people while it is unacceptable to a particular member of the group who may experience a more intense negative consequence of the hazardous exposure. For instance, the incident rate of a facility may be acceptable to all hands, but not to the person who next suffers the consequences of an accident. When this situation occurs, we generally have a very frustrated and irritated individual due to the fact that the group is unwilling to spend time and money to further reduce the incident rate.

Risk assessment is the characterization of the potential for adverse health effects of human exposure to chemical, physical, and other environmental hazards. A risk assessment has four parts. For the risk assessment of a chemical these are: hazard identification; dose-response curve; exposure assessment; and risk characterization. Risk assessment of physical agents has the same format. Some sort of dose-response curve is needed, even for physical hazards. Sometimes the dose-response of a physical hazard is assessed in a qualitative, rather than quantitative, fashion. But quantitative dose-response information is available. For instance, OSHA has a dose-response chart for noise exposure. A dose-response chart is available for illumination (light level) and other physical hazards as well.

The need for risk assessments arises from a source that releases hazardous energy that disperses to the workplace, leading to an exposure of a receptor. With exposure comes the probability of toxicity, injury, or other adverse health effects.

So, you walk through the plant and identify a list of three hundred potential safety problems. How serious are they? Which ones need to be fixed first? The problem, as Sheridan points out, is that an adequate amount of scientific data for assessing risk is rarely available. Inevitably some assumptions must be made in lieu of missing data. Hence, we are never quite sure what a safe exposure is. In the case of exposure to potential carcinogens, prudent engineers, scientists, and medical practitioners hold the opinion that no safe level of exposure exists.

Risk assessments, then, are necessarily subjective and based on individual judgment to some degree. Since risk is dependent on the human mind, it is not always based on the same reality. That is why many engineers and scientists differ from some of the public in opinions about nuclear power, radioactive waste storage and disposal, pesticides, and chemicals in general.

The presentation of risk assessments to management or compliance authorities must provide three things. First, the presentation must be understandable yet comprehensive. It must be consistent with both its purpose and the extent of available data. Secondly, the presentation must be directly usable in decision-making in regards to public policy. Its usefulness must be readily apparent or you can forget about anyone using it. Finally, the presentation must be both credible and fully defensible. If not, it will not survive the scrutiny of either the engineering-scientific world or the public as represented by compliance authorities. In some cases the public may represent themselves—for instance at a public hearing used by environmental compliance officials.

Additionally, it is highly desirable that the risk assessment include a clear, simple, and brief executive summary. The summary must present a balanced treatment of the relevant issues that are contended. The assessment report should also present the basis for critical assumptions. How were they chosen? What is the logic behind them? A clear presentation of your logic minimizes second guessing by Monday morning quarterbacks.

A periodic risk review of the facility serves three purposes. Such a review potentially identifies important contributing factors to the incident rate. These factors are situations that lead to accidents, which cause loss of production, threaten employee safety and health, and potentially threaten the public and the environment. A periodic facility review also gives you a chance to develop recommendations for reducing risk. Finally, it gives you an opportunity to identify risky processes and work areas that need more detailed study in order to determine sufficient measures to reduce risk.

Manage Change

If our plants were static, and nothing ever changed, we could all work hard for a few years, then sit back with a cup of coffee and enjoy our low incident rates until we retired. At least two things are wrong with this scenario. On a light note, why would our companies keep paying us our fat paychecks for doing nothing? On a more serious note, though, industrial plants tend to be dynamic places where change occurs often, planned or unplanned. A real part of our job as safety professionals will always be to help control changes that might give rise to new hazards. Figure 4-4 lists some changes to be aware of in your facility.

Figure 4-4
Facility Changes to Be Aware Of

1. Installation of new equipment, piping, electrical, or control systems
2. Modification of existing facilities
3. Addition or deletion of valves or piping
4. Changes in pipe size or material of construction
5. Introduction of a new chemical or microorganism
6. Reintroduction of a chemical that has not been used for two years or more
7. Changes in raw materials or waste streams
8. Revision of hardware/software control systems
9. Revision of a control scheme or alarm limit
10. Changes in the ranges of instrument transmitters
11. Installation of new buildings or structures
12. Modification of existing buildings or structures
13. Change in storage tank contents
14. Change in storage tank use
15. Generation of new solid or liquid waste streams
16. Installation of equipment emitting air pollution
17. Repair of equipment emitting air pollution
18. Replacement of equipment emitting air pollution
19. Installation of air pollution control equipment

After Jakubowski

It is important to have a handle on these changes, otherwise all your efforts to reduce the incident rate will be just spinning your wheels.

Identify and Control Particular Losses

Without walking through your plant it is hard to identify all the particular losses you could be faced with. However, we all face some common losses. Or, at least, some common categories of losses.

Workers' Compensation

Workers' compensation comes directly off your company's bottom line—straight out of profits. You (or someone in your plant) had better be interested in workers' compensation and have a handle on it or it could end up costing a lot of money. Payments for medical expenses, other expenses reimbursable to the employee, and lost wages are all called *loss,* which is technically the negative result of an incident that becomes the basis for submission of a claim by a worker and the subsequent payment.

In the early days of the industrial revolution workers were being maimed and killed by the scores. It was too difficult to win a law suit against your company (still is) for reasons we need not explore. However, the net result was that the surviving families of industrial accident victims were starving and in dire poverty because they had lost their breadwinner. Women generally did not work outside the home in those days. Workers' compensation was a development to provide for these workers and their families while offering protection from lawsuits to the companies who had to hand out the payments. In effect, workers' compensation is a no-fault insurance system. The worker gets compensated in return for waiving his or her rights to sue the company. Today, many states allow workers to waive their compensation rights in order to pursue litigation against their employer and all states allow workers to sue third parties.

Objectives of Workers' Compensation

The objectives of workers' compensation laws, in general, are five fold. First, the laws provide the injured employees with income replacement. This is usually not full pay, but enough to survive on. Some states have not increased their payments in years, however, and pay poverty-level wages. Second, these laws provide disabled workers with restoration of earnings. Again, remember that, in many cases, these earnings are at poverty level. By making the employer responsible for these payments, the third objective of these laws is to put economic pressure on employers to practice accident prevention. Fourth, the compensation laws generally aim to assure proper allocation of societal costs by making the employer, not the taxpayer, responsible for them. The final hope of the drafters of these laws is the efficient achievement of the first four objectives.

Workers' compensation insurance rates are based on accident frequency, loss rate in terms of dollars, and severity. The mathematical combination of these factors is called an *experience modifier,* an expression which is *weighted to frequency.*

$$EM = f \times LR \times S \qquad [4.1]$$

where

EM = experience modifier
f = frequency
LR = loss rate in dollars
S = severity

Therefore, you can actually save money by reducing the number of accidents in your plant.

Two generic types of workers' compensation laws are on the books today. *Elective workers' compensation* laws allow the employer to select whether or not to carry workers' compensation insurance. If the employer does not carry it, employees still have the right to litigate for damage recovery and punitive damages. States with *compulsory workers' compensation* do not give employers the option: they must carry workers' compensation insurance.

The laws typically provide for two types of payment systems. *Uncontested cases* receive direct payment in due course. An agreement system may be in place in which the employee can negotiate payments to a certain extent, but for the most part these are set by law with little leeway for negotiation. In *contested cases* a hearing is conducted in which the employee and employer may present opposing sides of the issue. An administrative panel hears the contested cases and decides on the final payment terms.

Workers' Comprehensive Risk Insurance

Workers' compensation insurers come in three forms. *Commercial insurers* are similar to other insurance companies except that they specialize in workers' compensation risk management. *State funds* are managed by a low-bidding insurance company or a pool of companies for high-risk employers and employers who cannot afford to pay for commercial coverage. The latter companies are usually not risk free, but must contribute to the operating cost of the fund. *Self-insureds* are typically larger companies that can afford to set aside a pool of money to cover possible workers' compensation losses.

Many companies are practicing loss control today; that is, someone, either an internal staff or an insurer, is seeking to minimize injuries and associated losses. As a program, loss control may include safety and ergonomics programs as well as environmental health analyses.

The chance of loss is called *risk.* The resources set aside to cover projected workers' compensation claims or losses are called the *loss reserve.* An employer can come up with the loss reserve in any number of ways. The most common way is through workers' compensation insurance. A pooling arrangement occurs when an organization of insurers reinsures among themselves on particular types of risks. Employers who are covered by pooling arrangements are referred to as the High-Risk Pool. The primary carrier is the insurance company that originated the policy. A purchasing group is an organization of high-risk companies banded together to be more attractive to an insurer. Purchasing groups do not share in the monetary risks and benefits as the self-insurance groups do. A self-insurance group is a form of mutual insurance in which employers group together to provide workers' compensation and employers' liability coverage. A self-insured situation assumes direct financial responsibility for potential losses. However, the group affords them the resources that they may not have individually.

When insurance companies offer coverage for sale the market is voluntary. A residual market provides employers the opportunity to purchase insurance not otherwise available. A retro plan is an insurance plan that provides for the reevaluation of a company's rate at the end of the policy period based on claims experience. Specialty carriers are insurance companies that offer very few or unusual types of insurance.

State funds are available in two types. Noncompetitive state funds are established by a state agency as the sole insurer for workers'-comp coverage. Only six states administer this type fund: Nevada, North Dakota, Ohio, Washington, West Virginia, and Wyoming. Other states establish competitive state funds, competing directly with other insurers. A second-injury fund is a state-administered fund for employees hired with existing conditions acquired at a previous job.

Insurance companies stand to lose large sums of money, too. To offset this risk, reinsurance is the strategy. Reinsurance occurs when one insurance company, called the reinsurer, assumes all or part of a risk originally undertaken by another insurance company, the ceding company.

A maximum weekly benefit is paid to a worker deemed to be temporarily, total disabled, as determined by the state workers' compensation agency. The amount varies considerably, based on either the average weekly wage in the state or the individual's average weekly wage.

Compensation is awarded to employees for work-related injury, disease, and death without regard to fault or liability of the employee or employer. The Notice of Loss is a formal report of a claim by the employer to the insurer. Mediation is one form of alternative dispute resolution that encourages parties to arrive at a compromise. Mediation may take place in or out of a judicial setting.

Success at managing workers' compensation depends on predictability. The insurer should have the ability to anticipate the number and size of claims. Predictability diminishes quickly when long-tail claims are awarded on an inconsistent basis.

Long-tail claims are those claims made or paid after the insured's policy has expired. This might happen, for instance, if a production worker filed a claim in 1999 for an injury that occurred in 1989.

Return-To-Work

A return-to-work process brings an injured employee back to work as quickly as possible without risk to the employee's health and well-being. This benefits the company by reducing workers' compensation indemnity costs by as much as twenty to forty percent. Employees can be productive while recovering from their disability. The obstacles of reintegrating the employee after full recovery are removed. Return-to-work provides the employer with an experienced worker who can help meet quality and productivity goals. Long-term disability benefits are eliminated. There is need to hire and train a replacement. On-the-job training of a new employee with its impact on productivity is eliminated. The employee is made to feel positive about his or her contributions to the organization, even with limited capacity. Emotional problems, such as depression, that are associated with long-term disability are also prevented.

Return-to-work takes place in four phases. The case management team is established in Phase I. Time is of vital importance. As the number of lost workdays increases for an injured employee, the chances of him or her returning to work diminish, according to Figure 4-5.

Figure 4-5
Probability of Return-to-Work Based on Number of Lost Work Days

Lost Work Days	Probability to Return-to-Work
6 months	50%
1 year	25%
2 years or more	0%

Establish communications with the attending physician and the injured employee as soon as possible. At least by the day after the injury, visit the employee in the hospital or at his or her home. Obtain information from the physician about the exact nature of the injury and any restrictions on activity due to the injury. The employee's immediate supervisor should also contact him or her in an empathetic manner, wishing for a speedy recovery. The supervisor may also hint about the possibility of return-to-work, just so the subject is broached. However, do not pressure the employee. The important thing at the moment is recovery. Social conventions such as sending flowers and get well cards are also appropriate.

Inform the workers' compensation administrator promptly. Experience has proven that reporting delays substantially increases costs. A delay of thirty days or more typically increases costs by one-third.

A team approach is vital to a successful return-to-work program. The team leader should be a human resources specialist. The leader has the responsibility to establish communication lines among the team members, coordinate activities, disseminate information, and for overall management of the process. The remainder of the team is composed of the injured employee, his or her supervisor, the health care provider (either the attending physician or a rehabilitation therapist), the ergonomist, an engineer who has authority to plan and implement changes, a representative of the bargaining unit, and the safety manager.

The team should educate itself about the type of injury or illness the employee is experiencing. The team leader must take the initiative to obtain the relevant information from a physician or a medical handbook. The vital information includes:

- Damage done to the body structure, tissues and/or organs
- Type of disability due to the injury/illness
- Physical, physiological, and psychological problems associated with the disability
- Brief description of the therapeutic and rehabilitative measures recommended by the physician
- Prognosis for such a disability
- Type of disability.

Disability

The type of disability has standard definitions.

Temporary partial (TP) disability anticipates the employee's full recovery. Such disability, prior to maximum medical improvement, results in reduced earnings but not in total incapacity. A partially disabled patient who shows rapid signs of clinical improvement falls under this category and has a very good chance of returning to work very soon.

Temporary total (TT) disability renders the employee incapable of any work for a limited time. Such disability, prior to maximum medical improvement, is of a nature that prevents return to work. However, full recovery is eventually anticipated.

Permanent partial (PP) disability implies the injured employee will not fully recover. However, the employee has some potential to do work. The disability is some form of permanent anatomical impairment that has an effect on subsequent employment. PP disabilities include loss of a finger or disfigurement.

Permanent total (PT) disabilities preclude the employee from returning to work ever. Some ambiguity exists between impairment and disability, which are not the same. *Impairment* is an anatomic or functional abnormality. For example, a facial disfiguration is an impairment. Also, a restricted range of motion for the hand is an impairment. A *disability,* on the other hand, is a limitation in performance, which may be caused by an impairment. One is disabled, for example, if unable to dress oneself. The inability to assemble nuts and bolts is also a disability.

Until recent years, the management of prolonged disability was as simple as not allowing the worker to return to the job until totally—that is, one hundred percent fit. Under scrutiny, this is an extremely costly process compared to providing a modified workplace to accommodate the injured employee or giving him or her part-time work.

Today, post-injury disability management by many companies includes efforts to reintroduce the recovering employee to the same workplace in some capacity. Other companies would like to manage disabilities this way, but are restricted by bargaining unit agreements that are too rigid to allow the employee to work outside his or her original classification (job description). Nevertheless, for those companies that can practice modern post-injury disability management, the timing of returning the employee to work is critical. Bringing the employee back too soon could lead to re-injury or the inability of the employee to do meaningful work. The latter situation can be very discouraging and has been observed to negatively impact the employee's morale. On the other hand, as shown above, if you wait too long to bring them back, the chances of successfully doing so in any capacity diminish greatly.

Health Care

When the injury or illness occurs, the most important thing is to provide acute care by a health care provider. Treatment by a physician, surgeon, chiropractor, or rehabilitation specialist ensures that the individual gets the required rest from function with a controlled amount of mobility. Controlled mobility is therapy, too. Next, a functional capacity assessment and job analysis are performed as Phase II of return-to-work.

The functional capacity assessment is made by the health care provider. This analysis is made to determine the available capacity in the injured employee's body parts as a whole. Strength, range of motion, and endurance, (called aerobic capacity), are all examined. The assessment allows the health care provider to determine the degree of disability and to make decisions regarding rehabilitation goals.

The job analysis, on the other hand, is performed by someone at your plant, perhaps even you. The purpose of this analysis is to match the rehabilitating employee to his or her functional capacities. The employer quantifies the strength, duration, repetitiveness, and range of motion required in a particular job. If this matches with what the health care provider determined in the functional capacity assessment, the employee can return to work. If not, we still have the options of work hardening, work conditioning, and job modification.

The health care provider applies *work hardening* to gradually strengthen and condition the worker to his or her maximum possible (restored) function. Part of work hardening is performed by simulating job activities. Work hardening is necessarily individualized for each employee, based on the results of the functional capacity assessment, particularly the gap between the assessment and the job analysis. A second assessment, called the exit assessment, is made after a period of work hardening to determine if the worker has rehabilitated sufficiently. If so, it is determined whether he or she can return to work.

Another process, *work conditioning,* is coordinated between the health care provider and the employer. Work conditioning places more emphasis on conditioning strength, muscle tone, and range of motion. The employee is educated about good work methods. Nutrition counseling is provided. Psychological counseling is used to work through discouragement and to attempt to keep morale up until the worker settles into his or her new task. The worker is also conditioned by performing the job in a graded manner. This means to work only restricted hours until he or she attains full capacity.

Job Modification

Job modification is done by facility personnel. This amounts to modification of the existing job or inventing a different (but meaningful and productive) job to enable early return to work.

The return-to-work team leader needs to keep in touch with the injured employee on a weekly basis to check his or her progress. The contact must be honest and sincere. The employee should not be made to feel annoyed or feel that the team leader is prying into personal matters or putting psychological pressure on him or her. If appropriate, the team leader should personally visit the injured employee. Also keep in touch with the health care provider to decide the timing of return-to-work, which typically coincides with work hardening.

When the employee is out of acute care, and the functional capacity assessment is completed, hold a meeting of the return-to-work team. The goal of this meeting is to make the following decisions.

- How to accommodate the worker's restrictions
- Identification of an existing or modified job for the returning employee

- How to provide psychological counseling as the employee reenters the workplace

Disability Syndrome

Psychological rehabilitation is as important as physical rehabilitation for employees returning to work after an injury. Returning employees often develop the *disability syndrome,* which is a discouraging state of mind that may ultimately prevent the worker from returning to employment in a job consistent with his or her level of physical recovery. The disability syndrome manifests itself for authentic and inauthentic reasons.

One authentic reason for this is the fear of re-injury. Another is a lack of confidence that he or she can perform the same job. Also, real and permanent residuals of the injury may affect performance. Previous problems, such as depression, drug abuse, or alcohol abuse may also interfere with the recovery.

Inauthentic reasons for the syndrome are associated with secondary gain factors. The injured employee may feel that he or she is entitled to a significant amount of money as a means of acknowledging his or her injury and to compensation for all of the pain and suffering. America is, after all, a litigious society. Hence, the disability syndrome may be nothing more than a form of revenge against the employer. It may also be an escape from responsibility. Either way, these things are too complicated for most of us, which is another reason to include psychological counseling as part of the return-to-work package. Although most of us would find it difficult to cope financially, some people find security in disability income. You will probably be unsuccessful in returning these individuals to work. Another inauthentic reason for disability syndrome is the effort to maintain the level of sympathy and attention from family and community. This person needing such special attention has problems beyond the typical safety manager's capabilities.

If the disability syndrome is not identified and controlled in the early stages, the chances that the employee will return to work diminish greatly. Psychological aspects of a compensable injury must be addressed without delay. If necessary, arrange for psychiatric care as well as general counseling.

If the disability syndrome is minor in nature, however, a qualified person at the plant, such as a human resources officer, can counsel the employee. The counselor must be familiar with the employee's work history before counseling begins. The main factors to be emphasized in counseling are:

- Empathize with the employee regarding the difficulty in making the transition.
- Motivate the employee by identifying his or her strengths.
- Identify for the employee the goals of the company.
- Provide an atmosphere of confidentiality.
- Demonstrate cases of successfully rehabilitated employees.

Normally, the health care provider should provide the team leader with a return-to-work recommendation record. However, the team leader is responsible for keeping in touch with the health care provider to identify the point of reentry for the injured employee.

Case Closure

The priority of return-to-work is to return the employee to the same job, if medically advisable. The second choice is to return the worker to accustomed but modified work in the same department. Job modification may include a temporarily reduced work schedule, light duty work, sharing the heavier aspects of the job with other employees, altering the way the work is performed, and use of specially adapted tools. Third in return-to-work priority is placement of the employee in a different position in a different department.

The recommended action is to take the advice given by the health care provider. However, if the return-to-work team has difficulty in modifying or identifying a suitable job at the facility to accommodate the employee, the team may identify the *best available job* that suits the circumstances. The health care provider's approval should be solicited before the employee is assigned the job. Modification of the workplace is Phase III of return-to-work and is covered thoroughly in Chapter 13.

Phase IV of return-to-work is case closure. A return-to-work effort terminates when one of the following happens:

- The injured employee is fully conditioned to assume his or her pre-injury work responsibilities
- The employee continues as a full-time operator at a modified or reassigned work station due to permanent partial disability

- The employer-employee relationship deteriorates beyond repair through litigation

Workers compensation carriers will make an attempt to recover medical expenses and compensation payments from culpable third parties. This practice is increasing in frequency and judgments are often favorable to recovery. Potential third parties include manufacturers or distributors of machinery or point of operation guards, toxic chemicals, instrumentation, or personal protective equipment as well as contractors and their subcontractors, consultants, or building owners.

Purchasing

Purchasing departments are vital to the minimization of losses. This has nothing to do with their primary function of ordering equipment, raw materials, and supplies for the company at economical prices. Purchasing agents can otherwise help or hinder the loss control effort. Purchase orders and contracts should include clauses that require the vendor to supply goods and services that comply with laws and regulations. Hazardous chemical orders should be preceded by Material Safety Data Sheets (MSDSs)and technical data sheets in order that a hazard analysis of the material may be completed. Purchasing agents should insist on no samples unless the supplier agrees to retain ownership of the material.

Enlist the assistance of your purchasing department in your loss control efforts.

Waste

Minimizing waste generation minimizes loss control. Many waste materials are flammable, or at least combustible. Some waste material contributes, at least potentially, to employee exposure. Operations personnel ought to be cooperative with your waste minimization efforts as waste represents lost production. This loss is either direct or indirect as a drain on operating income.

Environmental Concerns

More and more, safety managers are being saddled with environmental responsibilities. But even if your company does not require you to manage environmental matters, they are integral with your safety efforts. Environmental releases may impact the indoor air quality. Less obvious, outdoor releases may still represent an exposure to employees as well as to the general public and the environment.

Security

Plant security is another collateral responsibility often given to safety directors. In today's society security is becoming more linked with safety by the issue of workplace violence. Although I am leaving security for others to discuss, it deserves a mention here and workplace violence is discussed in more depth below.

Fire Protection

Upon notification of any fire, evacuate the building immediately. Do not play around with questions such as "How big is the fire?", "Can we put it out quickly?" and so on. Get everyone out of the building, except firefighters, and then sort out these questions later. Evacuation delays caused by hesitation to do the right thing has led to more than one plant disaster. Right or wrong, evacuate. Then you will not suffer the embarrassment of trying to explain your logic to a judge.

What sort of emergency response organization should you have? What emergencies are you willing to commit time and resources to? Figure 4-6 lists six strategies for dealing with fire emergencies.

Figure 4-6
Industrial Fire Fighting Strategies

Strategy 1:
Do nothing. Evacuate all personnel and allow local authorities to combat the fire. Recommended for extremely small facilities.

Strategy 2:
Assign and properly train all employees to fight incipient stage fires with portable fire extinguishers. Evacuate visitors. Recommended for facilities with few employees.

Figure 4-6 *(continued)*

Strategy 3:
Assign and properly train designated employees to fight incipient stage fires with portable fire extinguishers. Evacuate all other personnel. Recommended for most small manufacturing facilities.

Strategy 4:
Assign and properly train a fire brigade to combat incipient stage fires with available equipment. Evacuate all other personnel. Recommended for small to mid-size manufacturing facilities.

Strategy 5:
Assign and properly train a defensive, fully equipped, fire brigade to combat structural stage fires. Evacuate all other personnel. Recommended for large facilities of all types.

Strategy 6:
Assign and properly train an offensive, fully equipped, fire brigade to combat structural stage fires. Evacuate all other personnel. Recommended for large chemical processing and petroleum refining facilities.

The key words here are "properly trained." Figure 4-7 lists recommended training requirements.

As you see the training requirement becomes enormous as you plan to be more aggressive in your response.

Portable Fire Extinguishers

Portable fire extinguishers are designed to fight one or more of the four classes of fire. Class A fires involve the burning of ordinary combustible materials such as wood, paper, fabric, and some rubber or plastic materials. Class B fires involve flammable or combustible liquids, flammable gases, greases, and similar materials, and some rubber and plastic materials. Class C fires are energized electrical equipment. Fires that involve combustible metals such as magnesium, titanium, zirconium, sodium, lithium, or potas-

Figure 4-7
Training Requirements for Industrial Fire Emergencies

Strategy 1:
Fire prevention for all employees. Evacuation training for all employees upon employment and when changes are made to the evacuation procedure. Semi-annual drills recommended. Evacuation goal: 5 minutes from sounding of alarm to all hands accounted for.

Strategy 2:
Fire prevention and evacuation training as above. Fire chemistry. Familiarization with fire extinguisher types. Put out an incipient fire with a portable fire extinguisher.

Strategy 3:
Fire prevention and evacuation training as above. Fire chemistry. Familiarization with fire extinguisher types. Put out an incipient fire with a portable fire extinguisher.

Strategy 4:
Fire prevention and evacuation training as above. Fire chemistry. Familiarization with fire extinguisher types. Fire safety. Put out an incipient fire with available equipment twice per year.

Strategy 5:
Fire prevention and evacuation training as above. One week of fire academy training annually for entire fire brigade as a team. One timed response drill every six months.

Strategy 6:
Fire prevention and evacuation training as above. One week of fire academy training every six months for entire fire brigade as a team. One timed response drill every three months.

sium are considered Class D. Use approved fire extinguishers only. Replace any carbon tetrachloride or chlorobromomethane extinguishers.

Place fire extinguishers where required and where needed. Class A extinguishers are required to be placed every seventy-five feet, but a standpipe/hose combination may be substituted for a Class A extinguisher. Class B and C extinguishers are required every fifty feet. Class Ds are required every seventy-five feet, if needed. Be practical. Do not be stingy. Put additional fire extinguishers where you are most likely to have a fire.

Fire Extinguisher Inspection, Maintenance, and Testing

OSHA makes the employer responsible, while the National Fire Code makes the building owner responsible, for the routine inspection, maintenance, and testing of fire extinguishers, even if an outside service is contracted. You can delegate the work but not the responsibility. Fire extinguishers must always be charged, operable, and located where needed. Do not use soldered or riveted fire extinguishers nor the self-generating extinguishers that have to be turned upside down.

You must make a visual inspection of all extinguishers and hoses monthly. The inspector's job is to look for extinguishers that are out of place or missing (then replace them), or that have been used (then recharge them), or hose horns that have trash stashed in them (then clean them out). The inspection should be recorded and the best way is to hang a tag on each extinguisher so that the inspector may initial and date it as he or she gets to it. This provides a record of the inspection and allows you to tell immediately if one has been missed—by the absence of the initials and date.

Annual maintenance checks must also be logged. These are normally performed by a contractor who leaves a dated tag hanging on each extinguisher. In this case, you do not have to maintain a separate log. Sometimes these tags even have twelve spaces on the back for the monthly inspector to record his or her own inspection.

Advanced maintenance tests must be performed by qualified persons, meaning they have been trained by the manufacturer to perform the maintenance. Hydrostatic testing of cylinders is required every twelve years and can be contracted. Each portable fire extinguisher must be emptied and have maintenance tests performed on it every six years (every other time is the hydrostatic test). Carbon dioxide bottles must be tested every five years, which does not alternate well with the hydrostatic test. Keep some spare bottles on hand so that protection is provided while used or damaged fire extinguishers are being replenished, repaired, and/or replaced.

Fire Extinguisher Training Requirements

If portable fire extinguishers are provided for employee use, employees must be trained to use them. This training is required at initial employment and again as an annual refresher. The training must include the principles of fire extinguisher use and the hazards involved in incipient stage fire fighting. Designated fire fighting employees must be trained in actual use of the extinguishers—when assigned and again as an annual refresher.

Disasters and Emergency Preparedness

No one expects to be hit by a disaster. That is what makes them doubly dear in terms of losses. Some real probability exists ($p>0$) that a disaster may strike your facility. Yet that probability is less than a certainty ($p<1$) that a disaster will happen. How risky do you like to live? Part of the risk is covered by your insurance. Does that mean you ignore the remainder? I think not. The risk represents more than a building full of expensive equipment, processes, and products. The risk of disaster represents lives maimed or snuffed out. Friends gone. It also represents your personal and business reputation. No one will blame you for a disaster if you prepared as best you knew how to, but your detractors will have a field day if you did not prepare at all or made only a half-hearted effort.

Just when does an emergency exist? Figure 4-8 summarizes Callan on when an emergency exists.

If any one of the points in Figure 4-8 is true you have an emergency incident and need to treat the situation accordingly. Avoid wringing your hands and hoping the problem will go away despite your inaction. Good but indecisive men have gone to jail for criminal negligence where fatalities occurred during emergencies. In the case of fatalities, OSHA, public safety officials, survivors, and families of the deceased are going to ask some tough questions of you and they may not be patient with your answers. Be prepared.

By the way, if you have a hazardous substance release and one or more of the factors of Figure 4-8 apply, do not be shy in calling for assistance. Make timely reports to the area OSHA office, state environmental response and local emergency response authorities, and anyone else who may require such a report.

Figure 4-8
Signals that You Have an Emergency on Your Hands

- You need assistance from an outside source to control the incident
- You need to evacuate personnel
- The situation has the potential to create an immediate danger to life and health
- The situation has the potential to create a serious fire or explosion
- The situation requires immediate attention because the danger involved is imminent
- The situation may cause high level exposure to toxics
- You do not know if you can handle the situation given the equipment, personal protective gear, and trained personnel available
- The situation is unclear or information about important incident factors is unknown

Alarm Systems

The evacuation alarm must be "distinctive," although OSHA does not define what this means. For impaired personnel, use tactile alarms. OSHA leaves it to the employer to establish an alarm procedure. Every employee must be given training on the alarm system and the preferred means of reporting emergencies. The employer has the obligation to maintain the alarms in good operating condition. Test alarm systems every two months. A back-up for the alarm system is required. A supervised alarm should be tested annually. Alarm servicing and testing must be performed by qualified personnel.

Uncontrolled Release of Hazardous Substances

Callan asks the question: when is a release of a hazardous substance an emergency? After all, we tolerate some degree of air pollution inside and outside of our plants and most of these are not considered emergencies. OSHA has three categories of releases: incidental releases, releases requiring an emergency response, and releases that can go either way, depending on the circumstances.

Incidental releases are those that do not require an emergency response. Fugitive releases from a process line may be in this category. If we have measured or estimated the releases and know the air concentration, our response is guided according to Table 4-1.

Table 4-1
Response to Specific Air Concentrations

Concentration	Response
< ½ PEL (AL)	Routine work
< STEL	Work with time limitation
> ½ PEL, < PEL * RF	Work with cartridge respirator
> PEL * RF	Work with fresh air apparatus
> STEL	Work with fresh air apparatus
< IDLH	Escape immediately
> IDLH	Rescue victims

AL = action level
IDLH = immediately dangerous to life and health
PEL = permissible exposure limit
RF = respirator protection factor
STEL = short-term exposure limit

Fundamentals of an Emergency Response Plan

First, describe evacuation routes in a document and use it to train employees. While technically you may choose to only evacuate the involved areas of your plants, my experience has been that it is safer to bite the bullet and order a general evacuation in the case of fires, especially where an explosion hazard exists. If factors in your particular situation make a general evacuation hazardous in itself then use your better judgment, but be sure to document your decisions however.

Assign an evacuation warden for every twenty persons. If you do not do so, the goal of getting a head count of evacuees within five minutes will not be achievable. Also, give each group of twenty an assigned location to meet once they exit the plant. Designate an area for visitors and assign a warden to list their names, which should be compared to the visitors log. In this effort, the receptionist should bring the visitor's log to the place where the head count will be done.

Assign individuals to perform rescue operations and first aid functions, unless you choose to wait for the arrival of public fire fighters and emergency medical responders. This latter option is only recommended for small plants where the outsiders' response time is expected to take less than six minutes.

Assign a person to alert the public fire department. Assign either individual personnel or a fire brigade to use portable fire extinguishers, as appropriate to your situation.

Assign a person to check utilities, such as electricity and water. An electrician should stand by to shut down all power if needed. Someone should also be prepared to shut off all fuel supplies to the plant. Other persons should be assigned to shut down critical operations. Another to shut down ventilation systems. In very small facilities, one or two persons can handle all these assignments. In very large facilities, you may need to assign more than one person to each task.

Hazard Identification and Preparation

The next task, hopefully completed before your first emergency, is to identify and list fire hazards and other potential disasters in the plant. The presence of any fuel material constitutes a potential fire hazard. The presence of any finely divided material (dust, powder, aerosols) also constitutes a potential explosion hazard. Identify and list all ignition sources. Be particularly mindful of static electricity, and do not forget static generated by clothing in areas where finely divided materials are located. Correct any defective handling and storage of hazardous materials.

Equipment Maintenance. Designate a person or persons to be maintain fire protective equipment. Set up a monthly inspection and testing schedule for this equipment. Develop a written procedure that tells how and when preventive maintenance is to be performed on this equipment. Keep records of inspections, tests, and preventive maintenance of fire protection equipment.

Physical Training. For plants that have structural fire brigades, whether defensive or offensive, provide routine physical training for involved personnel.

Forty-five minutes of calisthenics three times per week is recommended. An annual physical examination is also recommended. Fire academies typically require proof of a recent physical examination and conduct strenuous exercises for the duration. Provide protective equipment for structural fire brigade members.

An evacuation alarm is required for your plant facility. It must be distinctive and tested on a regular basis. Testing the alarm in conjunction with an evacuation drill is recommended so that the sound made will be recognizable. Always announce drills ahead of time. People may panic and get hurt if they think the alarm signal is for a real emergency. The purpose of the drill is to get them to walk through the evacuation often enough that it becomes second nature to them. That kind of training discipline is what holds back panic. When untrained people are surprised, panic is invited.

OSHA requires fire extinguishers and standpipe hoses—unless they will not be used and the entire facility will be evacuated instead. However, remember that the standpipe hoses are merely window dressing unless you field and train a fire brigade to use them. Unless a team is trained as a unit at the fire academy, I recommend you leave the standpipe hoses alone. Standpipe hoses can cause electrocution, and poorly trained persons on hoses tend to stay too long in a losing battle. A loose hose can also break bones, yield concussions, render unconsciousness, or damage equipment. If you cannot commit to quality fire ground training, do no more than make available portable fire extinguishers and teach your personnel to know when to call it quits and evacuate.

Inspect portable fire extinguishers monthly for proper location and usability. The agent indicator gauge should register full. For CO_2 bottles with no gauge, weigh the extinguisher. Place a tag on each extinguisher and have the inspector initial and date it each month. In this way you will know your extinguishers are being taken care of. If an outside contractor comes in for the annual inspection and maintenance of fire extinguishers, show him or her the back of the tag to record monthly inspections.

How will you proceed if your fire protection systems are impaired? How will you protect the plant until the systems are restored? Your property insurance carrier can provide a procedure for this eventuality. Follow it as closely as practical.

Provide training. Provide initial instruction for assigned personnel in use and care of equipment and other duties. Conduct retraining if job assignments change or equipment is replaced. The more vigorously you decide to combat fire, the more frequent and aggressive training should be.

Medical/First Aid

Prompt delivery of medical services and first aid to the injured prevent bad situations from becoming worse. When accident prevention does not work for any reason, medical response is necessary. OSHA requires ready availability of medical personnel for advice and consultation. Many plants elect to have an occupational nurse on site. A few exceptionally large facilities have a medical staff including one or more doctors. OSHA does not specifically require onsite medical care; it simply requires that medical care be available within a reasonable amount of time. Unfortunately, what is "reasonable" is easier to interpret in retrospect,. However, the OSHA standard specifically calls for one or more persons to be trained in first aid in the absence of an infirmary, clinic, or hospital in near proximity. In addition, basic first aid supplies must be maintained on the premises and must be made available to persons trained to render first aid, regardless of whether they were assigned or volunteered.

Another requirement is that the company must have medical personnel readily available for advice and consultation. You need a written agreement with a hospital or physician to provide advice and consultation. Also, ambulance services need to be prearranged. Even if the local government provides ambulance services, maintain a written agreement or documentation that you tried to get a written agreement in your files.

OSHA also requires written emergency medical procedures. Sandler recommends that policies and procedures cover the items in Figure 4-9.

With respect to outlining the injuries and illnesses to be treated, Sandler recommends the American Medical Association's First Aid Guide as a comprehensive, yet do-able, list.

How many first aid-trained employees are needed? According to Goforth, medical experts suggest that outside emergency medical response should be no more than four minutes away. Two employees should be trained for every two hundred employees working per shift. A minimum of three first aid providers should be on duty each shift. Therefore, you must have enough people trained to render first aid to cover vacations, illness, and other normal absences.

Figure 4-9
Medical Care Policies and Procedures

Outline injuries and illnesses to be treated.

Provide training, certification, and recertification.

Integrate recordkeeping into medical files.

Specify the level of care to be provided by each first aid provider or other responder.

Include a bloodborne pathogen program.

Integrate local medical gatekeeper, emergency care providers, and medical specialists.

Document all procedures.

Familiarize the health care delivery team with workplace hazards.

Correlate event and exposure with incident/injuries.

Post emergency telephone numbers in conspicuous places at or near telephone sets.

Medical Surveillance

A medical surveillance program is part of a secondary loss control strategy. *Primary loss control,* with respect to worker well being and health, consists of matching the worker to the job, properly designing the workplace, training workers, and practicing industrial hygiene. *Secondary loss control* consists of *medical surveillance* for bodily health and an *employee assistance program* (EAP) for mental health. Medical surveillance, per Hoffman, is the systematic evaluation of employee health to monitor the early occurrence of disease.

Four types of hazardous exposures in need of medical surveillance are hazardous substances, repetitive motion, manual lifting, and physical hazards, such as noise, vibration, and light level. Biological monitoring determines the inner dose of an exposed worker as opposed to the outer possibility of dose, called exposure. By assessing blood, urine, and hair samples we can know if the worker is receiving an exposure to hazardous material. We may not monitor for the material directly in all cases, but may monitor for a *metabolite,* one of the residue compounds after the body's metabolic system has broken down the material. The American Conference of Government Industrial Hygienists (ACGIH) has developed Biological Exposure Indices (BEI) as guidelines to make judgments about biological monitoring results.

Biological monitoring takes into account absorption into the body via routes of exposure other than inhalation. Most workplace environmental monitoring is aimed at determining inhalation exposure only. Substances measured in body fluids during biological monitoring relate more directly to an adverse health effect than traditional environmental monitoring. Since personal hygiene habits—such as smoking, hand washing, and frequent touching of the face—vary greatly among individuals, the creditability of exposures determined by air monitoring as an indicator of dose can be questionable. Individual variation of physiological parameters, particularly respiration rate, affect the amount of exposure that becomes an internal dose. Environmental monitoring is not always feasible, either, making biological monitoring a good alternative strategy.

Exposures to physical hazards such as noise, vibration, light levels, radiation, and temperature extremes can also be part of the medical surveillance program. OSHA already requires hearing check ups as part of its Hearing Conservation Program. Likewise, radiation monitoring is required by OSHA and other regulatory agencies.

An employee assistance program monitors and intervenes in cases of stress, emotional disturbance, and substance abuse that affect the work performance of many of our employees. Injury rates and employee absenteeism are often directly related to these type mental health problems. EAP has been demonstrated to be an effective approach to intervene, prevent, and manage mental health problems. The EAP is a counseling service that typically is provided on an anonymous basis. The mental health counselor shows up for private one-on-one consultations or the telephone number of the counselor(s) is (are) posted in several prominent places throughout the facility so employees are able to make confidential off-site appointments.

Companies offering EAPs typically do not terminate employees the first time they are caught on the job with traces of substance usage in their blood or urine, nor for the first time their behavior announces possible substance abuse. Instead, these workers are referred to the EAP for assistance in gaining control over their problem. As long as they remain in the program and are not caught again, most often they are retained. All that the EAP counselor is required to report to the company is that the employee is actively seeking help, not the details of that assistance. The names of persons who voluntarily seek help from the EAP are not released to the company, only the numbers of persons who use the program.

Workplace Violence

Workplace violence will occur. Wilkinson cites some workplace statistics: In 1995, 17 percent of all workplace deaths were classified as homicides, the second leading cause of workplace fatalities. Homicide is the number one cause of death for women in the workplace. The annual cost of workplace violence to U.S. employers exceeds $4 billion dollars. One study found that more than two million Americans are physically attacked at work each year. Another six million are threatened with violence. In addition, sixteen million workers are harassed every year. That means that one in every four U.S. workers is attacked, threatened, or harassed during any twelve-month period on the job.

Men are more likely to be attacked by strangers while women are more likely to be attacked by somebody that they know. Employers who have effective conflict resolution, grievance, harassment, and security programs are the least likely to experience violence on their premises. Some employers are sued by survivors and victim's dependents for negligence and for not providing a workplace free from recognizable hazards.

Why does workplace violence occur? Robbery is the predominate motive for workplace homicides. Other factors also contribute. Diversity and change in the workplace are factors of frustration for some. Family problems can be another contributor. Shifts of responsibility at home, such as during a disputed divorce, mark potentially violent times. Layoffs, downsizing, rightsizing, wrongsizing, reengineering, and dismissals are also contributing factors. Individual employee conflicts are sometimes the causative agent of violence. Technological innovations can lead to violence, especially where there is lack of retraining. Substance abuse and mental illness are common contributing factors to potential violence. And even the prevalence of violence in our society in general plays a part as well, making the availability of weapons seem an easy solution to conflict resolution. Not only are violence and weapons such common elements, but the media tends to portray and exploit violence in an effort to better sell the news.

Potentially Violent Employees

Despite the uncertainty of the process, profiling a potentially violent co-worker is an important step to preventing workplace violence. Generalized information on the potentially violent employee comes from profiled behavioral characteristics of potentially violent individuals who have been studied by psychologists and criminologists.

Workers are usually quite familiar with their co-workers and have an excellent opportunity to make daily observations of behaviors that may later prove to be threatening. Supervisors and managers frequently know a good deal about their employees, too. They have an even better opportunity to measure the change of behavior over extended periods and often detect early warning signs if they are trained in the recognition process. Behavioral warning signs are the most reliable indicators of the escalating potential for violence.

Keep in mind that indicators are red flags, but not actual evidence. One indicator by itself is even suspicious, necessarily. Even the presence of several indicators suggesting violent tendencies does not mean the individual is definitely violent. Potentially violent people have these typical characteristics:

1. Male
2. White
3. 30-40 years old
4. Withdrawn or something of a loner (single, divorces, or widowed)
5. Has few interests outside of work
6. Relies heavily on his job for self-esteem
7. Probably owns weapons (typically fascinated with them)
8. Served in the military
9. May have a history of violence towards women, children, or animals
10. Has a history of substance abuse
11. Displaces aggression
12. Blames others

13. Has poor coping skills
14. Exhibits extreme mood swings
15. Does not seem to fit in
16. Prefers to remain anonymous, blend into the crowd
17. Makes off the cuff remarks
18. Makes implied or veiled threats

Certain risk factors may contribute to a violent behavior. The at-risk employee may have a:

1. History of conflict with co-workers or supervisors
2. History of unwelcome sexual comments
3. Hidden fear of losing his job
4. Sense of persecution or injustice; feels he gets blamed for everything
5. Decreased social connection or support; staying to himself
6. Recent stressful situation related to family, finances, or health
7. Fascination of previous workplace violence incidents
8. Difficult time accepting criticism
9. Tendency to hold grudges, especially against supervisors
10. Frequent display of paranoid behavior
11. Set of opinions and attitudes that are extreme
12. Sense of entitlement
13. History of physical or verbal intimidation
14. Fascination with weapons
15. History of substance abuse

These warning signs have often been expressed as a series of questions.

1. Does the employee have difficulty accepting authority or criticism?
2. Does the employee seem to hold grudges?
3. Has the employee expressed a desire to harm co-workers?
4. Has the employee been accused of any type of sexual harassment?
5. Has the employee threatened or tried to intimidate any co-workers?
6. Has the employee's conduct progressively gotten worse?
7. Is the employee argumentative, unreasonable, or uncooperative?
8. Does the employee have extreme opinions and attitudes?
9. Is the employee intrigued by previous workplace violence incidents?
10. Does the employee exhibit paranoid behavior and/or depression?
11. Does the employee have difficulty controlling his temper?

12. Is the employee a substance abuser?
13. Does the employee have irrational beliefs and ideas?
14. Does the employee have a fascination with weapons?
15. Are co-workers afraid of the employee?
16. Does the employee display unwarranted anger?
17. Does the employee have productivity or attendance problems?

In addition to the behavioral warning signs so often associated with a potentially violent employee, other indicators are available to supervisors and managers. These are often oriented to a manager's view of employee behavior, which may be more reliable than some of the warning indicators discussed above.

For instance, the violence-prone individual may also be prone to excessive tardiness or absences. Beyond simply missing any work, this employee may also reduce his or her workday be leaving early, departing the work site without authorization, or presenting numerous excuses for otherwise shortening the workday. This indicator is particularly significant if it occurs in an individual who has previously been consistently prompt and committed to a full work day.

The violent-prone individual may exhibit an increased need for supervision. Employees typically require less and less supervision as they get more proficient at their work. An employee who requires an increased need for supervision or with whom the supervisor must spend an inordinate amount of time may be signaling a need for help. Managers should be alert to such changes in behavior and consider offering professional intervention if the situation so warrants.

The at-risk employee may demonstrate a decline in productivity. If a previously efficient and productive employee experiences a sudden or sustained drop in performance, you should be concerned. This is a classic warning sign of dissatisfaction. His or her manager should meet with the employee to determine a mutually beneficial course of action.

An employee exhibiting inconsistent work habits may also need intervention. Normally employees are quite consistent in their work habits. Should a change occur without clear reasons, the manager has reason to suspect the individual needs assistance.

Many of the classic behavioral warning signs may be identified under the category of strained workplace relationships. Disruptive behavior requires immediate intervention. Take this seriously—a worker who exhibits disruptive behavior needs counseling or assistance as soon as such behavior exhibits itself.

The inability to concentrate may indicate a worker is distracted and in trouble. It may also be an indication of substance abuse.

Violation of safety procedures may be due to carelessness, insufficient training, or stress. If an employee who has traditionally adhered to safety rules is suddenly involved in accidents or safety violations, he may be operating under stressful conditions. Stress is a significant contributor to workplace violence.

An employee who suddenly disregards personal health or grooming may be signaling for help.

A sustained change in behavior, especially a change to unusual behavior, often indicates an employee is experiencing difficulty. Workers are typically quite familiar with the personalities of their peers and are often quick to notice significant changes. Manage the work environment in such a way as to protect trust and encourage open communication so that workers who are experiencing difficult times may be offered prompt assistance.

Fascination with weapons, as noted in each of the checklists above, is a classic behavioral warning sign that should be easily recognized by co-workers and managers and should always be taken seriously. The best company policy allows no firearms on the property, not even in the glove compartment of their car in the parking lot.

Substance abuse is another important warning sign. Every organization should have a methodology in place to identify and assist an employee who is experiencing drug or alcohol abuse.

Stress is a serious and widespread problem in the workplace. An organization should have procedures in place to identify victims of stress. If you identify stress victims, be sure to also have an intervention system in place for them.

The classic warning sign of potential violence that is most identified but also most ignored is the frequent use of excuses and blaming. A worker who engages in making excuses and transferring blame is often signaling for help.

Depression is a common ailment in America, but not all depressed people are prone to violence. However, a depression that has been sustained for a period of time needs professional intervention because a violent outcome is a real possibility.

Defusing a Violent Situation

Even relatively minor violent incidents in the workplace may indicate a serious problem is developing. The top four reasons workers give for not reporting incidents are: "I did not want to get involved;" "That guy was always threatening to kill someone;" "The company won't do anything about it anyway;" and "The company did not say we needed to report stuff like this." Let your people know what you expect of them. Make it misconduct not to report threats or violent incidents.

What are motives for attacks? Revenge, embarrassment, intimidation, desire for attention, entertainment, and boredom are a few motives for violence. What jobs are at risk? All jobs, really, but the most likely occupations to be attacked are taxi drivers, receptionists, human resource/personnel managers, executives, managers, supervisors, sales workers, money handlers, and service workers. People who commit violent acts are likely to be 1) current and former employees, 2) customers, clients, or patients, or 3) the general public, in that order.

The best way to reduce the potential violence of a threatening individual is to accept the premise that the individual needs help first. Stop what you are doing and give the person your full attention. This is critical because the individual often feels his or her grievances go unnoticed. Make eye contact with him. Attempt to get him or her to move to a quiet area where you can sit down and discuss the problem. Try to get the employee to sit down.

If confronted while standing, do not stand squarely in front of the individual. Turn your body slightly and stand at an angle. Do not cross your arms. In America, this is a classic threatening stance—square in front of your opponent with your arms crossed, as if daring him or her to cross the imaginary line between you. Use a less threatening stance instead.

Speak clearly in a calm, non-threatening manner. Never raise your voice or show emotion. Let the worker have his or her say. Listen attentively. Show concern and interest in what is being said. Ask for specific examples of grievances. Be careful to clearly define the problem. Explore all sides of the issue by offering clarifying feedback to ensure that you understand what is being said. Ask what can be done in his or her opinion. Ask specifically how you can help resolve the problem. Ask the individual why he or she feels that way. Ask him or her to name the real problem. Try not to become part of the problem. Ask how the problem should be resolved. Ask for an example of what is wanted. Encourage the individual to keep talking. What can be done to help solve the situation? How long has he or she felt this way? Never commit to or promise something that cannot be delivered. Make sure to follow-up. The purpose of this strategy is to let the person calm down and start focusing on facts, not opinions or personality dynamics. This allows you to diffuse the situation, calm the person, and increase intellectual control over an emotional reaction.

When a person is extremely angry and has nearly crossed that line separating civility from violence, special care must be taken to prevent a sudden escalation into violence. Be aware of the employee's behavior and physical gestures. The body reflects the intent of its owner. Make eye contact. Give the individual your whole attention and listen. Honor his or her anger. Do not try to "talk him or her out of it." Acknowledge this emotion and frustration. Do not pretend you do not know why he or she could possibly be angry.

Avoid an audience when attempting to diffuse anger. Establish appropriate boundaries for your own protection. Onlookers may make inappropriate comments or gestures that will thwart your efforts. Arrange yourself so that the employee cannot block your exit from the area.

Speak softly, slowly, and clearly. Project calmness, move and speak slowly, quietly, and confidently. Be sure to ask questions after he or she has finished talking. Be an empathetic listener. Encourage him or her to talk while you listen patiently.

Maintain a relaxed yet attentive posture and position yourself at a right angle rather than directly in front of him or her. Be reassuring and point out reasonable options and choices. Break big problems into smaller, more manageable ones.

Accept criticism in a positive way. Do not take it personally. Ask for recommendations. Repeat back what you feel is being requested of you. Establish ground rules if any unreasonable behavior persists. Avoid any physical contact whatsoever. If threatened with violence, calmly describe the consequences of any violent behavior.

When confronting an individual experiencing strong emotions, including anger, you should avoid doing certain things. First, avoid using any communication that generates hostility, such as apathy, brush off, coldness, condescension, going strictly by the rules, or giving the run-around. Do not reject all demands from the start. Do not pose in challenging stances such as standing directly in front of the individual, or putting arms on hips akimbo, or crossing your arms. Whatever you do, do not get into a shouting match, point your finger at, or stare at the potentially violent employee.

Do not make any sudden movements that may be seen as threatening. Be continually aware of your tone of voice, volume, and rate of speaking. Do not challenge, verbally threaten, or make a dare. Also, do not belittle or make him or her feel foolish. Do not be critical or show impatience. And since you are not an expert, do not attempt to bargain.

Especially do not clown around in an effort to make the situation seem less serious than it really is. Again, do not make false statements or promises that you cannot keep. When emotions are high it is not the time to impart technical knowledge or complicated information. Do not take sides, but do not agree with any distortions that you may hear. Do not suggest to the individual that he or she is overreacting.

Make certain not to invade his or her personal space in any manner. Leave a gap of at least three to six feet between you and the individual at all times. You do not in any way want this person to feel an invasion of personal space or that you will act in any way to restrain him or her.

Handling Threats

Of course, not every threat made in the workplace is going to lead to violence. However, adopt a zero tolerance regarding threats. Do you tolerate sexual threats, taunts, and innuendoes? Of course not. Neither should you tolerate threats. When a threat of violence has been made, record exactly what statements were made. Who made the threat? Against whom was the threat made? How was the threat communicated? Under what circumstances were the statements made? What was the time and place of the incident? Were weapons involved? Did the threatening party make any physical gestures that suggested follow through? Are other witnesses available? What information can they provide? Have any prior instances of threats or violent conduct been observed from the threatening employee? When? What were the circumstances? Were those threats reported? Have any significant changes occurred in this employee's behavior recently? Has his or her work performance changed significantly lately? Does this employee have a history of disciplinary problems? Has he or she recently been denied a promotion or transfer or had any reason to believe he or she might be laid off?

Unfortunately, violence in the workplace cannot be totally eliminated. By recognizing the potential for violence and the behavior signals of at-risk employees, however, we can take proactive steps to reduce the risk of an incident.

References

Callan, Michael J. "When Is a Hazardous Release an Emergency?" *NFPA Journal.* March/April 1994, pp. 18, 109.

Casada, M.L., J.Q. Kirkman, and H.M. Paula. "Facility Risk Review as an Approach to Prioritizing Loss Prevention Efforts." *Plant/Operations Progress.* October 1990, pp. 213-219.

Coletta, Gerard C. "Benefits from Your Risk Control Program." *Professional Safety.* March 1983, pp. 34-38.

Daugherty, Jack E. *Assessment of Chemical Exposures: Calculation Methods for Environmental Professionals.* Boca Raton, FL: Lewis Publishers, 1997.

Daugherty, Jack E. *Industrial Environmental Management: A Practical Handbook.* Rockville, MD: Government Institutes, Inc., 1996.

Factory Mutual System. *OSHA's Revised Fire Brigade Standards.* Pamphlet P8125, September 1989.

Gallagher, Vincent A. "Recognize Design Defects and Reduce Workers' Compensation Insurance." *Professional Safety.* September 1988, pp. 22-25.

Goforth, Candace. "The Critical Care." *Occupational Hazards.* September 1996, pp. 21-22.

Haddon, William, Jr. "On the Escape of Tigers: An Ecologic Note." *Technology Review.* May 1970.

Hoffman, Benjamin H. "Blueprint for Health Surveillance." *Occupational Health & Safety.* April 1993, pp. 50-55.

Jakubowski, Jake A. "Lessons Learned: Management of Change." *Professional Safety.* November 1996, pp. 29-31.

Keller, David B. "Putting Risk Assessment in Perspective." *Professional Safety.* February 1992, pp. 35-38.

Lark, James. "Risk Taking: Perspectives and Intervention." *Professional Safety.* November 1991, pp. 36-39.

Philley, Jack O. "Acceptable Risks?" *Professional Safety.* May 1991, pp. 32-36.

Philley, Jack O. "Acceptable Risk: An Overview." *Plant/Operations Progress.* October 1992, pp. 218-223.

Sandler, Howard M. "Fitting First Aid and Emergency Response into Your Occupational Health Program." *Occupational Hazards.* September 1996, pp. 43-45.

Sheridan, Peter J. "Risk Assessment: The Path of Uncertainty." *Occupational Hazards.* October 1990, pp. 68-72.

Smith, Thomas A. "A New Loss Control Management Theory." *Professional Safety.* February 1988, pp. 30-33.

Waddel, G. "A New Clinical Model for the Treatment of Low-Back Pain." *Spine.* December 1987, p. 12.

Wilkinson, Bruce S. "Preventing Violence, Threats, and Attacks at Work." Mississippi Manufacturers' Association. 3rd Annual Safety Conference and Exposition, May 7, 1997.

"Workers-Comp Update: Speak the Language (Part II)." *Safety + Health.* June 1997, pp. 24-5.

5

AUDITING, PERFORMANCE MEASUREMENTS, AND RISK

SAFETY AUDITS

Compliance of the safety process to laws, regulations, and company policies and procedures is measured effectively by audits. Self-audits can be effective, if done with objective honesty. According to Manuele, an effective audit is a systematic technique used to prepare 1.) a precise appraisal of safety process effectiveness, 2.) a diagnosis of safety problems, 3.) a description of where and when to expect trouble, and 4.) suggestions concerning what should be done about these problems. Unfortunately, most audits focus only on what employees are doing wrong, reinforcing a negative workplace culture.

External audits of the safety process are less likely to be self-serving and more likely to be objective than internal audits. An OSHA inspection is the true test of compliance, at least at the time of the inspection. Once the compliance officer leaves, you can no longer make a valid statement about your current compliance status. Also, with only a few exceptions, an OSHA inspection has a limited scope.

Some managers do not understand the value of a comprehensive audit or even self-inspection. The thinking is, "why discover things about yourself that will then need to be fixed?" "If we're getting by without this self-knowledge, don't fix it." "Don't change." "Let's not rock the boat." This type of manager sees inspection reports as a long list of defects or dirty laundry. Too many deficiencies are a potential threat to job security. Deficiencies make us look like we're incompetent. This type of manager feels that his or her daily presence should be enough to detect existing hazards and deter unsafe behaviors. Undetected hazards are no worry to them. Out of sight, out of mind, is the thinking. Other managers do not feel capable of making safety inspections. "Leave it to the Safety Department," they say.

Bad experiences with safety inspections or audits can have negative effects in an industrial facility. They:

- Can be construed as fault-finding;
- May produce only a superficial laundry list,
- May produce no remedial actions;
- May produce only inadequate remedial actions.

Safety audits are certainly a waste of time if the management leadership team does not wholeheartedly support them. If every maintenance work order you have generated from audits in the last three years is still stacked up on your desk awaiting completion, you may as well cease auditing. The purpose of auditing and inspecting is to *find and eliminate* hazards. If hazards are not getting eliminated, you may be doing more harm than good.

The other purpose of auditing is to assess the effectiveness of the safety program. If the findings of audits are merely stacking up, the safety program exists in your dreams. If this is not a reflection on you, perhaps more gainful employment may be found elsewhere.

Arnold suggests several principles for objectivity. First, an audit checklist should be designed. A checklist should help to reduce bias. Second, the auditors need to be familiar enough with the subject to delve into the spirit and intent of the audit rather than merely sticking to the audit instrument. Third, a prudent audit technique must also be followed. The auditors must establish that systems are in existence by verifying whether they are in place and operating properly. It is not enough for someone to show you a binder of procedures and claim that the safety program is "in there." In this case, it is a binder of paper and nothing more. Take a procedure from the binder and go out onto the plant floor and see if it works and is being used. Finally, auditors must avoid reaching hasty conclusions from the audit findings. Avoid forming first impressions. At least be willing to challenge your first impressions by testing them against reality.

Purpose and Scope

An audit or inspection is the monitoring function conducted in an industrial organization to locate and report existing and potential hazards, or management systems or conditions, that have the capacity to cause accidents or illnesses in the workplace. Audits can be viewed negatively or positively. They can be seen as *faultfinding,* with emphasis on criticism, or *fact-finding,* with the emphasis on locating potential hazards.

A safety audit should influence the culture of the audited facility with respect to its attitudes about safety. The scope of the audit must not only be the current operations but also design and engineering. As Manuele points out, the proof of management's commitment to the safety process lies in its insistence on safety review and safety by design at the planning stage. Design and engineering practices may far more about your safety process than how your employees behave.

Safety and health inspections of the workplace should cover all areas. That is not to say that you must do a wall-to-wall inspection each and every time. However, over a period of time—perhaps a year for large plants—your inspections should cover every department and general condition as discussed below.

As far as functional departments go, inspect manufacturing processes, receiving, shipping, material storage, and maintenance. Include housekeeping plus buildings and grounds conditions. Check electrical equipment, lighting, heating, ventilation, and air conditioning. Take a good look at manufacturing machinery and vessels, as well as hand and power tools and ancillary equipment. Look at the handling, storage, and transportation of chemicals. Examine personal protective equipment. Check the fire prevention program. Do not forget the status of training and the experience level of personnel.

Audit Procedures

Two types of inspections are the *continuous* and the *planned.* The supervisor and his or her employees do continuous inspecting everyday, all day. They should be constantly vigilant for conditions and behaviors that are risky. The planned inspection is the type of audit you normally think of, discussed here in detail.

The three types of planned audits are *periodic, intermittent,* and *general.* The periodic inspection is one that is made on a regular basis and with an announced schedule. The intermittent inspection is generally made on an irregular basis and it is always unannounced. A general inspection is a scheduled inspection that focuses on the out-of-the-way places, such as air ducts or storage areas.

Audit reports should be written on the assumption that they will become public documents. Therefore, anything that you do not wish to be made public has no place in the audit report. Certainly any statement made in an audit report could be admissible as evidence in court proceedings. Identify and describe hazards, violations of company policy, or violation of OSHA regulations very carefully. Mitigating factors should be carefully documented.

The report should not exaggerate, overstate, or oversimplify safety and health concerns.

Also, the report should be carefully analyzed for causes of defects, not just for quick fixes. Information obtained from inspection reports should not be used punitively. Otherwise you can forget cooperation in the future.

The report should always be accompanied by an action plan that addresses how all problems will be resolved within a reasonable time period. The audit should not be undertaken without a commitment to follow up on the concerns identified.

Audit Initiation

Senior management should explicitly ask corporate counsel, in writing, to undertake the audit. This establishes a potential for attorney-client privilege, which is a communication made in confidence to an attorney by a client, even a potential client, for the purpose of seeking legal advice. The privilege has four elements: 1) use of an attorney to provide legal advice, 2) communication between a client and an attorney, 3) confidentiality of the communication, and 4) absence of a waiver.

The request by management made to the attorney to initiate the audit should expressively seek his or her talents as a legal advisor. The attorney's function does not include technical or non-legal advice even if he or she is capable of rendering such advice. Purely technical advice from the attorney will not be privileged. The attorney cannot be just an audit team member who looks for uncontrolled hazards and procedural deficits, but must act as an attorney over and above anything else he or she may do.

The initial request should empower the attorney to interview employees at all levels, review documents, and retain outside consultants. The parties privy to the information are, generally, the company, its employees, the attorney, and consultants who assist during the audit. The client-attorney privilege may extend to the outside consultant's communications, if he or she is an agent of the attorney's.

The desire for confidentiality must be mentioned in the communication with the attorney. The attorney can allow others to do the legwork, but he or she must remain in charge of the process. All communications must be made in confidence, and that confidentiality must be protected by all. If a confidential report is given to someone who is not in the audit loop, the chances of a judge upholding the attorney-client privilege become diminished. A privilege is waived as soon as the information is disclosed to a party who was not privy to the information, unless to another attorney representing a client in a matter of common interest. The communication must, in that case, facilitate the rendering of legal services to both of the clients involved. Once claimed or privileged, the audit must not be disclosed in any context. No waiver will be granted.

An impediment to attorney-client privilege is that the protection extends only to the communication, not to any underlying facts. Also, courts may not apply this rule to regularly conducted audits.

As long as the attorney manages the audit in anticipation of a potential litigation, the audit report may also be protected by the work product doctrine. Some very strict rules apply here. You need to consult with your attorney about both the work-product doctrine and client-attorney confidentiality with respect to an audit. The work-product protects from discovery an attorney's work product in anticipation of litigation. Underlying facts may still be discoverable. Also, audits are not typically conducted in anticipation of litigation, so use this protection wisely.

Audits are performed the same way inspections are: with the use of checklists. You need to be extremely careful with checklists, especially if you are inexperienced in this area. The checklist in Figure 5-1 is adapted from Emery and Savely. from MSDSs, container labels, and storage signs whether cyanide is present.

Tailor the checklist to the abilities of your auditors or the information gleaned by the checklist may be meaningless, or at least defective. Checklists are merely aids, not ends in themselves. The advantage of using a checklist is that you have an organized way of looking at the workplace, you know where you are at any given time in the inspection process. Keep in mind that checklists can assist the inexperienced and unfamiliar with the inspection but you should be aware of their limitations.

Because it is a canned inspection program, the chief disadvantage of a checklist is that it is limited in scope. This is especially true for the inexperienced safety manager. Checklists in the hands of the novice are a temptation to commit *gundecking,* an old Navy term for making a false paper trail. A final disadvantage of using checklists is that a list invariably draws the user to the quick fix as opposed to the root or underlying cause. Checklists, to be of maximum use, need to be devised, at least selected, by a wise and experienced member of the staff and carefully worded at that.

Figure 5-1
Emery and Savely's Inspection Checklist

Physical Safety

Check for the presence of the following:

Electrical Hazards	Non-ionizing Radiation
Fire Extinguisher	Ionizing Radiation
Confined Spaces	Infrared Devices
Fall Hazards/Egress	Lasers
Heat	Intense Visible Light
Noise	Magnetic Fields
Vibration	Radio Frequency Waves
Repetitive Motion Hazards	Microwaves
Lifting Hazards	X-Ray Producing Machines

Chemical Safety

Check for presence/storage of the following chemicals:

Acids	Cleaners/Solvents	Halogens
Adhesives	Compressed Gases	Hazardous Waste
Alcohols	Cryogenics	Inks/Dyes
Aldehydes	Cyanides (*sic*)	Isocyanates
Alkalis	Epoxy Resins	Metal Fumes
Amines	Esters	Metals (heavy)
Ammonia	Explosives	Lubricants
Aromatics	Flammables	Oxidizers
Asbestos	Freons (CFCs)	Paints
Chromates	Fuels	Pesticides

Biological Safety

Check for potential presence of the following:

Bloodborne Pathogens	Allergens
Foodborne Pathogens	Fungus
Other Pathogens	Insects

Inspection Personnel

Who should do the inspection? Management has the ultimate responsibility for auditing, but should get as many people involved as practical in the auditing process. Self-inspection of the facility is important. Foremen, supervisors, safety committee members, safety engineer, and plant managers are all potential inspectors. This list is not intended to be all-inclusive, either. If not experienced, the inspector should be under the guidance and tutelage of an experienced inspector. Supervisors are important inspectors because their job is to require conformance with regulations, standards, and procedures. In cases of finding imminent danger, the supervisor has more credibility for shutting down a dangerous process or task than anyone else. Although any inspector should be able to do so, normally, the inspector's job is merely to inspect and report, leaving the operations alone.

A follow-up or deficiency tracking system should be devised so that appropriate personnel may be held accountable. Assign a tracking number to each deficiency and record the date it was found. Include a brief description and the location of where it was found. What is the appropriate OSHA standard for corrective measures? Record its hazard rating by potential consequence and probability. List the corrective action required, step-by-step. What is the estimated cost of correction? When was it corrected? How much did it actually cost to correct? Keeping track of inspection findings in this manner establishes clear proof of your good faith effort to make the workplace safer.

PERFORMANCE MEASUREMENT

Some companies are now using performance scores to determine efficiency of safety and health processes. If you cannot measure it, the thinking goes, you cannot control it. Measurement permits identification and prioritization. It also identifies which efforts are most rewarding and allows prioritization of company-wide issues.

The purpose of performance measurement is to continually improve. For great aims we must dare great things (Clausewitz, 1812). Zero accidents is the only worthwhile goal of a safety process, even though it is not achievable. Accidents are the best measure of safety. Illnesses are the measure of occupational health.

Pro

The benefit of safety metrics is that it arms you with information to identify where problems exist. Measurement of selected critical activities can help you identify areas to focus additional resources that will improve safety and health and reduce workers' compensation and other costs. Safety metrics allows you to make decisions based on facts, not merely as a reaction to an accident.

Con

Be careful how you use performance statistics. For instance, the incident rate measurement may be misleading. A common belief among safety professionals is that low incident rates indicate the safety process is working well. As Mathis points out, however, that is true only of the long term. If you keep monthly statistics, all the short-term incident rate tells you is that few or a lot of accidents happened during that month alone, depending on the rate. Short-term incident rates cannot be used to predict longer-term trends.

The same holds for year-to-date incident rates. This statistic will not give you a clue about the development of trends until late in the year when the year-to-date IR starts approaching the annual IR. In the interim, the year-to-date IR is noise on the information highway. If you wish to spot trends as they develop, use a twelve-month rolling IR. Anytime this number increases from the previous month, find out why. If a good reason does not exist, you have a problem.

Another concern about using a performance measure is that negative results put the onus on your company to take immediate measures to safeguard employees. If you do not take these steps, you could be liable for criminal instead of civil penalties. Therefore, it is crucial that performance measurement data be handled carefully. To minimize your potential liability, consult with corporate legal counsel about relevant policies for protecting your data from discovery and whether collecting the data triggers any regulatory requirements.

Shivers points out three potential flaws in such an evaluation process, when humans are the object, that he calls 1) the halo effect, 2) the horns effect, and 3) the Hawthorne effect. The first two terms he attributes to D.L. Kirkpatrick, *How to Improve Performance through Appraisal and Coaching* (1982). The latter term was coined in the 1920s when E. Mayo tested observations made by a group of Western Electric engineers in the Hawthorne experiments.

The Halo effect is a tendency to overrate a person being observed. Figure 5-2 list seven reasons for the Halo Effect.

Though these factors describe flaws in the human resource process of performance evaluation, they apply to the observation of safety performance in the workplace and explain why some people get disciplined for violating safety rules while others do not.

The Horns Effect is the opposite of the Halo Effect, as you might have guessed. Eight causes for this are listed in Figure 5-3.

The results of the Hawthorne experiment were that productivity, and hence quality and safety, improved no matter what physical changes were introduced to the workplace. Even negative physical changes produced positive results. Therefore, when measuring the performance of the safety process, you must be careful that the mere process of evaluation does not cause a positive change from baseline conditions. Such a change is welcome, of course, from the accident prevention standpoint. But if you draw erroneous conclusions, you may get a nasty surprise later.

Figure 5-2
Reasons for the Halo Effect

1. Effect of Past Record—Past good performance leads to the expectation of continued good performance, which carries over to future evaluations.
2. A person who has a pleasing personality and who is agreeable and compatible is typically overrated.
3. Effect of Recency—Recent outstanding behavior overshadows a preceding lengthy period of poor performance.
4. A person with an asset deemed important by an observer, even though irrelevant, may be overrated.
5. Blind-Spot Effect—The observer may overlook a bad or undeveloped trait if he or she possesses the same trait.
6. High Potential Effect—A worker may be judged by his or her potential instead of actual measured performance.
7. A person who never complains tends to be evaluated in a positive light.

Figure 5-3
Reasons for the Horns Effect

1. The high expectations of the evaluator cannot be met.
2. Someone who is disagreeable and argumentative is typically underrated.
3. Oddball Effect—A nonconformist is typically underrated.
4. Poor group performance leads to a lower evaluation of all members of the group, even if one has outstanding performance.
5. Guilt-by-Association Effect—A person is evaluated the same as those whose company he or she keeps.
6. Dramatic-Incident Effect—A recent mistake may overshadow a long period of good performance.
7. Personality-Trait Effect—An evaluator may associate aggressiveness, arrogance, passivity, and other character traits with poor performance.
8. Self-Comparison Effect—An evaluator may underrate a person who performs a task differently than he or she would.

Safety Metrics

Safety metrics is the measurement of key processes in order to permit process control. The purpose of safety metrics is to provide an objective basis to determine process performance. From the corporate level, metrics identifies divisions or facilities that need to improve within the safety process. At the plant level, metrics identifies departments and work areas that need to improve. Metrics can be used to determine whether an effective, basic safety process is in place at the facility. Finally, metrics can be used to identify and assist in the analysis of safety issues.

Safety metrics just does not happen overnight. Someone must establish a safety metric system and someone must be made accountable for recording the measurements and reporting the results. Also, a company's corporate or division headquarters must review these numbers, but must also review the facility reporting the numbers in order to be able to analyze and interpret them. Finally, someone must educate facility safety managers and plant managers on the new system of measurements so that they can understand the information they receive from and submit to corporate headquarters.

OSHA's strategic goals (Figure 5-4) as summarized by LaBar are worthy of imitation.

Figure 5-4
OSHA's Interdependent and Complementary Strategic Goals

1. Improve safety and health for all employees measured as fewer hazards, reduced exposures, and fewer injuries, illnesses, and no fatalities.
2. Change the workplace culture to increase management and employee awareness of, commitment to, and involvement in the safety and health process.
3. Secure public confidence through excellence in the development and delivery of OSHA's programs and services.

Figure 5-5, adapted from LaBar, summarizes some individual goals worthy of consideration.

Figure 5-5
Individual Safety and Health Goals

- Reduce most prevalent injury types by 10 percent per year.
- Increase direct employee involvement in safety-related activities by 50 percent, including goal setting committees, teams, task forces, and training and education.
- Continually improve compliance with company requirements for personal protective equipment and safe work practices.
- Build general safety and health awareness with attention to off the job and family safety.

If accidents were purely random events, then we could predict the number of accidents in our plant by using the Poisson distribution. Assuming that everyone has a 50-50 chance of injury:

$$P(x) = \frac{e^{-0.5}(0.5)^x}{X!} N \qquad [5.1]$$

where

$P(x)$ = probability of event occurring x times
N is the size of the group.

The expected number of injuries in groups of one thousand workers with x injuries is:

Number of injuries, x	Number of individuals injured, P(x)	% Workers	% Injuries
0	606	60.5	0
1	303	30.5	61
2	76	7.5	30
3	13	1.5	9
4	1-2	40	100

The significance of your incident rate is the difference between the actual incident rate and the number of accidents that may have occurred quite randomly. But the raw incident rate may be a smaller number than predicted here. Does that mean your plant does not have a significant problem? No, because only a portion of accidents are reported. Not only that but, for the sake of this statistical exercise, near misses are accidents, too. If one tenth of accidents and near misses get reported, then multiply the incident rate by ten. Now, does it look like you have a significance problem?

Introduction

The first metric requirement is to create comparable measurements and objective standards. Comparable measurements are those that are *apple to apple.* When comparing company components, remember that divisions and facilities vary in size. Departments within one facility are not equal in size.

Sometimes it is unfair to compare measurements of departments in the same plant unless you build in equalizers. Besides that, an objective standard must be created for all measured organizational units to achieve. To do this you must define the desired level of safety in terms that can be measured. Then you must identify the level of performance that is acceptable. Target resources at the division and facility level so that the worst performing units can improve the quickest. Also define measurements that allow you to track and control the safety process.

Measurables

Performance measurements are grouped into two categories: qualitative and quantitative. *Qualitative* performance measurements, such as checklists, are commonly used to assess the status of safety and health programs. *Quantitative,* or objective, performance measurements are necessary for precision, consistency, and reproducibility. Figure 5-6 compares these two types of measurements.

Quantitative performance measurements are either outcome or process oriented. *Outcome* measurements assess the *effectiveness* of a program by measuring the prevention of worker injury and illness as the reduction in number of cases. Outcome measures are innately after-the-fact measurements. Improvements in your safety and health program verified by current measurements may reflect preventive actions taken months or years before. Outcome measures are also nonspecific. That is, the injury/illness incident rate does not illustrate root causes of either improvement or decline. Additional analysis is required to fathom this information.

The *efficiency* of a program is determined by *process* measurements. Efficiency, in this case, is how well, how promptly, or how completely a program has been conducted in pursuit of the outcome desired. Process measures can reflect the efficiency of individual program elements in real time expressed as percentages on schedule or percentage completed. However, process measures do not necessarily predict a program's outcome, even though a carefully selected set of process variables can yield a valid forecast of program outcome.

Table 5-1 compares process and outcome measures.

Many measurements are available, far more than I can describe here. Not every measurement fits every workplace. On a fundamental level, measure-

Figure 5-6 Comparison of Performance Measurements	
Qualitative	**Quantitative**
Evaluates progress in implementation of management practices.	Incident rate
Surveys perception of employees and public.	Lost work day incident rate

ments are indicators of performance. On the whole, they provide a barometer that indicates the status of a program. They also serve as a meaningful continuous improvement tool. Two or three deliberately chosen measurements fitted to your situation can be a great help to you.

Statistical Analysis. Many organizations use the OSHA 200 reporting system on which to base their safety measurements. After all, these records need to be kept anyway. Use of the 200 log allows you to develop a measurement of the rate of injury. In fact, that is what it was designed to do.

Table 5-1
Comparison of Outcome and Process Measures

	Outcome	Process
Strengths	• Comprehensive • Direct measure of effect	• Selective • Early feedback allows predictions • Adjustable
Weaknesses	• Nonspecific • Retrospective • Delay between program changes and data	• Not comprehensive • Does not assess current outcome

Accident records can be examined statistically to help you make these performance measurements. They are a means of making an objective evaluation of the effectiveness of your safety process. The measurement variables allow you to identify high-rate areas. They also allow you to identify specific circumstances of occurrence. Shared appropriately for awareness, performance measurements can create interest in safety. They provide hard facts when presenting recommended hazard control to supervisors and the safety committee. They can also be used to measure the effects of individual hazard countermeasures.

Three measurements you can use are recordable incident rate (IR), Lost Workday Case (LWDC) rate, and Restricted Workday Case (RWDC) rate.

Incident rate is the number of incidents per two hundred thousand hours worked. By standardizing the measurement in this fashion, no matter how small or large a company may be, the frequency of accidents can be compared. The calculation of IR is as follows:

$$IR = \frac{\text{No. injuries \& illness} \times 200{,}000}{\text{total hours worked}} \quad [5.2]$$

The LWDC is similar

$$LWDC = \frac{\text{No. lost workdays} \times 200{,}000}{\text{total hours worked}} \quad [5.3]$$

Finally, the RWDC is also similar.

$$RWDC = \frac{\text{No. restricted workdays} \times 200{,}000}{\text{total hours worked}} \quad [5.4]$$

There are several and varied reasons for conducting statistical analysis of accident data. First, measurements allow you to identify and locate principle sources of accidents. They disclose the nature and size of the accident problem, may indicate need for engineering revision or disclose inefficiencies, and they may uncover unsafe practices that inspections have failed to find. Measurements also provide supervisors with useful information about the safety performance of their departments. Ultimately, they permit an objective evaluation of the safety process.

A standard you can set for your company is Best-in-Class. Determine your facility's SIC (Standard Industrial Classification) code to the 4-digit level. Using the annual Bureau of Labor Statistics report on occupational accidents, determine the weighted average of each of the above measurements (IR, LWDC, and RWDC) for your SIC code. Let the level of performance be based on how the unit compares with the SIC code average. Those units that meet the SIC code average are just adequate. Those that are above the average need to improve. Units that measure less than half their SIC code average are to be recognized.

The IR taken from the OSHA 200 log includes fatalities, diagnosed work related illness, and occupational injuries. The latter are recordable if they require medical treatment, involve loss of consciousness, result in restriction of work or motion, or lead to a job transfer. The rate is the number of injuries per one hundred full-time workers.

Metric Toolbox

Figure 5-7 lists some suggestions for safety metrics.

Incident Rate. The incident rate (IR) is discussed elsewhere in this chapter. The advantage of using this measure is that it is the one OSHA will look at

Figure 5-7
OSH Performance Measures

Outcome Measures
Incident Rate
Injury Rate
Illness Rate
Workers' Compensation Cost
Near Miss (Early Warning) Rate
Preventable Accident Index
Regulatory Visits
Citations Issued
Penalties
Lost Day Severity Rate
OSH Training

Figure 5-7 *(continued)*

Process Measures
Compliance with Company OSH Requirements
OSH Evaluation Follow-up
Exposure Assessment Strategies
Safety Hazard Assessments
Exposures above Action Level

when it inspects your facility. Therefore, it is smart to be aware of your IR at any given time. Post it somewhere that it can be seen often so it is always on your mind. If your IR is less than or equal to one-half the average IR for your industrial SIC code, OSHA looks at you as successful. If the IR exceeds one-half of the average, OSHA takes a closer look. If your IR is greater than the average IR, OSHA will typically conduct a wall-to-wall inspection of your facility.

Injury Incident Rate. Another performance measure is the rate of recordable injuries ($I_{inj}R$) as separate from occupational illnesses. While the IR is a total recordable incident rate, the $I_{inj}R$ looks at only the physical injuries on the job. You need to familiarize yourself with the Blue Book guidance on recording accidents and occupational illnesses, because if you are confused about what constitutes an injury as opposed to an illness the rate will be affected accordingly.

As in the IR, the $I_{inj}R$ is indexed to the number of employees and hours worked. One hundred employees or 200,000 hours worked is the standard. This measure allows you to assess the quality and performance of your safety program. Take your injury numbers from the OSHA log.

Measurement of occupational injuries provides a database from which you can target improvement for OSH performance. Analysis of injury cases helps to identify trend situations that require immediate hazard evaluation and/or employee training in order to achieve improved performance. Analysis of injuries allows you to identify jobs or tasks that require modification in order to improve not only safety but often productivity as well.

Inconsistencies in recordkeeping may cause injuries to be classified as illnesses, or vice versa. In some cases, a recordable injury may not even be recorded. This error can cause faulty interpretation of the injury rate. Often occupational injuries are difficult to classify due to off-the-job factors. In addition, the injury rate is only as good as the cooperation of employees to report all injuries, no matter how minor. Incentive programs are, in fact, sometimes devastating to safety metrics for this very reason. Also, minor injury data may sometimes skew the interpretation of safety performance. Large incidents may also skew results.

The $I_{inj}R$ is calculated in the same manner as the IR:

$$I_{inj}R = \frac{N \times 200,000}{H} \quad [5.5]$$

where

N = number of injury cases, and
H = actual hours worked.

Illness Incident Rate. The illness incident rate, $I_{ill}R$, is the other side of the incident coin. Occupational illnesses tend to be under-recorded. Chronic occupational illnesses, for one thing, are not easy to detect. They are also typically difficult to diagnose as work related. Employees often do not report occupational illnesses because they do not recognize the symptoms as work related, or they think the illness is not serious enough, or the symptoms too mild to even note. As in the case of injuries,—sometimes even more so—off-the-job factors can confound the classification of occupational illnesses.

The incident rate of occupational illness ($I_{ill}R$), highly dependent on accurate reporting, is calculated the same as for total incidents (IR) and injuries ($I_{inj}R$):

$$I_{ill}R = \frac{N \times 200,000}{H} \quad [5.6]$$

Workers Compensation Costs. Paid costs and reserved funds for workers compensation benefits allow us to measure the more direct costs associated with occupational injuries and illnesses. Managers tend to understand this measurement best, because it is given in terms of dollars off the bottom line.

More than any other measurement it relates to something they inherently understand: their pocketbook. This is a good measurement, then, for loosening up funds to get the more pressing safety problems addressed.

A direct relationship exists between workers' compensation costs and the accrued medical cost performance experience for your industry. Unfortunately, workers' compensation costs represent historical information. This measurement cannot, therefore, be used as an early warning indicator. Comparison of workers' compensation costs from site to site may be impaired by the fact that each state has its own workers' compensation laws and regulations and that they vary a good bit. Also, workers' compensation, unfortunately, is susceptible to fraud. Insurance companies and states are beginning to tighten the clamps on fraudulent claims, but at any given time you cannot be sure what percentage of your costs are due to fraud. Authorities have not been successful in estimating the cost of fraud. At least they have not shared this information with us, if it is indeed available.

Then, too, workers' compensation data does not correlate well with OSHA injury and illness recordability criteria. Most states, for instance, classify back injuries as "injuries," whereas OSHA classifies them as "chronic illness." Also, workers' compensation costs indicate nothing about near miss incidents or first aid cases.

Workers' compensation data is not readily available in all states. Most states use insurance companies to administer the system with a state commission to oversee and referee the process. Some states (Idaho, Maine, Nevada, North Dakota, Ohio, Washington, West Virginia, and Wyoming) administer the program fully, with no insurance company involvement. Consequently, it can be an exhausting task to get reliable workers' compensation data in a timely manner in these states. Also, no matter where your plant is located, workers' compensation costs are not the full cost to you when an accident occurs. Among other things to factor in are the loss of productivity, loss of product in the same accident, training of replacement employees, overtime to catch up, and many other indirect costs.

If you decide to use this measure you can get *loss run reports* from your insurance carrier or state. The National Safety Council prints the *Average Workers Compensation* report from which you can obtain cost data. Industry average data can be obtained from the National Council on Compensation Insurance, Inc., in its latest *Issues Report.* Some graphics that may

assist you are suggested below:
Compare measures for your company to data available for your industry. In the Employee Modification Rate (a number from the *Loss Run Report*) compare your rate to that of your industry and to that of your contractors. This will tell you whether you need to bear down on contractor safety or whether you need to concentrate on mending your own fences first.

Near Misses. The measurement of near miss incidents is based on identifiable events or conditions that have the potential to cause injury or illness. By using this measurement you can identify and respond to potentially hazardous situations before they result in injuries or illnesses. By identifying and correcting the causes of injuries and illnesses before they occur you can have a tremendous impact on accident prevention.

Since accidents and illnesses are, hopefully, few in number, statistics are not always meaningful with incident rates and workers' compensation costs. However, near misses are closer to the base of the traditional safety pyramid, giving more statistically meaningful data for analysis.

A near miss rate is easy to calculate and compare with medical treatment and lost workday cases, adding meaningful information to trend analysis. The near miss rate is a good tool to get employees involved in safety metrics. It also helps operating personnel to internalize proprietorship of their OSH program.

A good near miss reporting system can be used to integrate other reporting schemes into a single process. First aid cases, recordable OSH cases, environmental releases, and security problems can all be accomplished through a near miss reporting system.

Unfortunately, no databases for comparison of data are available. Nevertheless, near miss data can be used effectively in trend analysis. Near miss measurement does require a forward-looking attitude and an authentic resolve to prevent accidents and illnesses. Smaller companies may not have the resources and some companies, regardless of size, will not be willing to administer a near miss program. These companies would prefer to remain ignorant rather than know their weaknesses.

Near misses must be addressed promptly and with whatever resources are needed to correct any problems that are uncovered. Sources of near miss data include reports of unsafe conditions, observations of unsafe behaviors, and first aid cases.

Preventable Accident Index. The Preventable Accident Index (PAI) was developed by the Chemical Manufacturers Association (CMA) to identify

Type Graph	Y-Axis	X-Axis
Line	Average cost per claim	Year
Line	Total claim cost per total employees	Year
Bar	Claims per total employees	Year
Line	Claims per 100 employees	Year
Line	Employee Modification Rate for Industry Segment	Year

and assign a predetermined value to near misses based on the potential injury involved. The values are totaled and tracked monthly and then converted to an exposure hour rate for comparison with other safety metrics. This also allows a ranking by accident type based on the incident value. The PAI provides a method for measuring and improving the quality of safety performance by focusing on the base of the accident/incident pyramid.

The PAI, then, provides managers and employees with an objective measure for identifying and preventing non-injury events. It encourages prefatory investment of resources in order to improve safety performance. The PAI helps employees and managers to recognize the potential severity of unsafe acts or conditions without having someone suffer actual consequences. As such, the PAI provides a valuable measure for prioritizing investments in workplace safety and health.

Since the PAI is not commonly used, little data is available for comparison. Outside the chemical manufacturing industry it is almost unheard of. Unless managers understand the need for near miss reporting, and give it their full support, the PAI measure will be elusive. The danger is that the

PAI, and near miss rate, will always be perceived as a negative outcome, since other incident rates are reported in much lower amplitudes. Another limitation of the PAI is that it requires a lost time investment to train workers to identify potentially hazardous conditions. The near miss reporting and tracking itself requires a time investment that may be difficult to justify to operating personnel.

Each near miss accident is evaluated by accident type, identified in Table 5-2. Additional accident types can used to meet your needs. The CMA recommends that the exact number for the PAI be assigned locally to meet user needs. The range of 0-500 is suggested. In Table 5-2, I have suggested my own point assignment, but you may certainly adjust this for your facility as you deem appropriate.

If the exposure hours are constant, the points can be summed monthly based on the types of accidents or illnesses that occurred during the preceding month. The total PAI can be plotted on a linear graph on a monthly basis.

If exposure hours fluctuate, or if you wish to compare your total PAI to other facilities, you can weight the PAI as an incident rate:

$$\Sigma = \frac{PAI \times 200{,}000}{H} \qquad [5.7]$$

or, you may compare experience by accident type:

$$\Sigma_{AT} = \frac{PAI_{AT} \times E_{AT}}{N_{NM}} \qquad [5.8]$$

where

Σ_{AT} = total PAI based on accident type
PAI_{AT} = accident type points
E_{AT} = number of near misses for accident type
N_{NM} = total number of near misses

Another way PAI is used is to sum the points and divide this number by all the incidents as a monthly index over time and across organizations. This index is calculated as

$$\Sigma_I = \frac{PAI}{N_{NM}} \qquad [5.9]$$

Table 5-2 Preventable Accident Index (PAI) Points

Accident Type	PAI Points
Bodily reaction	400
Caught in	400
Chemical exposure preventative/precautionary	250
Chemical/radiation exposure	500
Contact by electrical current	500
Fall, different level, <5′	250
Fall, different level, >5′	500
Fall, same level, 0-1′	100
Foreign body in eye caused by work task	450
Foreign body in eye not related to work task	350
Laceration, puncture with sharp instrument	150
Motor vehicle accident, <30 mph	250
Motor vehicle accident, >30 mph	500
Not otherwise classified	100
Overexertion, back or knee strain, or hernia	500
Precautionary medical visit, but no injury	50
Rubbing, abrasion, or scratch	50
Struck against	150
Struck by	250
Thermal burn	500

Regulatory Visits, Citations, Penalties. OSHA or state agency regulatory visits (compliance inspections), the number of citations issued, and penalties assessed can be tracked as a measure of performance. Also, the related costs of these compliance inspections, such as lost productivity, may also be tracked. Fines and penalties from violations of regulatory standards reflect a direct cost of inadequate OSH performance.

These numbers are easily measured and communicate performance as a direct benefit of avoided costs. Compliance inspections are the most objective, and therefore most accurate, indication of a facility's OSH performance.

However, the lack of compliance inspections does not imply that a facility has a quality OSH program. Another limitation of these measures is that compliance inspections often focus on one small aspect of the overall OSH program. Thus, the performance of the overall program is not evaluated. At the same time, a compliance inspection, in and of itself, is not an accurate indicator of poor OSH performance, as the agency may have selected the facility for inspection based on a disgruntled employee complaint or for some other reason not apparent to the facility.

Lost Work Day Severity Rate. The Lost Work Day Severity Rate (LWDR) is a quantitative measure of consequences from an occupational injury or illness expressed as the number of days away from work. LWDR indicates the success of an OSH program in terms of severity.

Reporting the days away from work for occupational injuries and illnesses is an OSHA recordkeeping requirement. Measurement of LWDR provides a database upon which program goals and targeted emphasis can be focused for performance improvement.

As a direct cost, LWDR can be measured in terms of labor dollars lost. Other direct costs that result from each day away from work, such as the cost of replacement labor or overtime, can also be added.

LWDR is a productivity measure and a direct cost to operations.

A severity index may be a measurement of days away from work per case. This index allows for comparison with other cases, comparison between departments, and trend analysis. The severity index is an excellent indicator of case management of lost time accidents.

Keep in mind, single severe incidents can skew the index. Unfortunately, a good database does not exist by which you may compare yourself to others.

A graphic you may wish to use is:

Type Graph	Y-Axis	X-Axis
Bar	Days away from work (median for industry and your company)	Accident Type

The analysis of lost work days by accident type compared to industry median provides an indication of where to focus your resources. Severity rate may be analyzed for work centers, type of injury, task, incident rate, and many other groupings, limited only by your imagination.

OSH Training. Another potential performance measure is the percentage of employees who have completed all scheduled OSH training. This measure allows you to assess the degree with which you have completed required training.

OSH training requirements can be determined from individual OSHA safety standards and regulations that apply to your facility. (See Chapter 1.) Related EPA requirements may be found in regulations and consent orders. Some trade associations have their own list of recommended training. Your own corporate headquarters may also have specific training requirements.

A training measurement identifies clearly your compliance with requirements. It also clearly indicates your commitment to a quality OSH program. Historical records of the training measurement provide evidence that you have tried to meet your training obligations. These records also enable you to compare your efforts to sister facilities and even to other companies. They are an easy measurement to make.

The percentage of training completed does not address the quality of the training. Nor does the percentage of training completed indicate how well your employees have retained the material, assuming it was well presented in the first place. Finally, these records can be a potential liability to your company if it is slow to respond to training needs or makes half-hearted attempts at training precisely because they can be used as evidence against you.

An OSH training index, I_{Tr}, can be calculated as

$$I_{Tr} = \frac{W_{Tr}}{N_t \times W_{tgt}} \qquad [5.10]$$

where

W_{Tr} = number of workers who have been trained

N_t = number of training topics

W_{tgt} = number of workers who needed to be trained

Count each employee for each topic. One employee, for instance, may need to be trained in several topics, so count him or her as an employee needing that topic for the W_{tgt} count. Count him or her again for each training topic attended, as part of the W_{Tr} count. Consider a factory that has need to conduct training in seven OSH topics, N_t=7. The number of employees needing the training is 25, 40, 30, 12, 2, 40, and 100, respectively. W_{tgt}=249. After the training sessions and makeup sessions are complete your head count is 25, 39, 30, 10, 2, 40, and 88, respectively. W_{Tr}=224. Since not all the employees in the manufacturing unit have to be trained on every topic, the number of topics is not a factor. The training index, I_{Tr}, is 0.9—or you could say the training is 90 percent complete. If every employee, or the same number of employees must attend each of the topics, include the number of topics in the equation.

Compliance with Company OSH Requirements. The number of action items uncovered during internal OSH compliance audits can be used as a safety metric. Typically, the measure is the percentage of action items completed out of the total number identified in the audit. The objective is to identify the quality of performance within a company.

Such a measure helps to verify that the OSH program is effective. The measure provides an assessment of internal controls compared to internal and external standards of OSH management and administration. This internal measure provides objectives for continuous improvement of OSH performance. It ensures that OSH programs are being reviewed, verified, updated, and improved on a regular basis. It also provides a comparison between individual facilities within a larger corporation so that needed funds can be channeled to the most pressing needs. The corporate safety department can also identify the need for additional resources or the need for an increased level of commitment. The measure can be an opportunity for friendly competition among facilities. By minimizing potential OSH liabilities, the measure provides assurances to stakeholders.

This measure can be compared with other performance measures to develop links between OSH program management and accident prevention. This link, however, may not be obvious. Also, unless an active and vital audit program is in place, the measure is meaningless. Additional resources may be required to get an audit program up and going before the measure can be implemented. Attorneys are concerned that the compliance officer may request findings from internal audits as part of an inspection. To make valid comparisons of completed audit items from year-to-year, the business unit under scrutiny must be evaluated by similar criteria.

OSH Evaluation Follow-up. The metric is based on the percentage of OSH recommendations communicated and action plan received on schedule. Alternatively, it is the percentage of OSH recommended action plan fully implemented. The action items are identified in corporate policy documents or routine inspections. The difference between this and the last safety metric is that a full-blown internal audit is not involved. The action items may come from company-wide recommendations or specific inspections.

Exposure Assessment Strategies. The completion of exposure assessment strategies is another safety metric. Qualitative exposure assessment is a risk prioritization method for chemical, biological, and physical environmental hazards, such as radioactivity, electromagnetic fields, or other nonionizing radiation exposure. As a specific measure, the qualitative exposure assessment index is the percentage of all potential exposures that have either been qualitatively assessed or estimated. Quantitative exposure assessment consists of monitoring or modeling exposure. As a specific measure, the quantitative exposure assessment index is the percentage of an industrial hygiene monitoring plan or quantitative exposure modeling plan completed on schedule.

This metric allows an employer to assess and prioritize potential health risks faced by many or all workers. With this metric, the employer can decide which unacceptable exposures to control first after having differentiated between acceptable and unacceptable exposure. The remainder require PPE and/or administrative controls in the meanwhile.

The exposure assessment index also documents a record of exposure levels for the workforce and is an easy way to communicate exposure monitoring results to individuals. A historical file demonstrates compliance with exposure limits. It also provides physicians with data needed to make informed medical surveillance decisions. The file can also be used to support epidemiological studies.

While recommended approaches to exposure assessments exist, no standardized methods for either qualitative or quantitative exposure assessments strategies have been formulated. Although professionals agree in general on approaches, no consensus exists on methods for data interpretation. Therefore, data cannot always be readily compared between companies. Exposure assessment strategies are largely left to the industrial hygienist's discretion. Most companies do not have the internal resources to conduct a comprehensive exposure assessment.

The qualitative exposure assessment index, I_{qual}, is determined as

$$I_{qual} = \frac{N_{EA}}{N_{Haz} \times N_u} \quad [5.11]$$

where

N_{EA} = number of exposure assessments of homogenous groups
N_{Haz} = number of hazardous exposures to be assessed
N_u = number of business homogeneous groups to be assessed

A manufacturing company is concerned about twelve (12) chemical, biological, and physical hazards in its six departments. Last year the company conducted qualitative assessments of six (6) chemicals in three (3) departments, three (3) physical agents in two (2) departments, and one (1) biological agent in one (1) department. The Qualitative Exposure Assessment Index is

$$\frac{(6\times 3)+(3\times 2)+(1\times 1)}{12\times 6} = 0.35$$

or, the assessment strategy is 35% complete. A simpler way to track qualitative exposure assessments is to use a comprehensive "go or no go" approach. Either the assessments are 100% complete or they are not (0%).

Safety Hazard Assessments. This metric is based on observations conducted to identify hazards in the workplace. Completed safety assessments provide information about opportunities for preventing occupational injuries. The measure can be used to plan safety activities.

The safety hazard assessments measure demonstrates the company commitment to improving employee safety and health, improves understanding of hazards in the workplace by identifying potential concerns, and communicates the need for PPE by highlighting the existing conditions under which PPE should be used. These assessments verify the functioning of engineering controls. The development of proper practices is another benefit. Observations of work practices can lead to improved safety and productivity. Finally, it provides an excellent forum for employee involvement in the OSH program.

Different types of hazards, however, require different levels of expertise. Knowledgeable and skilled leaders are needed to supervise the safety hazard assessments. Participation of all affected employees is essential or the assessment findings and recommendations may not be well understood or effectively implemented. Liability may result from findings improperly addressed or not addressed in a timely fashion. Records either need to document the timely closure of each finding or document some follow-on decision.

Exposures above Action Level. Corrective action should be taken anytime a regulated exposure exceeds the Action Level (AL). OSHA defines the AL as ½ of the established PEL. Some companies opt to set the AL as low as 0.1 of the PEL or some other exposure limit. Anything between 0.1 and 0.5 is acceptable. This metric is the percentage of industrial hygiene samples indicating exposures above the AL adopted by the company. It is used to compare workplace exposure to chemical, biological, and physical agents with either the regulatory or stricter company requirements.

Implementation

Successful establishment of safety metrics requires initial planning and analysis of work processes. At the very beginning, identify the key stakeholders who need to understand and support the data collection and who need to receive the report of results.

Then you need to identify your company's key work processes relative to occupational safety and health (OSH). What activities and chain of events must occur in order to deliver the desired OSH products—the services and results expected? Figure 5-8 lists some OSH work processes that may affect you.

Separate the work process activities from the outcomes when you set up performance measures. Identify activities that can be measured and that

Figure 5-8
Key OSH Work Processes

OSH Program Management
Hazard Communication
Hazard Controls
Hazard Evaluation
Hazard Recognition

influence the outcome. Not all activities can be measured beyond the qualitative "Yes, it happened" or "No, it did not." Secondly, not all activities that can be measured have a significant influence on the desired results. Pay attention to detail and document how you establish your measurement process.

A well-defined process for data collection will assure the success of your safety metrics program. Figure 5-9 lists the steps of establishing data collection.

The painstaking identification of good reasons to collect data is the important initial step in data collection. We do not collect data for something to do or to justify our jobs, or just to impress our bosses. We collect data to improve OSH work processes. We must first identify opportunities to improve a work process before we can verify a problem exists, must less analyze the problem.

What are you going to measure? Focus for improvement begins by identifying a problem to address. Give your priority to OSH work processes and measurements that have some potential for improvement. Don't be discouraged by variations in observed data. Statistical variations are a fact of life. If you do not understand how to deal with data from a statistical standpoint, get someone on your data collection and analysis team who does.

Develop clear procedures that anyone can follow when collecting data. Carefully identify the concept you are going to evaluate—for instance—training efficiency, training effectiveness, or customer satisfaction. Define the concept for describing the standards against which the measures will be compared.

How will the measurements be obtained? What factors produce incon-

Figure 5-9
OSH Metrics Data Collection Process

Identify the reason for data collection.
Identify the measurables.
Determine the procedures for data collection.
Write instructions for data collection.
Collect data.
Analyze, summarize, and communicate results.
Address opportunities for improvement.
Document progress.

sistent data? Eliminate those factors. A pilot program that tests your data collection procedure is recommended. This allows you to make changes to the data collection process as necessary to minimize unnecessary confusion about variations in data.

After that follow your procedure and instructions religiously. Never collect data you have no intention of using. Fix the obvious problems immediately. Address data collection problems that are uncovered as part of continuous improvement.

Don't throw numbers around to managers and employees that they cannot digest easily. Summarize your data after evaluating it. Histograms and other charts should be used liberally to communicate interpreted results because most people understand numbers better when they have something to help them visualize. Identify causes of variation and make corrections, if needed, during the next data collection period. Factor out variations, when it makes sense to do so, but be careful not to fool yourself.

Remember always: your goal is to improve OSH work processes, not collect numbers. The numbers merely help you identify and prioritize problems. The real improvement comes when action is taken to correct a problem. The numbers of themselves have no value and are not sacred. Do not be offended if others do not accept your numbers, but be sensitive to the end results of your efforts.

Document activities to improve OSH work processes. This demonstrates the responsiveness of your company to hazard abatement. Documentation also helps to build credibility with OSHA and your own employees.

The Chemical Manufacturers Association lists some rules of thumb for safety metrics, which I have summarized in Figure 5-10.

Qualitative Risk Evaluation

A hazard evaluation is the major part of a risk evaluation. Once the hazard evaluation is completed, the remaining task is to assign risk. A typical hazard and risk evaluation is conducted as a sequence of steps as shown in Figure 5-11.

Quantitative risk evaluations, discussed below, add a tenth step: complete a quantitative hazard evaluation to reduce the uncertainty in the estimates of probabilities and consequences and to identify optimum investments for obtaining a level of risk deemed acceptable.

Figure 5-10
Safety Metrics Heuristics

- Keep it simple.
- Use existing data as much as possible.
- Involve stakeholders in the process.
- Failure to act in a timely fashion on findings makes data collection a wasted effort and leaves your company potentially exposed.

Probability Index

The probability of a hazard causing undesirable consequences is something between "it will never happen" (p=0) and "it's bound to happen" (p=1). However, the extreme probabilities are highly unlikely. If a hazard has an improbable chance of causing harm it is hardly hazardous. If a hazard continually causes harm, how are you going to be able to abate it?

Figure 5-11
Order of Qualitative Risk Evaluation

1. Identify hazards.
2. Estimate the consequences.
3. Identify opportunities to reduce the consequences.
4. Identify initiating events.
5. Estimate the probabilities of initiating events.
6. Identify opportunities to reduce the probabilities of the initiating events.
7. Identify the event sequences of accidents that may lead to the consequences anticipated.
8. Calculate the combined consequences and probabilities of the accident event sequences.
9. Identify opportunities to reduce the probabilities and/or consequences of the accident event sequences.

Table 5-3
Qualitative Hazard Probability

Probability	Designator	Definition
Frequent	A	Likely to occur immediately or within one year or less.
Probable	B	Probable to occur within 1 to 10 years.
Occasional	C	Possible to occur 10 to 100 years.
Remote	D	Unlikely to occur in less than 100 to 1,000 years.

In the real world $0<p<1$. The probabilities p=o and p=1 are just not realistic. However, as safety professionals we rarely have the luxury of knowing the true probability of an undesirable incident being caused by a known hazard. Therefore we assign a qualitative probability level to hazards we discover: remote, occasional, probable, and frequent. Table 5-3 explains these hazard probability levels and assigns a designator to them. This is not original, the National Safety Council's *Accident Prevention Manual for Industrial Operations—,I. Administration and Programs* has this information on page 70, with minor variations. The Australians, as reported by Mansdorf, use a slightly different system, shown in Figure 5-12.

Consequence Index

Consequence is the harm that a hazard may cause. The consequence of any particular hazard is something between insignificant and catastrophic. Table 5-4 outlines potential consequences of a hazard.

Mansdorf reports on a five by five system used in Australia that may be substituted, especially if it seems clearer to the user. Table 5-5 shows the Australian consequence chart.

Figure 5-12
Australian Probability System

Probability	Description
A	Common or repeating
B	Occurs or has occurred
C	Could occur
D	Not likely to occur
E	Practically impossible to occur

Table 5-4
Qualitative Hazard Consequence Categories

Consequence	Category	Definition
Catastrophic	1	Causing injury or illness resulting in permanent total disability, chronic or irreversible illness, or death.
Critical	2	Causing severe injury or illness resulting in permanent partial disability or temporary total disability lasting more than three months.
Marginal	3	Causing minor injury or illness resulting in hospitalization or in a temporary reversible illness with a limited period of disability (less than three months). Recordable medical treatment or lost time restricted.
Negligible	4	Causing less than minor injury or illness that does not result in hospitalization. Causing temporary reversible illness but requiring minor supportive treatment such as first aid.

Table 5-5
Australian Consequence Categories

Category	People	Equipment Damage/ Production Delay
1	Fatality or permanent disability	>$500K
2	Serious lost time	$100K-500K
3	Moderate lost time	$50K-100K
4	Minor lost time	$5K-50K
5	No lost time	<$5K

Risk Index

The risk index is the combination of probability and consequence. Table 5-6 shows the American risk index.

The Australian risk index is shown in Table 5-7.

Whichever system you use, the American 4X4 or the Australian 5X5, you can determine priorities for hazard abatement projects by one or the

Table 5-6
American Risk Index

	Probability			
Consequences	**A**	**B**	**C**	**D**
1	1	2	4	7
2	3	5	8	1
3	6	9	12	13
4	10	14	15	16

Table 5-7
Australian Risk Index

Consequences	Probability				
	A	**B**	**C**	**D**	**E**
1	1	2	4	7	11
2	3	5	8	12	16
3	6	9	13	17	20
4	10	14	18	21	23
5	15	19	22	24	25

other. Between any two projects, the one having the lowest risk index should get your attention first.

Accident Frequency Estimation

It is important that the estimation of accident frequencies be as conservative as available data, procedures, and circumstances warrant. Worst case estimates are often justified for decision making. What-If methods and HAZOP can provide the context of estimating frequencies. The primary purpose of Fault Tree Analysis, Cause Consequence Analysis, and Human Error Analysis is to estimate frequency of initiating events. The primary purpose of Fault Tree Analysis, Event Tree Analysis, and Cause Consequence Analysis is to estimate frequency of event sequences.

Accident Base Frequencies

The probability, hence frequency, of initiating events is obtained from failure rate data or human error probabilities. Also, a fault tree analysis can be used to relate the initiating event to its causes for which a failure rate or human error data is available. For complex systems, it is necessary to identify the entire event sequence. Combine the probabilities of the events in order to estimate the frequency of the accident.

Event Trees

Event trees appraise likely accident consequences resulting from an *initiating event.* The failure of a specific piece of equipment, a system, or a human error may be initiating events. The event tree analysis considers either operator or safety system response to the initiating event. How could a chain of responses develop into an accident and its subsequent consequences? Finding an answer to this question produces a sequence of events or event tree. When complete, the tree is a chronological set of failures or errors that define the accident under study.

Event trees are appropriate for analysis of systems that have safety systems or emergency procedures already in place. In design, event trees may be used to assess potential accidents resulting from assumed initiating events. Event tree analysis is also useful in operations, where they may be used to evaluate the adequacy of existing safety features or to examine the likely effect of equipment failure.

Outcome Frequencies from Event Trees

Figure 5-13 outlines the general procedure for event tree analysis.

The initiating event is a system or equipment failure, human error, or process upset that can result in a variety of effects, depending on how well the system or operators respond to the event. Event tree analysis is a forward thinking process, starting with the initiating event and proceeding forward in time to an accident. If the selected event is the direct cause an accident, the fault tree analysis is more accurate.

Figure 5-13
Procedure for Event Tree Analysis

1. Identify initiating event.
2. Identify safety functions designed to deal with the initiating event.
3. Construct the event tree.
4. Describe the resulting accident event sequence.

Safety functions that respond to an initiating event are the plant's defense against event. Such functions include safety systems, alarms, barriers, containment methods, operator actions, standard operating procedures, and safety procedures. The final consequences of any accident attributable to the initiating event are influenced by these safety functions.

The event tree itself is the chronological model of an accident. It begins with the initiating event and proceeds through successes and failures of various safety functions that respond to the initiating event.

You can construct the tree on a piece of notebook paper. Turn it long ways in the landscape fashion of computer printers. List the initiating event about midway down the left side. Across the top of the page, list from left to right the safety functions that respond, in chronological order. Underline each safety function and place an underline beneath the initiating event. The latter line represents the progression of the accident on a path from the initiating event to the first safety function.

Evaluate the first safety function. Did it succeed, or did it fail? (Assuming the initiating event has occurred.) A branch is drawn to the right of the initiating event to represent the two possibilities. Success is normally given the upper branch and failure is assigned to the lower one. If the safety function does not affect the course of the accident, it is neutral, a single line is drawn instead of a branch.

Going on to the next safety function listed at the top of the page, analysis will produce another series of branches or a single line to the next function. When the event tree is complete, prepare a narrative description of the accident event sequences using the tree. The sequences of importance are those that lead to an accident. Some sequences may end in a safe recovery, a return to normal operations, or to an orderly shutdown.

Assign each event in the tree a probability. By computing the probability of combinations of events, you arrive at the overall probability for the accident. A different probability is computed for each event path leading to an accident. The frequency, f, is

$$f = p \times n \qquad [5.12]$$

where

p = overall probability for the accident path
n = number of times the initiating event occurs per time period

Fault Tree Analysis

Fault tree analysis (FTA) uses deductive reasoning to focus on one particular accident event. FTA provides, therefore, a method for determining causes of that accident event. The fault tree models various sequences of equipment, faults, and failures that combine to result in the accident event. Contributing human errors are included, along with equipment failures. FTA is used in the design phase to expose the not-so-obvious failure modes that result from combinations of equipment failures. In operational situations, FTA is used to identify potential combinations of failures for specific accidents.

Fault trees are backward-looking models that start with an accident and work backwards to the initiating event. The fault tree is a graphical representation of the interrelationship of equipment failures, human errors, and a specific accident.

Three classes of equipment faults and failures are studied: primary, secondary, and command. Primary faults and failures occur in the environment for which the equipment was intended. Secondary faults and failures occur in an environment for which the equipment was not intended. Command malfunctions occur with proper operations but at inappropriate times or places.

You can draw a fault tree on a sheet of notebook paper. A top event is selected. This is the accident event and is the subject of the analysis. Since fault tree analysis is quite complicated and uses special symbols, we'll leave further discussion to experts. Suffice it to say, however, that it is worthwhile taking a course in fault tree analysis because you are quite capable of analyzing all but the most complex accidents. With the help of a task team, you should be able to analyze even the worst of accidents. Special templates can be purchased in office supply stores and drafting supply stores to assist you in drawing the tree.

Quantitative Risk Evaluation

Several options for evaluating hazards and quantifying the risks they present are available to safety professionals. Figure 5-14 lists some of the available hazard evaluation techniques for completing risk evaluations. Some of these are discussed above and some are discussed below. Information that may be needed to complete such an evaluation is summarized in Figure 5-15.

Figure 5-14
Hazard Evaluation Techniques

Cause Consequence Analysis Checklist Event Tree Analysis Failure Modes & Effects Analysis (FMEA) Fault Tree Analysis Hazard & Operability Analysis (HAZOP)	Human Reliability Analysis Preliminary Hazard Analysis Relative Ranking Safety Review What-If Analysis What-If Analysis with Checklist

Figure 5-15
Information for Hazard Evaluation

• Process chemistry • Process physics • Process limits—temperature, pressure, stored energy, contaminants • Process flow diagrams • Material balances • Energy balances • Material inventories • Material properties • Piping and instrumentation diagrams (P&ID) • Codes, standards, regulations, and corporate policies • Control system information	• Plot plans • Facility layouts • Topographical maps • Demographic data of surrounding population • Weather data • Utilities • Emergency systems • Equipment details • Standard operating procedures • Maintenance procedures • Maintenance schedules • Emergency procedures

Figure 5-16
The Hazard Evaluation Report

- The personnel who conducted the study and their credentials
- Scope of the study
- Hazard evaluation methodology
- Process information
- Findings
- Conclusions
- Recommendations

The hazard evaluation report is outlined in Figure 5-16.

Risk Perception

We have already talked about the perception of risk. This perception is very subjective and does not follow a logical sequence. It is difficult to explain to an angry or frightened person that his or her opinions and/or perceptions are not facts, and that the facts show that the facility is safe even when he or she doesn't feel so. Those who are not trained in the scientific method hold their opinions and perceptions very dear. Heck, even those of us who have been trained in scientific method can be pig-headed. The point is, do not be too disappointed if your best, most logical argument does not convince someone that he or she is safe, when he or she thinks otherwise. Nor can you expect to convince someone that he or she is at risk, if he or she feels safe.

Risk Acceptability

How do you get people to accept risk? First of all, be patient with them. Once you lose your cool you can forget about convincing them. Then, sometimes you can draw analogies. But sometimes that does work, either. Once, I was trying to draw analogies to explain why a cold cleaning solvent was relatively safe, but that gloves were needed to prevent defatting of skin, with subsequent dermatitis. I drew an analogy to dishpan hands and dishwashing detergent. That was the end of the training. An angry lady declared me "a complete idiot" and my credibility with the rest of the group was immediately blown. I don't know how to convince people to accept risk except to be honest and straightforward. You have to win their trust and confidence. Perhaps some personalities are better suited to this.

Failure Mode and Effect Analysis

Failure mode and effect analysis (FMEA) is an inductive method of analyzing hazards. In FMEA, the failure or malfunction of each component of a system is considered, including the *mode of failure.* Effects of hazards that led to the failure are traced through the system. The ultimate effect on task performance is evaluated. Because only one failure is considered at a time in FMEA, some possibilities may be overlooked. Once the FMEA is complete and the critical failures are detected, then a fault tree analysis facilitates an inspection checklist.

This procedure begins with a determination of the level of outcome necessary. At the plant level, the FMEA focuses on individual systems or subsystems. The study examines the failure modes and effects of these systems with respect to the hazard. At the system level, the FMEA addresses individual equipment and subsystems that make up the system.

Secondly, a consistent format for the study must be developed. The level of resolution desired must be considered in developing the format.

Next, a problem definition is developed. Part of this document includes a definition of the boundary conditions. The plant and systems that are the subject of the study are described. The physical system boundaries that encompass the system are also described. Recent reference information is collected to identify the equipment and its functional relationship to the remainder of the plant or system. A criticality ranking system that addresses the potential effects of equipment failure must be provided and it must be consistent with everything accomplished so far. The criticality ranking is defined in terms of the probability of the failure, the severity of the resulting accident, or a combination of these two factors. Facility- or process-specific assumptions are outlined. Table 5-8 suggests some criticality ranking definitions.

A FMEA table is completed. It identifies and describes the equipment or system. The table lists failure modes and effects and assigns a criticality rating. This document is used to assign resources and design safety systems.

Table 5-8
Criticality Ranking

Effects	Criticality Ranking
No effects	1
Minor process upset; small hazard to facilities and personnel; shutdown not required	2
Major process upset; significant hazard to facilities and personnel; orderly process shutdown required	3
Immediate hazard to facilities and personnel; emergency shutdown required	4

Fault Tree and Event Tree Analysis

Trees were discussed above. Deductive methods use combined events analysis, typically in the form of a tree. Positive trees easily transform into lists of DOs and DON'Ts. Fault trees demonstrate ways in which trouble develops. In fault trees, an undesired event is selected and all possible subordinate events that can lead to the designated event are diagrammed in the form of a tree. The branches of the tree are continued until independent initial events are reached. Probabilities are then estimated for each of these independent initiating events.

Trees have three advantages: they accomplish rigorous, thorough analysis without wordiness, they make analytical processes visible, permitting rapid and clear transfer of information, and they can be used as investigative tools by tracing backwards from the accident. In this case, the investigator may be able to reconstruct the system in graphic form and pinpoint the responsible elements as causes.

Consequence Analysis

This type of analysis combines the forward-looking event tree with the backward-looking fault tree. The approach connects specific accident consequences to their many possible causes. The work product of this study is a cause-consequence graphic that unveils the relationships between accident consequences and causes.

Dow Fire and Explosion Index

The Dow Index provides a direct, easy method for ranking risks relative to one another. Penalties and credits are assigned to various plant features. Penalties are assigned to process materials and conditions that are hazardous. Plant safety features are assigned credits if they can mitigate the effects of an accident. The algebraically combined penalties and credits yield a relative ranking of plant risk. Also included in the evaluation are estimates of consequences in terms of cost and outage time.

Designers can use the Dow Index to identify vulnerable areas and specify plant protective measures. Operations can use the index to provide relative information on plant hazards. The Dow Index can also be used by operations to determine how safety upgrades may be beneficial.

Using a plot plan of the plant, identify process units that contribute the most to a fire or explosion hazard, or a release of a toxic material. From a guide provided by Dow for some three hundred chemicals, determine the material factor (MF) for each unit. Instructions are provided for determining a MF for materials not included in the Dow Index.

Two types of hazards contribute to an incident: general process hazards and special process hazards. General process hazards contribute to the magnitude of the incident. Designated F_1, general hazards include exothermic reactions, endothermic reactions, material handling, material transfer, enclosed process units, inadequate access, and drainage. Special process hazards increase the probability the event will happen. They are designated F_2 and include process temperature, low pressure, operation in or near flammable range, dust explosion, relief pressure, low temperature, quantity of flammable material, corrosion, erosion, leakage in joints and packing, fired heaters, hot oil heat exchange, and rotating equipment.

A unit hazard factor, F_3, is calculated as follows:

$$F_3 = F_1 \times F_2 \qquad [5.13]$$

Next, the damage factor, D, is calculated, which represents the probable relative damage exposure.

$$D = MF \times F_3 \qquad [5.14]$$

The damage factor is also called the fire and explosion index, F&E. Fires and explosions also have blast effects, so an area of exposure is determined as the circular area around the process unit where damage may occur. The Dow F&E generally specifies the degree of hazard at the unit as shown in Table 5-9.

Next, the maximum probable property damage (MPPD) is determined. A base MPPD is calculated from the replacement value of equipment within the area of exposure. An actual MPPD is calculated by applying loss control credit factors to the base MPPD.

$$MPPD_A = MPPD_B \times Credit_{total} \qquad [5.15]$$

Other useful numbers include the maximum probable days outage (MPDO) and business interruption (BI) costs. BI considers cost of repairing or replacing damaged equipment and other property damage. Loss of production is included in BI also.

Communicating Risk to Employees

None of these quantitative determinations will ease the fears of employees. Tell them whether the hazard will hurt them. In the case of a chemical,

Table 5-9
Degree of Hazard vs. Fire and Explosion Index

Degree of Hazard	DOW F&E
Light	1 – 60
Moderate	61 – 96
Intermediate	97 – 127
Heavy	128 – 158
Severe	159+

will it hurt them if they can smell, taste, or see it? Even then, they may not trust the information. Risk communication is mostly a matter of trust. So, if you want to be credible, you need to do things that build trust.

Communicating Risk to the Public

Communicating risk to your employees is one thing, but communicating risk to the public can be quite another matter. When OSHA evaluates risk, it must conduct a cost-benefit study before issuing a new safety standard affecting the regulated community. OSHA is restricted by Section 3(8) of OSHAct to adopting practices that are reasonably necessary or appropriate. Yet in Section 6(b)(5) of the Act, OSHA is admonished to require that these practices be implemented to the extent feasible. Hence, a running battle between industry, OSHA, and the courts is created. Lately, the very body that charged OSHA to do its task, the U.S. Congress, has attacked OSHA. Nevertheless, because of public sympathy for workers maimed or killed on the job, industry will never be free of safety and health regulations.

In dealing with the public, you must be ready to face a certain amount of outrage that is not entirely rational. This is to be expected anytime that fear is a driving force in the communications process.

An assessment of the community is needed in order to understand the audience you will face. In Figure 5-17 Bruening lists these cardinal rules for risk communications:

Figure 5-17
Bruening's Cardinal Rules for Risk Communications

1. Accept and involve the general public as a legitimate planning partner.
2. Plan carefully; evaluate performance.
3. Listen to your audience.
4. Be honest, frank, and open.
5. Coordinate and collaborate with other credible sources.
6. Meet media needs.
7. Speak clearly; be compassionate.

References

Allison, William W. "Safety Statistics: A Perspective for Statistics in Accident Prevention." *Professional Safety.* October 1988, pp. 18-20.

Arnold, Robert M., Jr. "Measuring the Health of Your Safety Audit System." *Professional Safety.* April 1992, pp. 46-49.

Dreux, Mark S. and Frank A. White. "Protecting Your Audits from Compelled Disclosure." *Occupational Hazards.* February 1993, pp. 53-55.

Emery, Robert and Susanne M. Savely. "Soliciting Employee Concerns during Routine Safety Inspections." *Professional Safety.* July 1997, pp. 36-40.

Fiora, Glen and Paul G. Specht. "Cost-Benefit Analysis and Risk: In the Hands of the Supreme Court." *Professional Safety.* April 1992, pp. 24-28.

Hendershot, Dennis C. "Documentation and Utilization of the Results of Hazard Evaluation Studies." *Plant/ Operations Progress.* October 1992, pp. 256-263.

LaBar, Gregg. "My Point Exactly: OSHA Has a Strategic Plan; Do You?" *Occupational Hazards.* November 1997, p. 10.

Loud, James J. "Are Your Safety Inspections a Waste of Time?" *Professional Safety.* January 1989, pp. 30-32.

Manuele, Fred A. "Principles for the Practice of Safety." *Professional Safety.* July 1997, pp. 27-31.

Mansdorf, Zack. "Risk Assessment: Focusing on the Operator." *Occupational Hazards.* May 1997, pp. 93-98.

Mathis, Terry L. "Fallacies in the Safety Fable." *Occupational Hazards.* October 1997, pp. 155-156.

McElroy, Frank E., Ed. in Chief. *Accident Prevention Manual for Industrial Operations. Volume I: Administration and Programs.* 8th ed. Chicago: National Safety Council, 1981.

Milner, John E. *Protection of Environmental Compliance Audits from Discovery and Protection of Corporate Officers from Environmental Criminal Liability.* 1992 Environmental Seminar. Brunini, Grantham, Grower & Hewes, Jackson, Mississippi, May 27, 1992.

Petersen, Dan. "An Experiment in Positive Reinforcement." *Professional Safety.* May 1984, pp. 30-35.

Program Performance Measures: Resource Guide for Employee Health and Safety Code (Responsible Care®). Chemical Manufacturers Association, 1995.

Shivers, C. Herbert. "Halos, Horns, & Hawthorne: Potential Flaws in the Evaluation Process." *Professional Safety.* March 1998, pp. 38-41.

Witter, Ray E. "*Guidelines for Hazard Evaluation Procedures:* Second Edition." *Plant/Operations Progress.* April 1992, pp. 50-3.

6

SAFETY COMMUNICATIONS

Employers are responsible for keeping employees informed about matters concerning the Occupational Safety and Health Administration (OSHA) and their own safety and health. OSHA and states with their own occupational safety and health programs require that, at a minimum, each employer must post certain safety materials at a prominent location in the workplace. These are the four items discussed immediately below. Also, all employees have the right to examine any records kept by their employers regarding exposure to hazardous materials, or the results of medical surveillance (see Access to Records in Chapter 11).

The first basic information requirement is the display of the *Job Safety and Health Protection* workplace poster (OSHA 2203 or state equivalent). This poster informs employees of their rights and responsibilities under the Occupational Safety and Health Act (OSHAct). Besides displaying the poster in a prominent location, employers are required to make available copies of OSHAct and relevant OSHA rules to employees upon request. Obviously, the safety manager cannot supply this information to other employees if he or she does not have it themselves. Every workplace, no matter how small, needs access to these laws and to the OSHA rules. However, nothing states that a central location cannot provide this information for outlying locations.

Second, summaries of applications for variances from standards or petitions for variances for recordkeeping procedures must also be made available to employees in written form as well as being posted in a location where notices are normally posted. See Chapter 11 on variances and these requirements. This posting location will not necessarily be the same as the prominent place for the OSHA poster.

Third, copies of all OSHA citations for violations of standards must be posted. These must remain posted at or near the location of alleged violations for three days, or until the violations are corrected, whichever is longer.

In other words, you can not conceal from your employees this reminder that a law has been violated.

Finally, the OSHA 200 Log (see Chapter 11 under recordkeeping) must be posted from February 1 to March 1 of the year following the year of record.

Communications Meetings

As discussed briefly in the last chapter, the best way to communicate risk and to build confidence in hazard abatement is to have face-to-face meeting with employees. Of course, if you are insincere, or have built a reputation of being uncaring, do not expect a friendly audience. If you or your management leadership have a negative history, perhaps you need to prepare by first reestablishing confidence and trust between management and employees.

Effective communication, in fact, is one of the keys to a successful safety process. Communications links safety to productivity, quality, and customer service. Safety will not happen unless supervisors communicate with their employees. According to Pierce, communication is unnatural to humans. He goes on to say that communication is further impeded by our societal values. We teach our children "not speak until spoken to." We tell them "not to speak to strangers." Heaven forbid if you interrupt old Dad while he is speaking. We tell them "not to speak with their mouths full." "Silence is next to godliness," we say. "Don't talk." "Don't speak." "Be quiet." Then once they are grown we suddenly want them to communicate and we're surprised when they find all kinds of ways to mess it up. No wonder.

Communication is a two-way process and safe workplaces depend on the flow of information from the floor to the safety office and back down to the floor again. A memo from the safety office is not necessarily a communication. Likewise, a report or complaint of unsafe conditions from the floor is not necessarily a communication. Communication happens when the receiver reads and understands the document and acts on it. Otherwise a communication is merely a piece of paper.

The same holds true for safety policies and procedures. Previously, I said that a safety procedure binder is nothing more than a container of paper. Safety procedures mean absolutely nothing unless people are reading, understanding, and using them.

To have effective communication means safety is possible. Ineffective communication means it is not. To think otherwise is to believe in an illusion. Communication is so critical that it merits our entire attention.

Communications meetings should be brief. Keep them to fifteen or twenty minutes maximum. Add time for a video, if you have one that fits the subject well and will hold interest. You do not need a video for every meeting. I have witnessed some effective communications meetings that did not use videos at all. Each presenter used nothing more than a single overhead with a brief agenda outlined on it.

The presenter definitely needs to do his or her homework. An effective presenter reviews some background or resource material before the meeting. As the safety professional, assuming someone else is making the presentation, you should be there as a referee. As Bryan points out in his article on committee meetings, dominant personalities, articulate speakers, and grandstanders do not have a monopoly on intelligence and are not always right. Keep the meeting on an even keel, especially when a member of the audience tries to take over or if the presenter strays from the topic or gives incorrect information. If the latter is a real possibility, perhaps you should have a rehearsal presentation.

Do not shy away from controversial topics. The only way to handle these is to dive in and address them. Get all opinions and hurt feelings out in the open. Remember that opinions are not facts. At the same time, everybody is entitled to their opinions. A problem arises when some hold their opinions so holy and can not differentiate between their opinions and the facts. The best way to handle this is to get them out of the discussion. Allow an unbiased, full discussion to develop instead.

Do not be afraid to address the issue of responsibility either. Conscientious employees will take responsibility when it is clearly theirs to take. This issue is a two-edged sword, it cuts both ways. Make sure you have not dodged any of your own responsibilities before holding others accountable.

If your plant is typical, at least one hidden agenda will exist for each employee. More than once I have had a terrific communication meeting turn sour because I got blindsided by a hidden agenda. A useful byproduct of spending more time on the plant floor is that you pick up on these undercurrent attitudes and feelings. This is another reason why your line supervisors are your best allies—because they hear out these hidden agendas long before you do.

Another problem that will always plague you is irrational thinking. Be wary of what others consider "common sense": it is more often common irrationality. Common sense told an employee that if he could smell a certain solvent, his health was being affected. Air monitoring revealed that his exposure was almost nil. But, he disagreed—because he had common sense and I did not. He complained to OSHA. So I guess, they lacked common sense, too. They agreed with me. I have hundreds of "common sense" stories. My jaws clinch when I hear the words.

Figure 6-1 summarizes Pierce's rules of communication. Figure 6-2 lists some destructive business communication myths, also by Pierce.

Compliance by Rules

Unfortunately, we need an abundance of written rules and guidelines to get the safety job done. Much of the documentation OSHA expects to see when it visits your facility involves written rules. Many times specific documents are required explicitly by the OSHA standard. At other times, the requirement may be implied. Sometimes the requirement is implied by good business or engineering practice. But, if you think that you do not have to

Figure 6-1
Rules of Safety Communication

1. Communicate as much information as you can.
2. Communicate often.
3. Communicate honestly.
4. Communicate quickly.
5. Communicate in many forms.
6. Communicate on the worker's turf.
7. Invite open forums and questions.
8. Open your door.
9. Remember that too much information is better than too little.
10. Make communication two-way.

Figure 6-2
Destructive Communications Myths

1. Tell workers only what they need to know.
2. The workers cannot understand all this information.
3. Let's not tell them the whole story.
4. Let's sugar coat it.
5. Avoid open forums.
6. Communications is best done on management turf.

do something, such as devise a written document, because it is not called for by an OSHA standard, I'm sure a plaintiff attorney will have some tough questions for you one day. OSHA also recommends written plans and programs for all sorts of situations, not just those required for compliance.

Such written documents are formal, downward communication. Someone at the top—you—issues a directive or disseminates some safety information. At the next level of the hierarchy, someone passes the communication along to his or her subordinates. Such communication documents help to tie the hierarchy together.

OSHA requires, both explicitly and by implication, the development of certain programs to address specific safety issues. It is important to understand some terminology, specifically: plan, program, policy, standard, practice, and procedure. A *program,* whether required or not, is best thought of as a written description of a process. How are we going to make safety happen in our plant? Here, in this program (written process), is how. The difference between a program and a process is that the program has a start and stop point, while the process is continuous. A *plan* is a subset of a program or process. For instance, an Emergency Response Plan can be a subset of the Hazard Communication Program. A *policy* sets forth rules of conduct as expected by management. Policies outline general programs and assign responsibility.

Standards apply policies in detail. On the federal level, OSHA standards apply the policies or laws (specifically OSHAct) of Congress. Company standards outline what actions are required and who is to perform them. A good standard includes a timeline for the actions to be accomplished and provides performance measurements. A consensus standard is a set of guidelines issued by a body of professional practitioners, such as ASME, ASTM, ANSI, and NFPA. A *practice* is similar to a consensus standard in that it is merely a guideline and its methods may vary.

A *procedure,* on the other hand, is a nuts and bolts how-to-do-it document that outlines an exact step-by-step set of instructions. A procedure, unlike a practice, is mandatory and allows little, if any, variation from its methods. Figure 6-3 lists some required programs and plans required by OSHA.

When developing a written plan or program, research OSHA guidelines and standards first. Sometimes you can purchase a set of plans or a program through commercial publishers who specialize in such. However, make sure you customize the document to your facility.

Once I was called in to redo a Hazard Communication Program after an OSHA inspection had rejected the existing document. On the first page of the program was a statement that the reader was looking at a Hazard Communication Program for XYZ Company, a international chemical company. I asked my client if his company were an affiliate of XYZ. "No," he said, "I got this program from them and thought it was good. Should I have changed the name?"

I should say so! Look at other plans and programs and by all means use them extensively to develop your own. If the owners do not object to this adaptation, make sure the modified document reflects the conditions of your workplace.

Figure 6-3
OSHA Required Programs and Plans

Hazard Communication Program
Emergency Response Plan
Lockout-Tagout Program
Permit—Required Confined Space Program
Bloodborne Pathogens Program
Fire Prevention Program

Where do you start? Identify first whether you are subject to an OSHA regulation. If so, that is obviously where you want to begin. Evaluate your workplace with respect to the OSHA standard. Evaluate the activities in your workplace with respect to the OSHA standard. Use data from OSHA inspections.

Hazard analysis, discussed in the previous chapter, begs to be communicated to employees. An effective way to do this is to build risk communication into the standard operating procedures. The procedures must be easy to read and follow. The more like a recipe they read, the easier they are to implement. Define the process technology that the procedure will cover. Determine the tasks. Document the procedures.

Compliance by Understanding

The most solid compliance stance is built on an understanding of how hazards jeopardize our business. Understanding the importance of protecting property and human resources reinforces this position—not just lip service, but a real commitment to protection.

Employee Acceptance

Employees know instinctively when they are being cared for. The words and actions of management speak volumes about their commitment to safety. We have discussed this elsewhere, but it bears repeating because it is a vital point. Management leadership teams that pay only lip service to commitment remind me of an old Pogo comic strip. One time Pogo said, *We have met the enemy and they are us.* Or something to that effect, anyway.

Relationships between Supervisors and Employees

Familiarity and trust are the key elements of communication between supervisors and employees. Supervisors must be consistent and fair in order to earn trust. When supervisors walk out onto the floor of the plant, employees should see in them a trusted friend coming. They should see someone to whom they can turn for help and with whom they know they will get a fair hearing. But if the supervisor is the only member of management with whom this trust is shared, the battle is already lost. At a minimum, more than half the line supervisors should have this kind of relationship with their employees.

Line supervision is the level where your safety program will happen or not. Help your supervisors gain the skills they need to relate to employees. Training courses in the soft skills abound. Schedule one today.

Hazard Communication

OSHA requirements for hazard communications, typically called HazCom, were mandated in 1985. The standard requires manufacturers and distributors to list hazards and precautions for the hazardous products that they sell. The manufacturers and distributors were made responsible for preparing the Material Safety Data Sheets (MSDSs). Employers who use the hazardous products have the responsibility to obtain an MSDS for each material received on-site, to train employees on the hazards and precautions mentioned therein, and to maintain the documents.

Information on the risk to workers from chemical hazards can be obtained from the Material Safety Data Sheet which OSHA's Hazard Communication Standard requires be supplied by the manufacturer or importer to the purchaser of all hazardous materials. The MSDS is a summary of the important health, safety, and toxicological information on the chemical or the mixture's ingredients. Other provisions of the Hazard Communication Standard require that all containers of hazardous substances in the workplace have appropriate warning and identification labels.

The chief problem of MSDSs, as Altvater points out, is the failure to supply minimum data more than inaccuracy of data. Some inaccuracies do exist. How accurate is an MSDS that lists no hazardous ingredients but contains three pages of health warnings? Also, some MSDSs apparently switch back and forth in reporting properties for pure ingredients and mixtures.

Another problem with MSDSs is clarity. What does "use adequate ventilation" mean? How can you "dispose of in accordance with all applicable laws and regulations" when none of the ingredients are revealed? How about, "Dispose of properly"?—I love that one. Parella suggests using symbols in combination with words on the MSDS to ensure communication with all employees.

Figure 6-4 lists the informational requirements for MSDSs.

Figure 6-5 lists some important questions to ask about your hazard communication effort.

Keep in mind that the purpose of hazard communication is to get pertinent information about chemicals to the user. Presumably, he or she can then protect himself or herself. Therefore, demand complete, accurate, and clear MSDSs. Call the number on the form if any information is missing and ask for a new MSDS with the completed information. Make a record of your telephone call. Attach the record and a copy of any follow-up letters to the file copy of the MSDS. Do the same for MSDSs containing inaccurate data or unclear statements.

Make it a policy that labels are not to be removed, defaced, or destroyed from any container. Do not get into the labeling business yourself. Do you know the contents of the container? If you prepare labels yourself for containers of material that you did not manufacture, you may be participating in hazard miscommunication! Figure 6-6 lists major label requirements.

Hazardous Waste Operations and Emergency Response

Perhaps the most misunderstood of all OSHA's standards is HAZWOPER. Most facilities overkill the training aspect of compliance. HAZWOPER applies to specific situations, not to every industrial facility. In fact, HAZWOPER is mostly aimed at sites where hazardous waste operations are the routine. To which sites does HAZWOPER generally apply?

1. To sites where uncontrolled releases of hazardous substances are being cleaned up.
2. To sites where hazardous waste is routinely handled.
3. To other sites during emergency response to spills and leaks.
4. To emergency personnel entering a Category 1 or 2 site.

Sites where a Superfund cleanup is underway fit Category 1 as do RCRA corrective action sites. Category 2 pertains to active RCRA permitted sites, called Treatment, Storage, and Disposal (TSD) facilities. Your industrial plant would be a Category 2 site if you have a RCRA Part B Permit. Otherwise, if you are a RCRA generator who is allowed only ninety-day storage before shipment, you are neither Category 1 nor 2.

If neither 1 nor 2, you could still fall under the HAZWOPER umbrella if you have a team assigned to respond to a hazardous waste emergency, such as a spill or leak. Also, any outside emergency responders to your hazardous waste facility, whether public or contractor personnel, are required to be HAZWOPER trained, and you have a role to play in that.

Figure 6-4
Information in Material Safety Data Sheets

- Chemical identity of each hazardous component
- Physical and chemical characteristics
- Physical hazards (fire, explosion, reactivity)
- Primary route of entry
- OSHA Permissible Exposure Limit (PEL)
- Carcinogenicity
- Precautions for safe handling and use
- Control measures
- Emergency and first aid procedures
- Name, address, and telephone number of the manufacturer or distributor
- Name of preparer
- Date prepared or last modified

Figure 6-5
How Effective Is Your Program?

- What is the program's goal?
- Do you have a list of hazardous chemicals?
- Where are the MSDSs for these chemicals?
- How can an employee get a copy of an MSDS?
- How will you inform employees of the hazards of nonroutine tasks?
- How will you inform on-site contractors of the hazards their employees may be exposed to?
- Where is the written program available for inspection?
- What is your company policy on labeling containers?
- Who enforces the container labeling policy?
- Who will conduct hazard communication training, and when and who will receive this training?

Figure 6-6
Label Requirements

1. Identity of the hazardous components
2. Appropriate organ specific hazard warnings
3. Name and address of manufacturer, importer, or distributor

Otherwise, if a lone employee can handle any potential spill or leak of hazardous waste on your property, and that employee has been given hazard communication training on the specific materials involved (as hazardous waste), your facility does not need to implement HAZWOPER. A lot of facilities have one or more personnel who received HAZWOPER forty hour training who no more need it than the man in the moon. In the case of HAZWOPER, however, overkill is good.

Most industrial facilities, at least those which are generators of hazardous waste only and not permitted TSD facilities, fall under 29 CFR 1910.120(q) for emergency response. HAZOPER training requirements can be avoided as explained above if a lone employee with hazard communication training responds to spills and leaks. This does not preclude the ability to have any employee respond, only that one at a time may respond. Otherwise, a team of qualified employees must be trained to make an emergency response. They must receive training and recertification annually. Training must pertain to specific topics. Various levels of training are allowed under the standard. Written plans must be developed. Safety equipment must be provided.

For those who require HAZWOPER training, certain topics must be covered by a person qualified to teach them, as summarized in Table 6-1, found on the following two pages.

Table 6-1
Training Topics for HAZWOPER

Topic	Subtopics
Basic Chemistry	Element Compound Chemical Formula/Names Vapor Pressure Molecular Weight Vapor Density Flash Point Flammable/Combustible pH - Acid/Base Forms - Solid/Liquid/Vapor/Gas
Basic Physiology	Respiratory System Skin Gastrointestinal Cardiovascular Eyes Central Nervous System Liver Kidney
Basic Toxicology	Dose Response Toxicity—Acute/Chronic Resistant v. Hypersensitive LD_{50}/LC_{50} TLV-PEL-REL-IDLH IARC-NTP Mixture Route of Entry Carcinogenicity Target Organ Other Effects
Detection Methods	Calculations Senses (Odor, Visual) Direct Sampling Methods Indirect Sampling Methods Sampling Equipment Accuracy and Precision MSDS Signs and Labels Manufacturer Information

Table 6-1 *(continued)*

Personal Protective Equipment	EPA Levels of Protection
	Respiratory Protection
	    Air Purifying
	    Air Supplying
	    SCBA
	Skin Protection
	    Permeation
	    Penetration
	    Degradation
	Selection Methods
	Limitations

Terms	OSHA-NIOSH-ACGIH EPA-TSCA-FIFRA-CERCLA-RCRA DOT-IATA HMIS-NFPA MSDS

Risk Assessment	People-Environment-Property
	Quantity-Quality
	Physical-Toxic
	Conditions
	    Weather
	    Reactions
	    Fire
	Resource Limitations
	Critical Operations

Procedures	Prevention First Aid and Medical Plans (Emergency Action) Control-Confinement Decontamination Standard Operating Procedure (SOP) Termination Follow-up Training

References

All about OSHA. U.S. Department of Labor. OSHA Publication 2056. U.S. Government Printing Office, 1995.

Altvater, Thomas S. "Material Safety Data Sheets: A User Perspective." *Professional Safety*. October 1990, pp. 17-9.

Bryan, Leslie A., Jr. "Group Dynamics: toward More Effective Committee Meetings." *Professional Safety*. October 1991, pp. 28-31.

Daugherty, Jack E. "Hazard Communication for Small Plants." *Occupational Hazards*. February 1995, pp. 37-40.

Ford, Christopher M. "Building Safety into Operating Procedures." *Occupational Hazards*. February 1997, pp. 39-44.

Haimann, Theo, and Raymond L. Hilgert. *Supervision: Concepts and Practices of Management*. Cincinnati: Southwestern Publishing Company, 1977.

Parella, Stephen D. "Illiteracy and Right-to-Know—Bridging the Gap." *Occupational Hazards*. October 1990, pp. 113-5.

Pierce, F. David. "Ten Rules for Better Communication." *Occupational Hazards*. May 1996, pp. 78-80.

7

ACCIDENT/ILLNESS PREVENTION

CAUSES AND EFFECTS

Accidents are malfunctions of the management system as it relates to the following: 1) situational work factors, 2) human factors, and 3) environmental factors. Situational work factors are totally outside the worker's control, except for the responsibility to recognize and report hazards in the workplace. Situational work factors include defects in materials and design of processes and equipment. Human factors on the other hand are any unsafe acts. This unsafe behavior is within the worker's control and can, thus, be modified by his or her supervisor. Physical, chemical, and biological agents are environmental factors that may cause harm and, for the most part, are beyond the worker's control.

Situational Factors

Defects in the design of processes, equipment, or machinery are not always detectable by the worker until some accident takes place. For instance, one employee who worked on a machine that ratcheted around, taking parts to six different stations where various operations were performed, hurt her finger. Thinking that it was her own fault, and her finger was only slightly mashed, she did not report the accident to her supervisor. Several days later, however, her finger was cut and she reported this accident. The supervisor, thinking the accident was caused by an unsafe act, routinely reported it. While the accident report was still making its round, the woman nearly got her finger cut off. This was the third incident, the second reported. She and her supervisor did not relate the incidents to each other. However, upon investigation, it was discovered that the machine required a safety device that newer versions of the same machine already had.

Poor construction is another situational factor that workers have no control over and are unlikely to detect in time. Improper storage of hazardous materials is another situational factor. With proper training and education, a worker can participate in controlling these types of hazards, as well as identifying them. Inadequate planning, layout, and workplace design are beyond the workers' control unless you allow them to participate in the planning and design processes.

Human Factors

Human factors are important in accident causation. Anywhere from 6 to 15 percent of accidents are caused entirely by the behavior of employees. These human factors can be easily corrected when management has a good relationship with its workers. Training, leadership, discipline, and trust are required. Unauthorized removal of guards from a machine, for one, is a common hazard caused by a human factor. Operating machinery or vehicles at unsafe speeds is another example of a human factor. Knowingly using a defective tool is another example, but if the employee does not *know* the tool is defective, then it is a situational factor that is behind the potential accident.

Human beings are naturally limited in what they can do. One such limitation is vision. Bryan points out that visual acuity is a problem if it is not accounted for in machine design. Figure 7-1 lists some specific visual acuity problems. Additional sight problems are discussed in Chapter 14.

Auditory acuity is another human limitation. More discussion is provided on this in Chapter 14.

The sense of touch—tactile sense—is pronouncedly affected by environmental temperature. Obviously, any switches, buttons, and dials that depend on tactile sense for operation are potential sources of human error.

Figure 7-1
Visual Acuity Limits

- Discrimination or resolution of detail
- Stereoscopic acuity or depth perception
- Convergence or double vision in close work
- Color discrimination
- Dark adaptation

Motion, balance, and position are called the propiosenses. Fatigue of muscles, tendons, and the vestibular organs in the ear can throw these senses off, resulting in potential human error.

Human-Machine Relationship Errors

Eight types of errors are caused by the human-machine relationship: substitution error, selection error, reading error, forgetting error, reversal error, unintentional activation error, mental overload, and physical limitations.

Typically, substitution error is the result of habit patterns, causing the operator to choose incorrectly. For instance, an operator may throw the wrong switch simply because it is the same size and shape of a switch he or she has always flipped on another machine. Use standard shapes and sizes for selector switches to avoid this.

A selection error, though similar to a substitution error, is not caused by habit. Anytime the machine operator inadvertently selects the wrong switch, wrong switch position, or otherwise sets a machine up wrong, he or she has made a selection error, if it was not due to habit. Design in preselected settings as much as possible to avoid this error.

We have all made reading errors from time to time. How often do you misread the thermometer in the morning and dress too warmly or not warm enough? How often do you misread a watch? These sort of reading errors happen with analog readouts. Use digital readouts to lessen the probability of error.

Forgetting errors happen more often when a machine or operation is very complex. To avoid these types of errors, prepare a checklist for the operator to follow. Memory experts have found that most forgetfulness is merely mental laziness, not the lack of capacity for memory. Trainers should therefore challenge the memories of their students.

In a reversal error, something is operated in the wrong direction,—a switch is flipped on instead of off or vice versa. Stress, inattention, and design contribute to reversal error.

Unintentional activation is often caused when loose jewelry or baggy clothes snag a switch lever. Critical switches can be guarded from unintentional activation. Lockouts can be useful. Prohibit loose jewelry and clothing from being worn.

Mental overload is too common. Too many things happening at once leads to an inappropriate response at some point, unless you are lucky.

According to Bryan, more than one hundred alarms went off in the first few minutes of the Three Mile Island nuclear incident. No one can make decisions fast enough to respond appropriately to one hundred problems at once. I saw the aftermath of a machine wreck where the operator was startled when only three alarms sounded at once and she cycled the machine to the next operation when it was not ready. A simple limit switch could have prevented this accident. Thank goodness no one was hurt, but product and the machine were affected. Fatigue, poor training, and lack of experience lead to mental overload.

Physical limitation errors are caused when the operator's body cannot reach, fit, or operate, leading to non-response or incorrect response. Anthropometric data is available to the designers. With more women and handicapped people in the workplace, these physical limitations need to be allowed for in design. Otherwise a retrofit must be designed.

Table 7-1, adapted from the Bryan article, summarizes the abatement practices where human error is involved.

McClay proposed a lengthier list of human limitations, summarized in Figure 7-2.

Table 7-1
Abatement of Human Error

Error	Engineering	Administrative	Training		
			Knowledge	Skill	Attitude
Substitution	Separate Control/Guards	Warning Labels	Awareness	Practice	
Selection	Preset position or detents	Warning Labels	Awareness	Practice	
Reading	Digital read out			Practice	
Forgetting		Checklist	Drill memory	Practice	Acceptance of Checklist
Reversal	Design	Warning Labels	Awareness		Caution
Unintentional Activation	Guards	Clothing Rules	Awareness		Caution
Mental Overload	Design	Proper Placement	Drill memory	Practice	
Physical Limits	Design	Proper Placement		Compensating Skill	

Figure 7-2
McClay's List of Human Limitations

Force exerted by body member	Sound recognition or speech interpretation
Rate of force exertion	Tactile recognition
Endurance	Smell and taste
Speed of movement	Balance
Dexterity	Reaction time
Reach distance	Memory recall
Postural limits	Knowledge and understanding
Work load	Motivation
Visual acuity under normal, darkened, or glare conditions	Attention span
Aural acuity at various frequencies, or with background noise	Capacity for attending to simultaneous tasks

Environmental factors are physical, chemical, or biological agents. Physical agents include noise, light, vibration, ionizing radiation, heat, cold, and electro-magnetic forces. Chemical agents include potentially harmful gases, vapors, dusts, mists, aerosols, fumes, and smoke. Biological agents include viral and bacterial pathogens, fungi, and vectors such as mosquitoes, fleas, rats, mice, and insect pests.

Loss Causation Model

Manuele discusses a causation model based on the assumptions listed in Figure 7-3.

A general causation model is shown in Figure 7-4.

Cultural and poor management practice causal factors are summarized in Figure 7-5.

Figure 7-3
Manuele's Causation Model Assumptions

1. Occupational injury and illness risk levels are determined fundamentally by the culture of an organization.
2. Management commitment to safety is an extension of culture and is the source of decision-making affecting the avoidance, elimination, or control of hazards.
3. Safety practice that focuses on unsafe acts as principal causation factors is unproductive and unrealistic.
4. A causation model for hazards-related incidents should stress the genesis of decision making, rather than consequences, such as human error.
5. Most incident causal factors are systemic and evolve from the workplace and work methods created by management; responsibility for the remainder of the incident causal factors lies with the worker.
6. Causal factors derive from poor policies, standards, procedures, and accountability systems or their defective implementation, which impact on design management, operations management, and task performance.
7. Design management, operations management, and task performance are interdependent and mutually exclusive.
8. Energy releases or exposure to hazardous environments are essential to hazards-related incidents.
9. Hazards-related incidents, even ordinary or numerous ones, are complicated and have manifold interacting causal factors.

Figure 7-4
General Causation Model

Management leadership establishes and influences corporate culture, which is colored by the attitude of the employees.

The degree of management commitment to safety, which is an expression of culture, demonstrates the system of expected behavior.

Figure 7-4 *(continued)*

Causal factors derive from culture and management practices when safety policies, standards, procedures, or the accountability system, or the implementation of these are less than adequate.

Additional causal factors derive from inadequate design, operations, and task performance practices and complicate an already bad situation.

A hazard-related incident occurs.

Harm or damage occurs, or could have occurred.

Figure 7-5
Poor Management Causal Factors

Design Management inadequacies:	**Operations Management inadequacies:**
1. Hazard/risk assessment	1. Hazard/risk assessments
2. Facilities	2. Work methods
3. Hardware	3. Personnel selection
4. Equipment	4. Supervision
5. Tooling	5. Personnel motivation
6. Materials	6. Training
7. Layout/configuration	7. Work scheduling
8. Energy control/substitution	8. Management of change
9. Environmental concerns	9. Maintenance
	10. Incident investigations
Task Performance inadequacies:	11. Inspections
1. Originate from inadequate design or operations practices	12. Personal protective equipment
2. Errors of commission or omission	

Figure 7-6
Universal Causation Model

1. Find the point of irreversibility.
2. Find the first loss incident that produces effects or a near miss without subsequent events or conditions.
3. Connect these two points.
4. Add aggravating or mitigating factors between the two points.
5. Find the next loss incident.
6. Add aggravating or mitigating factors between the loss incidents.
7. Find all remaining loss incidents.
8. Identify and list all final effects.
9. Identify causal factors, those hazards that immediately precede the point of irreversibility.
10. Identify the causal factors that precede these hazards.
11. Show causal factors only if they are conditions or events that meet all these requirements:
    a. Occur in the same timeframe as the loss incidents.
    b. Occur in the same location as the loss incidents.
    c. Immediately precede the diagrammed hazards.
12. Do not include supervisory actions or omissions, programs, or policies.

McClay gives us the universal model shown in Figure 7-6.

A hazard-related incident results from unwanted energy flows or exposures to harmful environments. Another cause of hazard-related incidents is when either a person or thing in the system or both are stressed beyond the limits of endurance or recoverability. The incident begins with an initiating event, which triggers a series of causal events. Multiple interacting events occur sequentially or in parallel until the chain of events results in harm or damage. If harm or damage does not occur, these could have been the result if the situation or hazardous exposures had been slightly different in either degree or sequence.

Identify, Evaluate, Control

Six elements make up the entire process of hazard control: hazard identification; hazard evaluation; decision; prevention and correction; observation; and process evaluation.

Hazard Identification

Before hazards can be controlled they must be discovered. Two basic approaches to identifying hazards that contribute to accidents are *after-the-fact* and *before the fact.*

After-the-fact *hazard identification* relies on accident reports, insurance audits and loss control reports, and frequency and severity ratings. The problem with these methods is that someone is already hurt or property is already damaged before they are measured. However, that does not taint the information; it is still good and valid data.

Before-the-fact methods are preferable in an aggressive accident prevention process. One of these methods is the critical incident technique in which employees' opinions about their safety are solicited and analyzed. The plant safety committee can also conduct direct observations, in the form of audits, inspections, and safety sampling. Finally, the job safety analysis is another before-the-fact method of hazard identification.

Workplace monitoring is an effective means of acquiring hazard information before- or after-the-fact. The four principle methods of monitoring are hazard analysis, inspection, measurement, and accident investigation.

Hazard Analysis

Hazard analysis is an orderly process that acquires specific information pertinent to a given system. The hazard analysis can uncover hazards that may have been overlooked in the original design of a workplace or component. It can locate hazards that may have developed after a task or process was installed. A hazard analysis can determine the essential factors in and requirements of a specific job, process, operation, or task. It can indicate the need for modifying processes, operations, and individual tasks. Situational hazards, human factors, exposure factors, and physical factors can all be identified by conducting a hazard analysis. Finally, a hazard analysis can be used to determine workplace sampling and instrumental monitoring methods and the maintenance standards needed for safety.

The mathematical relationship between hazards and their effects can be expressed as a function:

$$X = f\{A,B,E\} \qquad [7.1]$$

where

X = adverse effects, and
A, B, E = hazardous conditions

Two formal methods are used to conduct a hazard analysis: inductive and deductive. The *inductive method* considers a system's operation from the standpoint of its components, the failure of those components in a particular operating condition, and the effect of that failure on the system as a whole. The job safety analysis (JSA) discussed in Chapter 9 is another inductive method of hazard analysis.

The *deductive method* postulates failure of an entire system and then identifies how the components could contribute to that failure.

The *cost effectiveness method* of hazard analysis can be used independently of the deductive or inductive method and may be used to support findings by these methods. The cost of system changes (price of compliance) is compared to the cost of serious failures (potential cost of noncompliance). Cost effectiveness is then used to decide among several alternative control solutions to abate hazards.

Who should participate in the hazard analysis? To be fully effective, hazard analysis should represent as many viewpoints as practical. Regardless, it is important to select the tasks to analyze with care. Some priority must be assumed, especially if many tasks have yet to be analyzed. Six risk factors that are considered in the selection process are listed in Figure 7-7.

Hazard Evaluation and Decision

Next, after identification, hazards must be evaluated to determine risk. This *hazard evaluation* involves making frequency and consequence deter-

Figure 7-7
Risk Factors in Order of Priority

1. Frequency
2. Potential for injury
3. Severity of injury
4. New processes, altered equipment
5. Excessive material, waste, or equipment damage
6. All jobs, eventually

minations to analyze relative risk, as discussed in Chapter 5. Hazards are then ranked by risk, the worst listed first.

When hazards are prioritized according to frequency and consequence, the next step is for management to make a decision. *Decision* means the safety professional must provide management with full and accurate information, including a brief discussion of all alternatives. This report should be made in such a way as to promote action. If management chooses not to take action it is usually for one of three reasons. Either management feels it cannot take the required action for some unforeseen reason or it has been presented with limited alternatives, none of which were appealing. Or, management may refuse to take action because it disagrees in principle either that a hazard exists or that the risk does not warrant action.

One way to expedite the decision making process is to present your findings in such a clear manner that management can easily understand: 1.) the nature of the hazards, 2.) their location, 3.) their importance, 4.) the risk, 5.) necessary corrective actions, and 6.) the estimated cost. Besides estimating actual cost of controlling the hazards, give management the amount of sales dollars that must be generated to cover the cost of a potential disaster. I call this the *potential cost of noncompliance.* This is sometimes a staggering figure and usually gets heads to nod in favor of spending money for control. I call this expenditure the *price of safety.* Comparing the price of safety to the potential cost of noncompliance will often sell the abatement project for you.

Hazard Control

Next you control the hazards. *Hazard control* is the function that is directed toward eliminating, or at least minimizing, the destructive effects of hazards emanating from human errors, situational conditions, or environmental conditions of the workplace. Hazard control is failure oriented. By that I do not mean that it is an unsuccessful effort, rather it focuses on systems, processes, and objects that could fail, thus causing harm to people or damage to property. The manner in which a system exhibits failure is called its *failure mode.*

Hazard control is implemented by installation of preventive and corrective measures. The two kinds of hazard control are *administrative controls* and *engineering controls.* The three places where hazards may be controlled are at the source, on the pathway, and at the receptor. The *source* is the place from where the hazard emanates. Control a hazard at the source by eliminating it entirely or substituting equipment or material or a process that is less hazardous. The *pathway* is how the danger gets from the source to a victim or object. The pathway for noise, light, and most chemicals is air. Control a hazard along the pathway by using a barrier of some kind. The *receptor* is the human victim of the hazard or the property that is damaged. Control a hazard at the receptor by using remote controls or personal protective equipment to isolate the receptor from the hazard. Figure 7-8 lists nine methods used to control exposure to hazards.

Observation

After hazard control comes the monitoring or observation process. Monitoring does not necessarily mean taking measurements of concentration—that kind of monitoring was probably accomplished in the hazard identification phase and provided data for the hazard evaluation phase. Monitoring in this instance means the steps taken to provide assurance that hazard controls are working properly. It may include the measurement of concentrations and other scientific measurements to validate the abatement. Monitoring may also be necessary to ensure that subsequent modifications to the workplace do not reduce the effectiveness of hazard controls. Finally, ongoing monitoring may be required in order to discover new hazards as they develop.

Figure 7-8
Control of Exposure to Hazards

1. Substitution
2. Minimize contact
3. Isolation or enclosure
4. Wet methods, chemistry
5. General or dilution ventilation
6. Local exhaust
7. Personal protective devices
8. Housekeeping
9. Training and education

Process Evaluation

The sixth step in the overall hazard control process is the evaluation of effectiveness of the process. How much is being spent to maintain abated hazard conditions? What benefits are being received? What impact are benefits of hazard abatement having on improving operational efficiency and effectiveness? Answers to these questions are needed in order to evaluate the process of hazard control. Figure 7-9 lists the criteria for measuring effectiveness of hazard control.

Property Damage

The risk of property damage is usually underwritten by property and casualty insurance. This financial assurance should not lure us into a lack of vigilance. Property damage is often accompanied by human losses, for one thing. For another, we should never make business decisions based on letting insurance take up the slack. Insurance is a safety net when all else fails. If we do not build in other safeguards through engineering and management controls, we may be disappointed in the long run.

Figure 7-9
Hazard Control Effectiveness Criteria

- Frequency and severity rates
- Cost of occupational medical care
- Material damage cost
- Equipment replacement cost (after accidents)
- Days of lost time
- Experience rating

Personnel at Risk

Preventing humans from being harmed or killed is the business of the safety professional. Personnel are placed at risk by engineering design, operational situations, environmental situations, and themselves. We have already discussed these. The safety professional's role is to oversee the elimination of the risk.

Waste

Waste is often related to hazards. Of course, hazardous waste is always hazardous, but other forms of waste are also risk related. For instance, waste oil can contribute to the fuel value of a fire. Scrap metal can be a liability when handled and disposed improperly. Poor housekeeping is a fire prevention nightmare.

Categories of Occupational Health and Safety Risks

Recognition of OSH risks, or at least the underlying potential hazards, is a learned skill and therefore an art. I know a machinist who can find ten to twenty accidents waiting to happen for every one I find. Some have the knack and some do not. I am getting better at it and if I did not believe you could also, I would not include this section. However, sometimes beginners are better at recognizing accident potential or OSH risk than experienced people are. The reason for this is that the experienced tend to specialize their perception over the years and they literally stop seeing the obvious unless it appears within their interest range.

Fire and Explosion

Fire and explosion have the potential for causing the most harm and incurring the highest losses in most industrial plants. Prevention and management techniques for fines and explosions are discussed in Chapter 20.

Chemical Burns

Acids and caustic materials can cause severe burns to the skin. Chemical burns are discussed in Chapter 12.

Chemical Toxicity

Harmful chemical compounds in the form of solids, liquids, gases, mists, dusts, fumes, and vapors exert toxic effects by inhalation (breathing), absorption (through direct contact with the skin), or ingestion (eating or drinking). Airborne chemical hazards exist as concentrations of mists, vapors, gases, fumes, or solids. Some are toxic through inhalation. Some of them irritate the skin on contact. Some can be toxic by absorption through the skin or through ingestion, and some are corrosive to living tissue.

The degree of worker risk from exposure to any given substance depends on the nature and potency of the toxic effects and the magnitude and duration of exposure. Toxicity is further discussed in Chapter 12.

Inhalation

Air contaminants are commonly classified as either particulate or gas and vapor contaminants. The most common particulate contaminants include dusts, fumes, mists, aerosols, and fibers. *Dusts* are solid particles generated by handling, crushing, grinding, colliding, exploding, and heating organic or inorganic materials such as rock, ore, metal, coal, wood, and grain. Any process that produces dust fine enough to remain in the air long enough to be inhaled or ingested should be regarded as hazardous until proven otherwise. *Fumes* are formed when material from a volatilized solid condenses in cool air. In most cases, the solid particles resulting from the condensation react with air to form an oxide. The term *mist* is applied to liquid suspended in the atmosphere. Mists are generated by liquids condensing from a vapor back to a liquid or by a liquid being dispersed by splashing or atomizing.

Aerosols are also a form of a mist characterized by being highly respirable, minute liquid particles. *Fibers* are solid particles whose length is several times greater than their diameter, such as asbestos. *Gases* are formless fluids that expand to occupy the space or enclosure in which they are confined. They are atomic, diatomic, or molecular in nature as opposed to droplets or particles, which are made up of millions of atoms or molecules. Through evaporation, liquids change into vapors and mix with the surrounding atmosphere. *Vapors* are the volatile form of substances that are normally in a solid or liquid state at room temperature and pressure. Vapors are gases in that true vapors are atomic or molecular in nature.

Dermal

Materials that irritate as opposed to causing burns can contact the skin through immersion or through wetting by aerosols. Some toxic materials are absorbed through the skin and cause systemic or organ damage.

Other Routes of Exposure

Subcutaneous injections are normally delivered through a syringe. However, high pressure spraying liquids can also inject materials under the skin.

Human Factors and Ergonomics

Chapter 13 will discuss human factors in safety and ergonomics.

Sight and Sound

The physical hazards of sight and sound include, among other things, excessive levels of ionizing and nonionizing electromagnetic radiation, noise, vibration, and illumination. In occupations where there is exposure to ionizing radiation, time, distance, and shielding are important tools in ensuring worker safety. Danger from radiation increases with the amount of time one is exposed to it. The shorter the time of exposure, the smaller the radiation danger. Distance is measured between the worker and the source. For example, at a point ten feet from a source, the radiation is 1/100th of the intensity at one foot from the source. Shielding is another way to protect against radiation. The greater the protective mass between a radioactive source and the worker, the lower the radiation exposure. Similarly, shielding workers from nonionizing radiation can also be an effective control method. In some instances, however, limiting exposure to or increasing distance from certain forms of nonionizing radiation, such as lasers, is not effective—for example, an exposure to laser radiation that is faster than the blinking eye. These are discussed further in Chapter 15.

Heat and Cold

Exposure to heat and cold, the injuries and illnesses caused, and effective means of abatement or treatment are discussed in Chapter 15.

Physical Injury

Physical injuries are those incidents where a victim suffers acute trauma to the body or one of its component parts.

Site and Structures

Sometimes our physical surroundings can present potential for accidents to happen. Conditions to look for include traffic patterns, landscaping and vegetation, utilities, neighborhood conditions, walking conditions, weather conditions, releases, and lighting.

Complex traffic patterns may set up the potential for vehicle accidents in and around the plant site. What are the patterns for motor vehicles, rail cars, ships and barges, or foot traffic? Consider the vehicle routing system that is adjacent to your site. How is traffic routed near storage tanks or sheds? How is traffic routed with respect to pedestrians? Look at intersections and crossings near the plant and on the plant grounds. Investigate rail sidings and piers. Take a look around your parking lot.

Undesirable landscaping and vegetation is more than an eyesore—it may be a potential fire hazard. Even attractive landscape vegetation can be a potential hazard. Does it conceal a prowler with violence on his mind? Are rape, armed robbery, or vengeful murder possibilities due to the nature of the landscaping? High grass and weeds are a definite problem. Heavy shrubbery requires some judgment. What about rabid wildlife coming out of your wooded areas? How about ticks and other pests and parasites? What is the potential for flooding on your property?

The ill-planned location of utilities has exacerbated more disasters than we have time to discuss. Where are your power transformers located? Do you have natural gas or other fuel in the vicinity of areas with high fire or explosion potential? If so, you have a multiple disaster in the making. Are your communications lines located where a disaster will easily put them out of order? Perhaps it does not matter, but perhaps it does. Do you know what to look for?

Do you have unusual neighborhood problems? What is your labor turnover like? Are the buildings run down and/or vacated? What issues are causing tension at your plant? Could these boil over into workplace violence? What is the potential for civil disturbance in your neighborhood? How about the history of bomb scares?

References

Bryan, Leslie A. "The Human Factor: Implications for Engineers and Managers." *Professional Safety.* November 1989, pp. 15-18.

Manuele, Fred A. "A Causation Model for Hazardous Incidents." *Occupational Hazards.* October 1997, pp. 160-165.

McClay, Robert E. "Toward a More Universal Model of Loss Incident Causation, Part I." *Professional Safety.* January, 1989, pp. 15-20.

McClay, Robert E. "Toward a More Universal Model of Loss Incident Causation, Part II." *Professional Safety.* February, 1989, pp. 34-39.

8

ACCIDENT/INCIDENT INVESTIGATION

The purpose of accident investigation is not to find fault and assign blame, but to determine what problems to correct so that accidents do not recur. In order to decide what to change and how, you need facts. Besides finding the root cause, accident investigations are used to uncover indirect causes and prevent similar accidents from recurring. By documenting facts and providing information of costs, safety is further ensured.

Who should conduct the accident investigation? Senecal and Burke suggest that the investigators should consist of the injured employee (this is not always possible), a peer who understands the work process involved in the accident, and the supervisor. In most plants only the supervisor conducts the accident investigation, though he or she interviews the injured employee and witnesses.

Figure 8-1 summarizes the ten elements of an effective accident investigation. This table assumes that the injured have been treated. Emergency care for the victims is the first priority. The investigation can wait.

Accident investigations have some common weaknesses. For one, inexperienced or uninformed investigators often settle for the quick and easy answer, or a non-answer that sounds good. For instance, a too common cause listed on accident investigations by the inexperienced, is: "She was careless." That statement has no meaning. Even if she were careless, how so? Was she in haste because her supervisor gave her too many tasks? Was she ever trained to perform the task "carefully"?

Another common weakness found in accident reports is the reluctance of the investigator to accept responsibility for his or her area. Instead of identifying the root cause, such a supervisor is too quick to find fault in the misbehavior of one or more employees. The investigator also often has a narrow interpretation of environmental factors. Placing erroneous emphasis

Figure 8-1
Elements of an Effective Accident Investigation

1. Secure the accident scene.
2. Obtain a description of the accident.
3. Document the sequence of events.
4. Record interviews with witnesses.
5. Recreate and control the same accident scene.
6. Determine existing engineering controls.
7. Review applicable employee/supervisor training.
8. Review existing enforcement training procedures and established safety rules.
9. Review supervisor accountability.
10. Determine most likely causes of the accident.

on a single cause out of many possibilities is another common fault. Another common error is to judge the effect of the accident to be the cause. For instance, stating that the cause of a damaged wall was the forklift that ran into it is not accurate enough. Did the brakes on the forklift fail? Was the driver alert? Was there oil on the floor? The forklift running into the wall was obviously the *effect,* not the *cause* of the accident.

> All accidents, no matter how small and insignificant they may seem, should be investigated promptly.

The inexperienced investigator may arrive at conclusions too early, before all the facts have been identified and considered. He or she also may lack good interviewing techniques, leaving much information uninvestigated. The final common weakness in industrial accident investigation is undue delay in initiating the investigation. Remember that once the scene is cold, the facts will start to disappear.

Figure 8-2
Accident Investigation Kit

Chalk	Graph paper	Clipboard
Pens	Plain paper	Labels
Pencils	Steno pad	Hang tags
Colored markers	Accident investigation forms	
China marker (for metal surfaces)		Protractor
Large manila envelopes		Ruler
Small manila envelopes		150-foot tape
Carpenter's rule	35mm camera	50mm lens
Film	75-100mm zoom lens	Macro lens
Photo log book	Autoflash	Flashlight
Cardboard arrows	Fluorescent flagging tape	
Lockout/tagout set	Copy of standards	Warning signs
Evidence log	Chain-of-custody forms	Hardhat
Safety glasses	Leather work gloves	Latex gloves
Plastic gloves	Micro-recorder	Video camera

In order to assist your investigators provide them with an accident investigation kit. Figure 8-2 lists the minimum kit items, which will fit into a modest size attaché case. Make the kit available in a central location for their convenience.

Recording the Accident Scene

The most critical skills you need as an accident investigator are observation and recognition, which are discussed in Figure 8-3. At the scene we focus our attention on the symptoms of the accident. Here is where we differentiate between unsafe acts and unsafe conditions. In Figure 8-4 some sources of unsafe acts and conditions are listed.

Figure 8-3
Accident Investigation Skills

Observation and Recognition Techniques

1. Understand the objective of the work activity you observe.
   - Identify insufficient actions/steps.
   - Increase efficiency.
2. Be familiar with the standard/accepted methodology for completing the task under observation.
   - Recognize deviations.
3. Look for attitudes, regardless of behavior.
   - Attitudes precede behavior, just as behavior precedes accidents.
4. Trust your initial impression.
   - Behavior may change once a worker knows he or she is being watched.
5. Know the accident history of the facility.
   - Know specifically what hazards to look for, and therefore what behaviors can lead to accidents.
6. Document your findings.
   - Create a paper trail for implementing changes in training, inspections, and investigations.
7. Have an immediate reaction.
   - Correct unsafe acts as soon as possible.
   - Speak to the employee about unsafe behavior.
   - Investigate the source of poor attitude.

What do we look for in an accident investigation? We look for what the injured employee was doing at the time of the accident. Was the injured employee authorized to perform the task that resulted in an accident? We also should find out what nearby employees were doing. Was the proper equipment in use? Look for new processes, equipment, and tasks. Where was the supervisor at the time? Did the injured employee receive hazard

Figure 8-4
Unsafe Acts and Unsafe Conditions

Unsafe Acts
An unsafe act is any hazard created as a result of human action or behavior. An unsafe act can be attributed to:

- lack of adequate training
- improper lifting technique
- poor attitude
- utilizing shortcuts to save time
- lack of proper equipment and tools
- poor leadership

Unsafe Conditions
An unsafe condition is any physical hazard related to equipment, materials, structures, or other physical elements of a worker's environment. Unsafe conditions may include:

- poor housekeeping
- lack of guarding
- poor maintenance
- defective equipment or tools
- improper material storage
- slip and fall hazards

recognition training before the accident? What was the physical condition of the area? What actions may have prevented the accident? What corrective action was recommended before the accident, but not adopted?

The root cause of every accident is some sort of system breakdown—find it.

That may be easier said than done. First, establish the facts. Then review the facts until you identify the primary systematic failure. Next, begin tracing cause and effect linkages forward and backwards. Select likely operational errors and investigate them further. The goal is to develop feasible courses of action for preventing such incidents in the future.

Scene Control and Security

The first thing you need to do is to stop any ongoing hazards. Your priority is the safety of those persons still on the scene. If necessary send for help from the emergency medical responders, the emergency response team, fire brigade, salvage team, or other help as needed. Rope off the area immediately after taking care of the injured and stopping all hazards. Enlist personnel to keep curious bystanders beyond the rope or tape barrier.

Once the area has been secured, note transient conditions such as: air quality, odors, evaporating pools, melting solids, loose material that may be blown away or moved, footprints, tire tracks, and any evidence of the accident that may disappear with time and wear. Note weather conditions, even if the accident is inside the facility. You never can tell how weather may affect a situation until you start reconstructing what happened. If you arrive more than fifteen minutes after the accident, ask witnesses to relate transient conditions and weather conditions at the time of the accident.

> The safety of people is prime.

Photography

Photographs are essential to reconstructing what happened. Keep a photograph log with a description of each photograph taken, date, time, your position with respect to the accident scene or the object photographed, distance from objects, type of film, shutter speed, and comments that may help validate the accuracy of the photograph. Date and sign the log for each roll of film.

Use indicators such as arrows, rulers, coins, or other familiar objects to depict scale.

Sketching

A scale drawing of the accident scene can be invaluable at a later date. Include topography, signs, objects, and location of witnesses. Measure everything with respect to the location of the victim and relative to other important object locations.

Mapping

If a plant layout or a topographic map of suitable scale is available, mark the scene on it. It makes no difference whether you map or sketch so long as six months after the accident people can understand what has happened. Also, a good map helps the investigator recall the scene better.

COLLECTING EVIDENCE

You are not a criminal investigator, but you do need to take care of evidence. Some helpful hints are discussed below.

Scene Management

During your investigation, nothing should be moved from the accident scene. No one should enter or leave without your approval. As mentioned above, enlist some people to help you control onlookers and senior managers who come to the scene out of curiosity. Maintain strict control over the scene until you have finished photographing, completed your sketch and/or map, and selected witnesses to be interviewed.

Looking for Evidence

Remember, you want to find causes, not fault. In non-safety cultures, supervisors often try to avoid their responsibilities by finding fault with employees for unsafe behavior. On the accident report the faultfinding supervisor will blame the injured employee for being lazy, careless, stupid, inattentive, or accident-prone. This seldom results in positive worker change and does nothing to solicit management change.

Supervisors in safety cultures focus on the process rather than the individual to identify true accident causes. These supervisors know that finding the facts is the only effective way to determine what management changes are necessary. Figure 8-5 lists some statements from supervisors' accident reports. Which are factual and which are faultfinding?

Three production factors involved in all operations must be thoroughly explored for accidents and near misses: equipment, materials, and people.

Figure 8-5
Sample Report Statements

Larry did not wear proper protective equipment.

The load was not properly tied down.

Sue was careless.

George fell over uncoiled hoses.

Mike was daydreaming.

Bill was not paying attention at the safety meeting.

The guard was not replaced over the belt.

Mary, like most workers, is too lazy to work safely.

Sometimes accidents, or near misses, occur from the improper use and improper selection of equipment. Equipment that has been improperly maintained can also cause an accident. Other incidents may result from contact with materials or from improper material handling. This is another category of production factors causing accidents. An exposure to toxic vapors is one example. A back injury caused by improper lifting technique is another.

Behavior of people is another production factor cause of accidents. Figures 8-6 and 8-7 list some areas of investigation for potential evidence.

Figure 8-6
Areas of Investigation

Equipment	Material	People
Selection	Selection	Selection
Use	Handling	Training
Maintenance	Processing	Leadership

Figure 8-7
Investigation Questions for Consideration

INVESTIGATION GUIDE If the symptom appears to be: **Unsafe Condition**	 **Unsafe Act**
Why did it exist? Why had no one noticed and corrected it?	Why was it done? Why was it done this way? Why was job or detail necessary?
What caused it to exist? What caused it to be involved?	What was its purpose? What other way could it have been done? What details could be eliminated? What instructions were not followed?
Where was it? Where was its source? Where else does it exist? Where can I find out?	Where should it be done? Where else is it done?
When did it occur? When do similar conditions occur?	When should it be done?
Who was responsible for it? Who can give me answers? Who should take corrective action?	Who is qualified to do it? Who can give me answers? Who can show me what was done?
How should it be corrected? How can it be avoided in the future?	What is the best way to do it? How can job or detail be improved?

Removal of Evidence

After observing the scene, select items that may be useful evidence for preservation. You may want some input from witnesses before selecting all your evidence and releasing the scene for cleaning and restoration of

Figure 8-8
Sample Accident Scenario

A crane operator is moving a 500 pound pump with a length of pipe loosely attached. The load shifts and the pipe falls and crushes his foot. A nearby worker trying to avoid the falling pipe slips on an oil spot made by a leaky machine and breaks her arm.

What equipment is involved in this accident?

What material is involved in this accident?

What people are involved in this accident?

function. Figure 8-8 gives an accident scenario. Answer the questions listed.

Collect operating logs, charts, records, identification numbers, maintenance records, and physical objects that have bearing on the accident. If you cannot have the original, make a copy. Do not forget personnel files and medical records.

Cataloging and Tagging

Keep a log of evidence that you select for preservation. Catalog where you found it. Indicate its location on a sketch or map. Describe its condition as you found it. What was its apparent relationship to the accident? Place hangtags on larger items. Record the date, time, location of the evidence when found, name of the incident, and your initials on the tag.

Packaging

Place smaller evidence in manila envelopes and mark the outside with date, time, location of the evidence when found, name of the incident, and your initials. Quart or gallon sized self-sealing bags are also useful for storing evidence.

Chain-of-Custody

Track the location and possession of evidence with a chain-of-custody form. Restrict access to the evidence and keep written records of people who are given access and when.

Storing Evidence

Store evidence in a controlled area where you alone have access. The environment of the storage area should be conducive to preserving the evidence.

INTERVIEWING TECHNIQUES

Not every investigation requires the taking of written witness statements. Written statements are appropriate and useful when

- A serious incident involves hospitalizing injuries or death, or illnesses to other employees, visitors, or contractors
- Termination of an employee is expected because of the accident
- The event is reasonably expected to lead to arbitration, administrative hearing, or litigation

A good interview is complete, correct, and pertinent. As the investigator, your goal is to hear and record all of the information given. Ask the six key questions over and over again: who, what, where, when, how, and why.

Have a plan and know where the interview is going to lead. If possible, prepare questions ahead of time. Make sure you understand the technology of the equipment or process involved in the accident. Avoid a hurried atmosphere. Hold the interview in private. Make the person being interviewed feel at ease. Do not be an inquisitor. Avoid asking questions that suggest answers. Also avoid questions that can be answered simply "yes" or "no." Instead, ask open-ended questions that require a statement. Allow the person to talk freely, but not to get off the topic. When necessary, interrupt and turn the conversation back to the germane subject.

Allow witnesses to first describe what happened with no questions from you. Let them tell their story uninterrupted, except to return to the subject, if necessary. Then ask your questions to get at the story. Write down both versions.

The contents of written statements and the manner in which they are obtained are important. The following guidelines should be observed.

1. Statements should be given voluntarily and without the offer of reward or the threat of punishment.
2. Preferably the witness will relate the information to a person who writes down the statement. When this is not feasible, a witness may write out the statement.
3. The information furnished should be based on facts, not opinions, and should be a true statement to the best of the witness's knowledge.
4. Specific information regarding who, what when, where, and how should be obtained.
5. Personal, first-hand knowledge is desired. Hearsay is acceptable only when identified as such.
6. A statement may be handwritten or typed.
7. The statement must be signed, witnessed, and dated.

Wilkinson states

> Accident investigations should always be directed at *Fact-Finding* if you want to find *Fault.* If investigations are directed at *Faultfinding,* you won't find *Facts.*

Witness interviews should always be conducted in private, preferably out of and away from the incident scene.

The person taking the statement should be a good listener and should allow the witness to talk. The statement writer should avoid cross-examination. Suggest that the witness begin with a chronological account of what happened. Tell the witness the purpose for taking the statement and at all times be open and honest with him or her.

As the witness tells key or important information, slow the narrative down and ask for specifics. Ask the witness to estimate distances, vantagepoints, noise levels, and any other relevant information as to the best of their ability. Some key questions are listed in Figure 8-9.

Once all of the obvious questions have been asked, start asking open-ended questions. For example, "Do you have anything to add?" "Is there something that you feel has been missed?"

Close the interview firmly. Encourage the witness to contact you if they remember some pertinent detail later. Immediately after the interview, record your impressions and judgments of the interview's strong and weak points.

Figure 8-9
Witness Questions for Accident Investigation

What did you observe at the time of the incident?

What was observed just before, during, and after the incident?

What equipment was being used at the time of the incident?

What deviations occurred in operating procedures or equipment functions?

Who else was in the area?

What actions were taken to alert management of the incident or to get medical treatment?

What conditions or procedures may have been contributing factors?

Witness Behavior

Take notes on the behavior of the witness during the interview. For instance, is the witness nervous? Defensive? These notes may be important later.

Quality of Testimony

Some witnesses do better than others at relating what they saw. Be careful about how you coach a poor witness—It is easy to put words into a person's mouth. It is better to take notes as best you can than to coach too much. Avoid forming an opinion of the witness based on the quality of his or her testimony. A poor witness may be more truthful than a sophisticated liar.

Acquiring Information on Near-Misses

Investigate near misses as if an accident actually happened. Make no compromises in finding the truth or sifting the facts. Remember, the next time it may not be a near miss, but a real accident.

EVALUATING WHAT HAPPENED

You have evidence, photos, sketches, maps, and interviews. Now what do you do with all of this information? Remember, we are not criminal investigators—whodunit is not nearly as important as why did it happen or what really occurred.

Causes and Effects

An accident is a system breakdown. The role of the investigator is to identify and describe the system and its failed components. One of the components of any system is the human or humans involved with it. All probable causes must be considered when investigating and analyzing an accident. Fault trees, event trees, universal accident models, or general accident models can be used to sift facts and assess the accident.

We know the effect. The effect is the accident. What happened to the victim? What happened to the machinery? The infrastructure? Those are the specific effects of the accident. A chain of events led to the accident. At the beginning of this chain is a root cause. That is our objective—once we know the root cause we have something that can be fixed.

In order to recognize accident causes, we must be able to differentiate between accident symptoms and underlying causes. Figure 8-10 discusses these two concepts. Figure 8-11 lists some examples of accident symptoms. Can you pick out probable causes for each?

Cause Analysis Methods

Types of root causes have been identified as procedural, hazard, facilities/ equipment, communication, haste, training, or miscellaneous factors.

Procedurally, the most common safety failure is lack of a procedure. Sometimes a procedure exists but the victim fails to follow it. Perhaps he or she is not trained to follow it. Or perhaps the victim does not understand the procedure adequately. Perhaps the procedure is inaccurate and does not work. On occasion, the victim does not follow the procedure for some other reason.

Hazards may be created by people, or by some external factor. Too often a hazard is documented but not abated in a timely manner. Most often a hazard goes unidentified until an accident happens. Sometimes a hazard will be identified but the employees involved have already accepted its risk as part of their routine. Sometimes the hazard can be abated but the effort to

Figure 8-10
Accident Symptoms and Underlying Causes

Symptoms - The unsafe acts and conditions that we can see that often result in accidents but are not necessarily the root cause.

If only symptoms are corrected, accidents can continue to occur. Examples of symptoms include:

- careless operation of a forklift
- oil on a floor
- climbing a storage rack
- improper lifting technique
- not wearing eye protection
- standing on the top rung of a ladder

Root Causes - The underlying reasons for accidents, which we cannot see, can only be identified by a thorough investigation. Examples of causes include:

- inadequate employee training
- ineffective employee motivation
- lack of accountability
- inadequate policies and procedures
- improper selection of equipment or material
- poor maintenance of facilities or equipment

Figure 8-11
Example Accident Symptoms

Worker standing on top rung of a ladder.

Worker killed when he fell.

Worker removed a guard from a machine and did not replace.

Worker slipped on oil from a forklift.

Worker's hand was amputated on unguarded machine.

Worker not wearing proper eye protection.

Small metal chip struck worker's forehead.

Worker repairs a machine without locking it out.

do so is inadequate. Or maybe the hazard is abated but some new conditions are not properly communicated, thus a new accidental possibility is opened up. A lack of documentation is another shortcoming of hazard abatement.

Equipment may be faulty by design or disrepair. Corrosion and wear and tear both lead to machinery breakdowns, which can have consequences for the employees involved. Ergonomic factors may also be involved.

Insufficient planning is in itself a breakdown of the communications system. Sometimes the breakdown in communications is between workers and at other times it is between workers and their supervisors. Communications breakdowns between work teams may lead to accidents. Sometimes confusion exists even after communications because people did not take the time to understand each other and thus the communications are ineffective.

"Haste makes waste" goes the adage. Haste also makes accidents. A supervisor may demand speed and workers respond by being hasty and taking risks. Or, an employee may have a perceived need for haste and take risks on his or her own. Friendly competition between employees or work groups may also lead to dangerous haste. External factors may also cause hasty work habits. If the workload is too heavy frequent shortcuts may be taken. A lack of teamwork can also create haste. Finally, some individuals naturally take shortcuts, and this practice inevitably leads to accidents.

Insufficient training can be the cause of an accident. Also, even sufficient training may not have covered the precise circumstances that lead to an accident. The incorrect use of tools should also be considered a training failure.

Other factors that contribute to accidents include weather and temperature, working long hours, physical overexertion, and failure, misuse, or non-use of personal protective equipment. Improper body positioning or the ill-suited workplace or tools to the physical needs of the worker, both ergonomic root causes, are other miscellaneous factors to be considered.

Cause and Effect Sequence Analysis

Accidents typically result from known factors in the workplace. The premise of behavior-based accident investigation is that the unchecked dangerous behavior of the worker is the sole cause of every accident. The behavior is considered unchecked because it has been observed, but not changed and therefore condoned and encouraged.

Systems Safety Analysis

Was it caused by human error or by system error? The debate rages on, but a little of both is usually the answer, not all of one and none of the other.

The human behavior school assumes that at any given moment the employee makes a choice to work safely or not. The systems school assumes that accidents are defects in the total system. People, then, are only part of this system. According to this thinking, even if they decide to work safely, they may still be involved in an accident.

The behavioral school assumes that accidents are caused by human error. The systems school assumes that multiple causes exist for accidents.

The main measurement of the behavioral school is incident rates. Systems proponents favor multiple measurements. Along with incident rates, they look at accident investigation reports, control charts, near-miss investigation reports, surveys of employee attitudes and perceptions, and process measurements.

To the behaviorist, the method for safety improvement relies on persuasion and the appeal to employees to be more careful. The method for improvement for the systems proponent, in comparison, involves identifying appropriate ways to improve the system.

Fault Tree Analysis

Analytical trees are merely systematic perceptiveness.

Two major types of trees are used: the objective or positive tree (Figure 8.12) and the fault tree. The *objective tree* emphasizes the necessary steps required to do a task properly. A *fault tree* (Figure 8.13) is based upon a specific failure and emphasizes those things that can go wrong to produce the failure. A fault tree fashioned for one job can be generalized to cover many jobs.

Management Oversight and Risk Tree (MORT)

The MORT diagram (Figure 8.14) describes an ideal situation in an orderly, logical manner. A MORT has three basic branches. The first branch deals with specific oversights and omissions at the workplace. The second branch deals with the management system that establishes policies and drives the overall organization. Finally, an assumed risk branch visually demonstrates that no activity is completely risk-free and that risk management functions must exist within a well-managed organization.

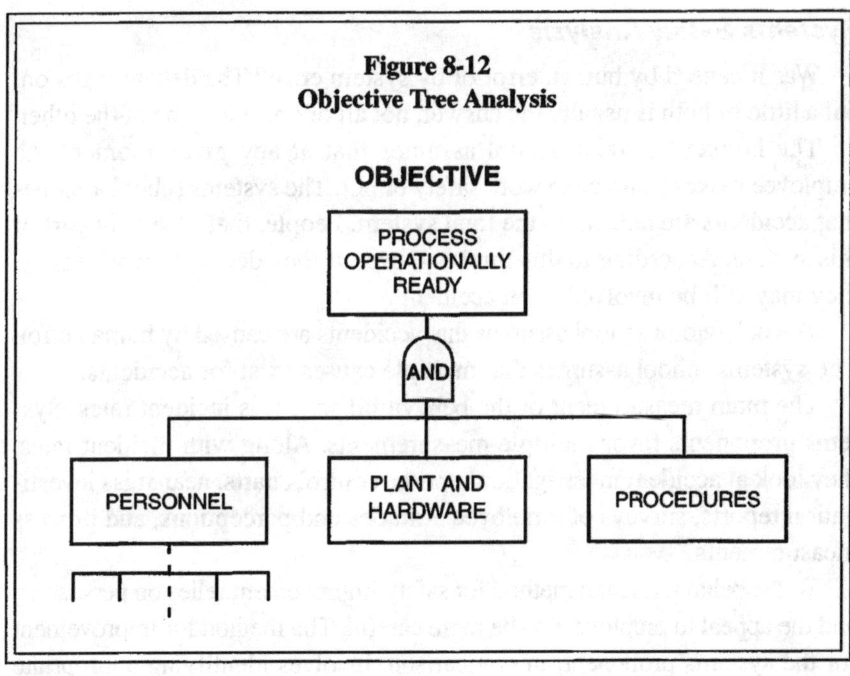

Figure 8-12
Objective Tree Analysis

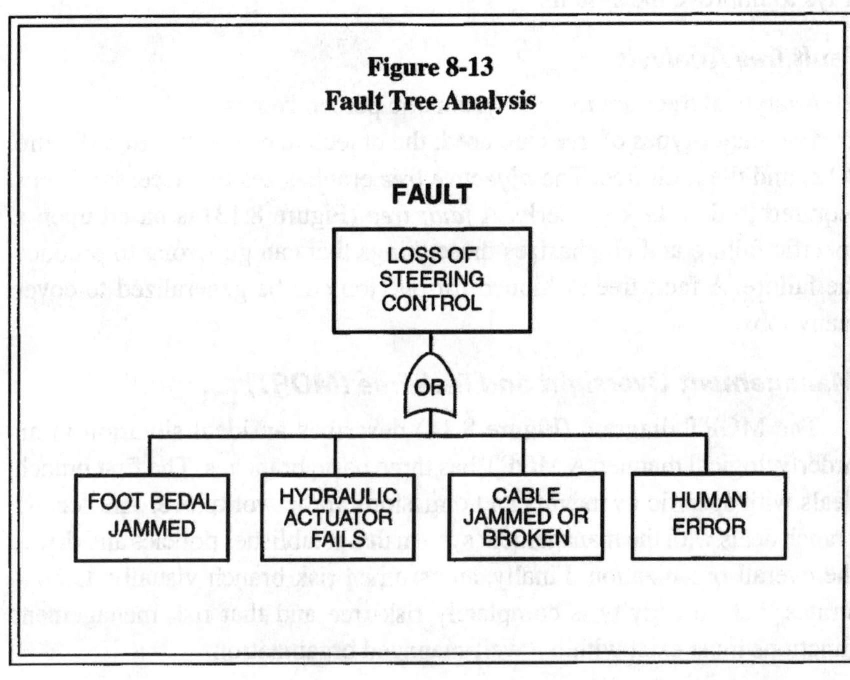

Figure 8-13
Fault Tree Analysis

Figure 8-14
MORT Diagram

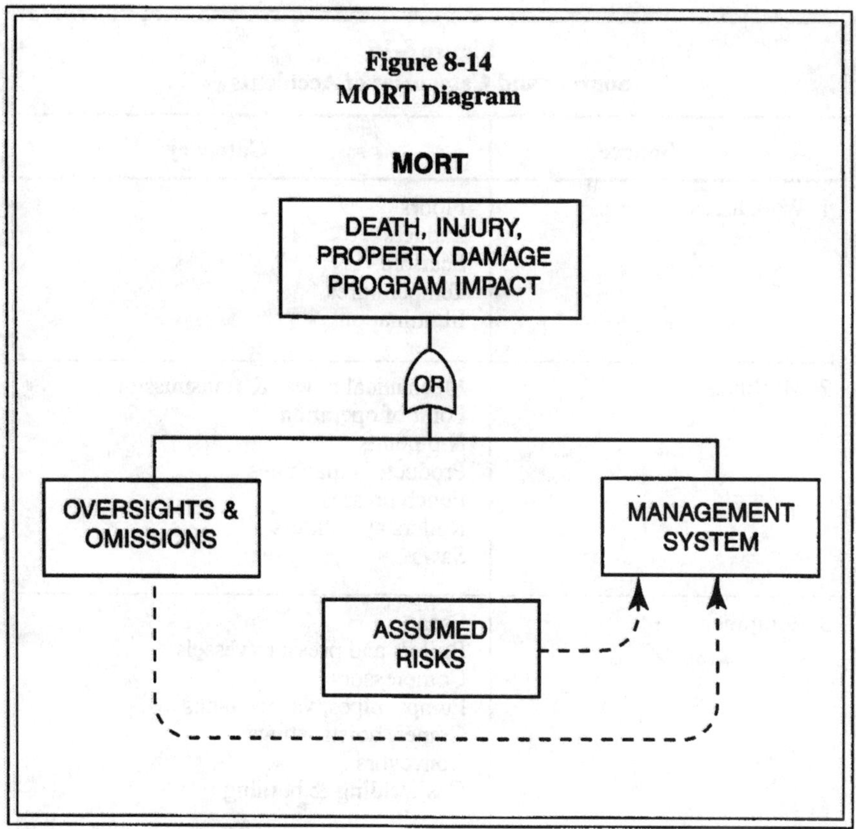

Identify Real Causes

The identification of root causes in any system is the key to problem solving and continuous improvement. Lack provides a checklist for accident sources, which he defines as the specific thing—the unsafe condition that is most closely related to a hazard and can produce an accident. Figure 8-15 summarizes Lack's sources and categories. The source named Multi-Category pertains to critical procedures that relate to more than one category.

Analysis of Damaged Parts, Property

You may have to spend some time studying the damaged parts or property in order to pin point the root cause. This is not always obvious at first.

Figure 8-15
Sources and Categories of Accidents

Source	Category
1. Work areas	Floors Ladders Scaffold Temperatures Illumination
2. Machines	Mechanical power & transmission Point of operation Nip points Production machines Punch presses Rollers and slitters Saws
3. Equipment	Fans Boilers and pressure vessels Compressors Pumps, pipes, valves, tanks Cranes, hoists, slings Conveyors Gas welding & burning
4. Portable tools	Hand tools, manual Hand tools, power (except electric)
5. Materials-health	Skin irritants Toxic materials
6. Materials-traumatic	Compressed gases and liquids Acids and caustics Flammables and explosives Molten metals Hot materials Flying particles Sharp edges Weight-shape stacking Ergonomic risk factors Radiation Noise

Figure 8-15 *(continued)*

7. Electrical	Motors and generators Transformers and rectifiers Wiring(conductors) Electrical hand tools Electrical welding & burning equipment
8. Vehicles	Manual Autos Power Trucks Forklifts Railroads Marine vessels Aircraft
9. Multi-Category	Confined space entry Lockout/Tagout Hazardous area entry

Accident Reconstruction

Using one of the techniques discussed above you may have to reconstruct what happened, testing one hypothesis at a time until you arrive a reasonably probable chain-of-events.

Examination and Testing by Experts

Occasionally experts must be called in or evidence sent to their laboratories for in-depth study and testing. This is not typically the case, but it does happen. Weigh the liability risk against the cost of expert analysis and decide if it is economically feasible.

Correct Real Problems

Recognition and investigation of unsafe acts and conditions does not accomplish anything unless corrective action is taken to change behaviors and correct conditions. Making the change permanent and effective is the safety professional's greatest challenge. Figure 8-16 gives some suggestions.

Figure 8-16
Making Permanent Behavior Changes

Worker Change

1. Clearly identify the behavior to be changed.
2. Explain the basis of that behavior and the benefits of the change.
3. Make the worker accountable for the change—and be consistent.
4. Monitor the degree and effectiveness of the change over time.
5. Avoid faultfinding—focus on the benefits.
6. Implement the change as soon as possible after recognition of its need.

Management Change

1. Retrain employees relative to any new equipment, materials, or procedures.
2. Relate safety changes to productivity enhancements, and quantify, if possible.
3. Obtain the approval and buy-in of upper management.
4. Monitor and quantify the results of the changes.

Report Development

The immediate supervisor of the injured employee should complete a preliminary investigation and submit a report by the end of the shift. The injured employee's statement can be postponed if medical attention is required.

Report Forms

Assign an incident report number to the accident report. Use of military time is recommended to avoid a.m./p.m. confusion. Clearly identify the date of the incident and date of report. If they differ, explain the reason for late reporting within the report. Give the detailed and precise location of the incident. Also give a precise description of the injury type and affected body part.

Remedial Action: Focal Point of Investigation

Summaries of vital information about recommended corrective action should be distributed to appropriate department heads. Real corrective action should be undertaken. Until action is taken, all reports are merely paper promises.

Upper Management Responsibilities

Accident reports do not exist for the entertainment of management. Reports should elicit a response in the form of a policy statement, an order to provide a new or modified procedure or training, or an order to install or modify safer equipment. This does not exhaust all the possibilities—but the point here is that management must do *something* with the accident report. Too often it ends up in limbo and employees feel that their safety is unimportant to management.

Trend Analysis

You may want to use some of the safety metric techniques discussed in Chapter 5 to see if a trend is developing. If a trend is detected, usually the circumstances leading to an accident can be remedied.

Preparation for Accidents

If we could really prepare for accidents would we have any? Of course not. What I mean is that, in the least, the accident investigation report should be used to prevent the next accident of its type from occurring.

Investigation and Reporting Administration

Accident investigation does not end with the completion of the report. As mentioned earlier, follow-up is required. To aid in follow-up, establish a tracking program similar to the one for follow-up of inspection deficiencies. Hold people accountable until corrective action is taken.

References

Hallock, Richard G. and D.A. Weaver. "Controlling Losses and Enhancing Management Systems with TOR Analysis." *Professional Safety.* October 1990, pp. 24-26.

Krause, Thomas R. and Larry R. Russell. "The Behavior-Based Approach to Proactive Accident Investigation." *Professional Safety.* March 1994, pp. 22-26.

Lack, Richard. "Get to the Source of Your Accident Problem—A Nine Point System for Classifying Accident Source Areas." *Management Newsletter.* ASSE Management Division.

Senecal, Patricia and Ellen Burke. "Root Cause Analysis: What Took Us So Long?" *Occupational Hazards.* March 1993, pp. 63-65.

Tiedt, Thao and Roger Kindley. "A Lawyer's Perspective on Accident Investigations." *Professional Safety.* August 1987, pp. 11-17.

Wilkinson, Bruce S. "Savings through Accident Investigation." A Paper given at the 1992 Mississippi Workers' Compensation Safety Conference, May 29, 1992.

9

Job Hazard Analysis and Safety Procedure Development

Planning Task Observance

A traditional way to prevent workplace injuries is to establish safe job procedures and then train employees to use them. The enhanced ability to create safe job procedures is a benefit of conducting job hazard analyses. The job hazard analysis (JHA) is a careful study of each step of a job, which is then written down as a permanent record. The purpose of this study is to identify existing or potential job hazards with the goal of determining the best way to eliminate or reduce hazards.

A JHA can be performed by anyone, but the supervisor is the ideal person to do so. The analysis can be performed for any job, whether special or routine. OSHA recommends reviewing the injury and illness reports to determine which jobs should be analyzed first. Of course, the jobs with the highest rates of accidents and disabling injuries should be analyzed first. Jobs where close calls have occurred should also be given priority. Analyses of new jobs and jobs where changes have been made in processes and procedures should follow. Eventually, all jobs should have a JHA on file. The standard at 29 CFR 1910.132 requires JHA.

Job hazard analyses are relatively cheap to do. Also, JHAs are fairly simple and get the supervisors involved in the safety process. They are easy to sell to management as a tool and JHAs are a good way to reduce the number and frequency of accidents.

With respect to OSHA compliance, JHAs document your good faith attempt to identify hazards. The JHA document is also a permanent record of the content of your training for that particular hazard—that is, if you used it as the basis for developing your training. Lastly, a JHA document can be used as a guide with which to review accidents for their causes. A JHA helps identify safety systems or devices left out in design. JHAs find issues that arise after startup. They also help identify unauthorized changes to equipment and processes done by employees.

Self-Audit of Task and Workplace

Select jobs for analysis. A job is a series of steps or activities necessary to complete a task. Jobs selected for JHAs are prioritized according to a hierarchy, as listed in Figure 9-1.

Observation of Employee(s)

Discuss the procedure with the employee whose job is being analyzed and explain its purpose. The JHA studies the job itself and not the performance of the employee. Involve the employee in all phases of the JHA—from reviewing job steps to discussing potential hazards—and recommended safe procedures. Include long-time workers who have performed the task for their experience, but also include the newest worker assigned to the task because a fresh viewpoint may provide the answer you seek.

Nearly every job can be broken down into steps. List each step in order of occurrence as you watch the employee performing the job. Select an experienced employee to perform the job. Record enough information to describe each job action, but do not make the steps too detailed. Then go

Figure 9-1
Prioritizing Jobs for JHAs

From highest priority to the least, select jobs for JHAs as follows:

1. Jobs with high accident frequencies
2. Jobs with severe injuries or high property losses
3. Jobs with potentially severe injuries or losses
4. Jobs where close calls have occurred
5. New jobs, including changes in operations
6. All other jobs

Figure 9-2
Breaking the Job into Steps

- List each job activity as it occurs.
- Describe each job activity.
- Go over your list of job activities with the employee.

over the job steps with the employee. Do not overlook preparations, startup, and shutdown activities. Figure 9-2 summarizes your activities to this point.

After you have recorded the job steps, examine each step to determine existing or potential hazards. Is the worker wearing clothing or jewelry that may get caught in machinery? Could fixed objects cause injury? (Sharp machine edges, for example?) Can the worker get caught in or between machine parts? Can reaching over moving machine parts or materials hurt the worker? Is the worker in an off position at any time? Is the worker positioned to the machine in a way that may be dangerous? Is the worker required to make movements that could cause hand or foot injuries, repetitive motion injuries, or strain from lifting? Can an object strike the worker, lean against or strike a machine part or object? Can the worker fall from one level to another? Can the worker be injured from lifting objects or from carrying heavy objects? The job observation is repeated as often as necessary until all the job hazards are identified. Figure 9-3 summarizes the eleven potential accident types. Figure 9-4 lists some general work conditions to look for.

Observation of Tools

Do hand or machine tools need repair? Do suspended loads or potential energy sources pose hazards? Keep in mind that potential energy is the force stored up in compressed springs, hydraulic reservoirs under pressure, weight lifted on jacks, and pneumatic reservoirs under pressure.

Observation of the Environment

Are there materials on the floor that could trip a worker? Is the lighting adequate? Are any live electrical hazards exposed? What explosive hazards

Figure 9-3
Eleven Potential Accident Types

1. Struck by (SB)
2. Struck against (SA)
3. Contact by (CB)
4. Contact with (CW)
5. Caught on (CO)
6. Caught in (CI)
7. Caught between (CBT)
8. Foot level fall (FLF)
9. Fall to below (FB)
10. Over exertion (OE)
11. Exposure (E)

Figure 9-4
General Work Conditions

- Slip, trip, fall hazards
- Lighting and other nonionizing radiation
- Noise
- Electrical hazards
- Tools and equipment
- Fire protection
- Emergency exits
- Trucks and motorized vehicles
- Personal protective equipment (PPE)
- Ventilation and air quality

are present or may develop? Does the noise level hinder worker communication? How accessible is fire protection equipment? Are emergency exits clearly marked? Are forklifts and other motorized industrial vehicles equipped with brakes, overhead guards, backup signals, horns, steering gear, and necessary identification? Are vehicle operators properly trained and authorized? Are employees wearing proper personal protective equipment for the jobs they perform? Do employees complain of headaches, breathing problems, dizziness, or strong odors at this workstation? Is ventilation adequate? Does the area contain a confined space that must be entered?

Each work site has its own requirements and environmental conditions. The key to a successful JHA is asking the right questions. OSHA suggests the ones above, but you should add your own questions. OSHA also suggests that you take photographs of the workplace for a more detailed analysis of the site.

Evaluation of Hazards

After observing worker, tools, and environment, the next step is to study what could cause these hazards. Think about events that could lead to an injury or illness for each hazard identified. What kind of preventive measures can you take to either eliminate (preferably) or control the hazard? Figure 9-5 lists the hierarchy of hazard control.

Figure 9-5
Hierarchy of Hazard Control

1. Eliminate the hazard.
2. Install safety equipment to mitigate the hazard.
3. Implement safer job activities.
4. Implement physical changes in the workplace or on the equipment.
5. Reduce the frequency of performing the hazardous task.
6. Implement corrective or mitigating suggestions of employees.

Is the worker wearing appropriate protective clothing, including safety belts or harnesses? Does the clothing fit properly? Has the worker been trained to use the appropriate PPE? Are work positions, machinery, pits or holes, and hazardous operations adequately guarded? Are lockout procedures used for machinery deactivation during maintenance procedures? Is the flow of work safely organized? For example, does that worker have to make movements that are too rapid? How are dusts and chemicals dispersed in the air? What are the sources of noise, radiation, and heat? What causes a worker to contact sharp surfaces? Why would a worker be tempted to reach into moving machine parts?

Recommendations should be based on the reliability of the solution. Generally, the most reliable protection is to eliminate the source or cause of the hazard. Methods of hazard elimination include redesigning equipment, changing tools, installing ventilation, or adding machine guards. If the hazard cannot be eliminated, the next preference is to reduce the danger as much as possible. Improving the work procedure or use of PPE are methods of hazard reduction. Training programs aimed at covering new or modified procedures and equipment in detail should accompany changes.

The JHA Report

A good job hazard analysis becomes a report—typically reported on a form similar to Figure 9-6, but the key elements are discussed here.

The job title and description are necessary for tracking and filing purposes. The name of the supervisor or department head is needed in order to be able to confer about the JHA at any future time. The job location completes the identification of the job to be analyzed.

Identify and record the materials or compounds to be used for each job task. Also, identify and record machines, tools, and equipment to be used to complete the job task. Itemize each step of the job from start to finish.

For each task, identify the potential hazards. What conditions or actions could cause an injury or affect health adversely? What safe practices do you recommend for each potential hazard listed? What protective clothing, personal protective devices, special clothing, or procedures are needed to perform the task safely? How, when, and to whom are problems to be reported? The Merritt Company has provided a list of potential hazards (See Figure 9-7.) that may not be all-inclusive but is a good starting checklist.

Figure 9-6
Sample Job Safety Analysis

JOB SAFETY ANALYSIS

page ___ of ___

JOB TITLE / DESCRIPTION	LOCATION / DEPARTMENT	PREPARED BY:	DATE:
		SUPERVISOR	

KEY JOB PROCESS STEPS	TOOLS / MATERIALS USED	POTENTIAL HAZARDS: CONDITIONS OR ACTIONS WHICH COULD CAUSE AN INJURY / AFFECT HEALTH	RECOMMENDED SAFE PRACTICES: PERSONAL PROTECTIVE DEVICES, SPECIAL CLOTHING, PROCEDURES

The job hazard analysis ought to be reviewed periodically. How often depends on your particular situation, but revise and update the JHA as the situation demands. Changes in workplace, equipment, or personnel suggest a review. Review any JHA for a job where an accident occurs. Be sure to retrain the involved employees any time a JHA is revised.

Accident History

When selecting jobs to be analyzed, look at the accident history of the plant. Where a trend has developed, a high priority needs to be assigned to conducting job hazard analysis. You may need to retrieve old accident

Figure 9-7
Potential Problem Behaviors and Conditions

Health Hazards

Dusts
Explosive chemicals
Flammable liquids
Fumes
Gases
Ignitions sources
Noxious odors
Reactive chemicals
Solvents
Toxic chemicals
Vapors

Housekeeping and Working Conditions

Care and condition of tools
Holding, handling objects
Lifting operations
Orderliness in work area
Position of body, arms, legs
Pushing materials, carts
Running, walking too fast
Sitting, stretching
Walking, carrying

Personal Protection

Eye protection
Face protection
Finger rings
Gloves
Hand creams
Head coverings
Long hair
Long sleeves
Respirators
Safety shoes
Ties, kerchiefs

Figure 9-7 *(continued)*

Potential Hazard Conditions

Abrasion, laceration, puncture wound
Availability of intermediate treatment for injuries
Contact with electrical conductor
Exposure to a corrosive (caustic or acid)
Exposure to extreme heat, hot objects
Exposure to radioactivity
Falls over objects
Falls to same or lower level
Finger, hand, body caught in or between objects
Sprain, strain
Struck by or against

Work Rules

Eye protection areas
Hard hat areas
Hearing protection areas
"No smoking" areas
Reporting procedures
Treatment of injuries

investigation reports to assure that every detail that could contribute to an accident has be considered in the JHA.

Identifying the Hazards

We have already spent a good deal of time in earlier chapters discussing hazard identification. Use these techniques to find hazards for the JHA.

Procedure Development

After each hazard has been identified and reviewed, can the job be performed in some other way to eliminate hazards? Can job steps be combined? Can the sequence be changed? Is additional equipment needed to make the job safe? Do precautions need to be posted or written into the job procedure? Is training needed to recognize hazards?

Where safer job steps are feasible, list each new step in a detailed description. List exactly what the employee needs to know to perform the job using a new method. General statements such as "be careful" are meaningless and should not be used. Written procedures should be as specific as feasible. If hazards are still present, try to reduce the necessity for performing the job or the frequency of performing it. Before finalizing, go over the new procedure with one or more employees who perform the job in order to get their suggestions. Their ideas may be invaluable. Napoleon used to do this with his battle plans. He would give them to a common soldier to read and then ask the man what the plans meant. If the soldier could not explain the emperor's plans, they were rewritten.

The best procedures are written as cookbook recipes. They identify who is responsible and list step-by-step what is to be done. The headings in a procedure might look something like Figure 9-8.

Procedure Implementation

Not going over new procedures with employees who will do the work is the downfall of most safety processes. Show me the finest safety manual in

Figure 9-8
Sample Procedure Requirements

Detailed Requirements

Responsibility	**Actions**
1.0 Operator	1.1 Get bucket of water.
	1.2 Add soap.
	1.3 Don rubber gloves.
	1.4 Stir soap and water.
	1.5 Don back support.
	1.6 Mop floor.

the world, and it's worthless if no one follows or knows the procedures. How can anyone be expected to follow procedures they know nothing about? They cannot. Use procedures to conduct procedural training and you have a practical application that will more likely be implemented.

REFERENCES

OSHA Manual. The Merritt Company, 1988.

Today's OSHA: A Compliance Update. Leawood, KS: American Management Association/Keye Productivity Center, 1996.

U.S. Department of Labor. Occupational Safety and Health Administration. *Job Hazard Analysis.* OSHA 3071, 1992.

10

Hazard and Operability (HAZOP) Analysis

What Is HAZOP and What Isn't It?

It is not the purpose of this chapter to make you a HAZOP expert. And although we briefly discussed HAZOP in Chapter 5, I think it deserves a chapter of its own. If you find something here you wish to use, however, I recommend that you get expert assistance first before implementing the HAZOP program.

HAZOP is one of several techniques that may be used to identify hazards, sort out root causes of potential accidents waiting to happen, and determine rational courses of corrective action. HAZOP is not the end of all your problems and it certainly is not an easy exercise for the uninitiated. It is, however, a powerful tool when used correctly.

The Need for HAZOP

If you have accidents that seem to elude corrective action, you may need HAZOP. If you have accidents that defy investigators' attempts at finding root causes, you may need HAZOP. If you are building a new plant where hazards will abound, you definitely need HAZOP. It was designed for just such usage.

Figure 10-1 lists some reasons for a HAZOP.

What Is HAZOP?

HAZOP was originally a study developed by chemical engineers to identify hazards in chemical process plants. Specifically, the desire was to find, in advance, those operability obstacles that might or might not be hazardous, but could possibly jeopardize the plant's capacity to achieve design productivity. The HAZOP was first developed to anticipate hazards and operability problems. It was to be used to analyze new or novel technology where operating experience was limited. HAZOP has been found to be an effective tool in other stages of a plant's life, not just in the design phase.

Figure 10-1
Reasons for a HAZOP Study

- Check the safety of a design.
- Decide whether to build a plant.
- Decide where to build a plant.
- Develop a list of questions to ask a supplier.
- Check operating/safety procedures.
- Improve safety in an existing facility.
- Verify safety instrumentation.
- Satisfy regulatory requirements (OSHA: Process Safety Management; EPA: Chemical Accident Prevention Program).

What HAZOP Isn't

A HAZOP can go way beyond hazard identification. It is not something you can perform with one person. Nor is it merely a committee activity. A HAZOP can be an expensive undertaking. From a cost viewpoint, however, it can be an effective program in the final design of a costly new plant with many hazards. It can also be used in existing facilities where accidents take a toll on the bottom line and OSHA is on your back.

How to Conduct a HAZOP

A multidisciplinary team is formed to conduct the HAZOP. The team works together to brainstorm ideas. This brainstorming identifies hazards and operability problems. The team, in effect, searches for ways in which the intent of the designer has been deviated from. An experienced team leader who can guide the team through the plant design should be appointed.

The ideal HAZOP team consists of five to seven members. In small plants, where the task is limited to one process or manufacturing line, a smaller team may suffice. Figure 10-2 lists some persons who should be enlisted on the team.

Several factors influence the success of the HAZOP. For one thing, drawings and other data used as the basis for the study must be complete and

Figure 10-2
Desirable HAZOP Team Members

- Design engineer
- Process engineer
- Operations supervisor
- Instrument design engineer
- Chemist
- Maintenance supervisor
- Safety engineer

accurate. The team must have the technical skills and insight to make the study successful. The HAZOP approach must aid the imaginations of the team in visualizing deviations, causes, and consequences. Finally, the team must have the ability to concentrate on serious hazards. Clearly, team selection is vital to the study. Also, outside influences must be minimized if the team is to complete the study successfully.

The study proceeds in several steps. First, define the purpose, objectives, and scope of the study. Next, select the team that has the best chance for achieving success. Give the team time to prepare for the study by doing background investigations and coming up on the learning curve. This is partly why HAZOPs are expensive. In the next step, the team carries out its review. Lastly, and no less importantly, the team records results. Some of these steps occur simultaneously.

The task of the team leader is to keep everyone focused. The primary task of the team is to identify problems, not solve them.

Guide Words

The team in brainstorming sessions uses guide words to systematically examine the plant design. Checklists or personal knowledge may also be used. Specific points in the design are labeled "study nodes." Guide words are applied to these study nodes to identify potential deviations of the plant process parameters. The team leader typically assigns study nodes before the team meeting.

Figure 10-3
HAZOP Guide Words

Guide Words	Meaning
No	Negation of the design intent
Less	Quantitative decrease
More	Quantitative increase
Part of	Qualitative decrease
As well as	Qualitative increase
Reverse	Logical opposite of the intent
Other than	Complete substitution

Figure 10-3 lists HAZOP guide words. In Figure 10-4, modifications to guide words are listed.

Deviations

Deviations are departures from the intention. They are discovered by systematic application of the guide words to the plant design. A guide word applied to a process parameter yields a potential deviation as shown in Figure 10-5.

Figure 10-4
Modifications to Guide Words

- SOONER or LATER for OTHER THAN when considering time.
- WHERE ELSE for OTHER THAN when considering position, sources, or destination.
- HIGHER and LOWER for MORE and LESS when considering elevations, temperatures, or pressures.

Figure 10-5
Use of Guide Words to Find Deviations

Guide Words	Parameter	Deviation
No	Pressure	No pressure
More	Temperature	High temperature
As well as	Two phase	One phase
Other than	Operation	Maintenance

Causes

Causes are the reason why deviations might occur. If a deviation has a believable cause, it is significant and warrants further study. Causes include hardware failures, human errors, changes in process states, or external interruptions.

Consequences

Consequences are the results of the deviations, when they occur. Drop the study of consequences that are inconsequential—it is a waste of valuable time.

Figure 10-6 lists specific consequences to be considered in the HAZOP study.

Actions

The potential actions or omissions of human beings are important factors in hazard analysis. The Navy calls this "sailor-proofing." In industry the term is usually "idiot-proofing." Bizarre errors or mistakes rarely cause accidents, however. More typically, human errors are commonplace actions that occur at inappropriate times.

It is nearsighted to point to a human error as the sole cause of any accident. More to the point, a human-machine or human-task mismatch has occurred. Focusing on the human error itself as the cause leads to a disciplinary matter. Discipline is not precluded, but it does not really solve your problem. Seek, rather, events and conditions that are sensitive to improvement.

Figure 10-6
Consequences to Consider

- Employee safety and health
- Loss of plant
- Loss of equipment
- Loss of production
- Liability
- Insurability
- Public safety
- Environmental deterioration

Even if not the root cause, such events and conditions can be improved and thus break the domino chain of sequential events leading to a new accident. Human errors are failed experiments in hostile environments.

Rasmussen quotes several industrial psychologists on human error:

> Knowledge and error flows [*sic*] from the same mental sources, only success can tell one from the other.—E. Mach (1905) *Knowledge and Error.*
>
> . . . errors in problem-solving are not stochastic events, but . . . results of solution trials with regard to a task, which is somewhat misconceived.—O. Selz (1922) *Zur Psychologie des Production Denkens und des Irrtums.*
>
> Systematic error and correct performance are two sides of the same coin.—J. Reason (1982) *Absent Minded?*

I would not say that it is never appropriate to discipline workers for unsafe behaviors. In fact, I believe that discipline is a crucial aspect of any successful safety process. But, clearly, discipline does not always solve our problem in accident prevention. When your disciplinary response to an accident is complete, ask yourself this question. Have I done what is necessary to prevent this accident from happening again?

A task analysis can be an effective means of identifying opportunities for human error in the workplace. (Chapter 9 discusses job hazard analysis.) The goal is to identify human errors that may affect the productivity, quality, and safety performance of the plant with emphasis on safety. Target those errors that may impact plant performance negatively.

What are the causes of human error? Complexity, stress, fatigue, environment, lack of training and experience are some.

The complexity of the system the worker must deal with, for one, is a major factor. My first supervisor, as a safety engineer, used to say to me repeatedly, "KISS!: Keep It Simple, Stupid!" Back then I bristled, but now I understand the wisdom of his offensive acronym. Keep it simple! Also, make it stress free. When a worker must perform a task on complex and dangerous machinery with quick precision while looking after his or her hide, too, that is stressful.

Lack of sleep, excessive overtime, and little time for rest between shifts bring about fatigue. Temperature, noise, humidity, lighting, and air contaminants can interact with complexity, stress, and fatigue to set a worker up for failure. So too does inadequate training or experience. Experience and training are integral parts of one another, so an inadequate level of either is a common source of human error.

Certain things complicate human errors. Complexity is a complicating factor as well as a cause. Even trained and experienced hands that are one day tired or stressed out can get into trouble in a complex operation. Poor judgment is another complicating factor that is related to inexperience and poor training. Collective decisions complicate human errors—mostly collective indecision.

Maintenance is another complicating factor. If machinery is malfunctioning or inoperative, even the experienced worker can be drawn into making a poor judgment call. Finally, the perception of risk, discussed in an earlier chapter, is a complicating factor. Some people, for various reasons valid or not, will be susceptible to taking more risk than others will.

Records

Records are a vital part of HAZOP. Not all that is said during a team meeting can or should be recorded, but important ideas ought to be recorded. This can be accomplished by writing the ideas on newsprint, flipchart paper, or a recordable dry erase board, which have a scanner that sweeps across the board when you are ready to record anything written on it.

The Role of Documentation

The purpose of a team meeting record is to present the best ideas to management.

What to Record

A HAZOP form is devised for each team meeting as shown in Figure 10-7.

How to Record

The team recorder does not necessarily have to be part of the team during brainstorming sessions. However, the recorder ought to be a technically competent person who can understand what is being said during the meetings in order to accurately record this information. The chart shown in Figure 10-7 can be used to record the meeting. A tape recorder or video recorder with audio may be used to augment the note taker's efforts.

PREPARING TO CONDUCT HAZOP

The time needed to prepare for a HAZOP is directly proportion to the size and complexity of the plant. Piping and instrumentation diagrams (P&Ids), process flowsheets, plant layouts, isometric drawings of piping systems, and fabrication drawings must be collected and studied. Operating

Figure 10-7
HAZOP Form

Guide Word	Deviation	Consequences	Causes	Suggested Action
No Less More Part of As well as Reverse Other than				

instructions, instrument sequence control charts, logic diagrams, and computer programs must be gathered and studied. Plant manuals and equipment manufacturers' manuals must also be collected. All this data must then be inspected to ascertain whether they pertain to the study and contain no discrepancies or ambiguities.

Next the data has to be converted to a form the team can study. A plan of the study sequence must be devised. Included in the plan is an arrangement for the necessary team meetings wherein the study nodes will be brainstormed relative to guide words.

Team Dynamics

Everyone on the team must thoroughly study the assigned study node before the next team meeting. Team meetings are not places to compete with each other, but each personality should draw the others out for open discussion. Listening is a skill that all members of the team need, especially the team leader. No one should be attacked or put on the defensive during a meeting. All ideas should be accepted as equally worthy to any other idea. Sorting of worthiness can be done later. A high level of thinking energy is desired during the meetings. Therefore, the team leader should call for frequent breaks.

References

Bell, Barbara Jean. "Evaluating the Contribution of Human Errors to Accidents." *Professional Safety.* October 1988, pp. 27-32.

The Center for Chemical Process Safety. *Guidelines for Hazard Evaluation Procedures.* New York: American Institute of Chemical Engineers, 1985.

Rasmussen, Jens. "Approaches to the Control of the Effects of Human Error on Chemical Plant Safety." *Professional Safety.* December 1988, pp. 23-29.

Welch, Niles T. "Human Error Risk Assessment: An Undefined Role in System Safety." *Professional Safety.* February 1998, pp. 18-22.

instructions, and inspect sequence control charts, logic diagrams, and computer programs must be gathered and studied. Plant manuals and equipment manufacturers' manuals must also be collected. All this data must then be inspected to ascertain whether they pertain to the study and to determine discrepancies or ambiguities.

Next the data has to be converted to a form the team can study. A plan of the study sequence must be devised. Included in the plan is an arrangement of the necessary team meetings when the study nodes will be [illegible] words.

Team Dynamics

Everyone on the team must thoroughly study the assigned study nodes before the next team meeting. Team meetings are not places to compete with each other, but each personality should draw the others out for open discussion. Listening is a skill that all members of the team need, especially the team leader. No one should be attacked or put on the defensive during a meeting. All ideas should be accepted as equally worthy in any [illegible]. Solving [illegible] problems can be done later. A high level of thinking energy is used during team meetings. Therefore, the team leader should call for frequent breaks.

References

Bell, B. J. [illegible] "Evaluating the contribution of human error to accidents," [illegible] 1983, [illegible].

Center for Chemical Process Safety. *Guidelines for Hazard Evaluation Procedures.* New York: American Institute of Chemical Engineers, 1985.

[illegible] "Recommendations for Control of the [illegible] Process Hazards [illegible]" [illegible] 1983.

[illegible]

11

The Occupational Safety and Health Act

A Capsule History of the Act

Each day more than ninety million Americans leave home and go to work. Their intended goal, of course, is to earn a living. Few, if any, consciously think that another goal is to simply return home after work with good health, life, and all their limbs. Industrial workplaces are inherently dangerous. In a way, when a person accepts employment in this setting, they are to some degree buying into the associated risks. For many years, especially during the years of the Industrial Revolution, injuries and illnesses on the job were blamed almost entirely on careless employees. An accident or incident of illness was just "tough luck." By not voting with their feet and leaving the job, employees were assumed to have accepted the dangers of the job and, therefore, the employer was not considered blameworthy.

While this logic is not entirely faulty, the employee has little control over his or her surroundings compared to the control the employer has. Strict justice, as I understand it, demands that the one who has control of a situation also is responsible for the situation and its consequences. As more people bought into this idea, two relief systems developed. First, in the early twentieth century, workers' compensation legislation was passed as discussed elsewhere. Workers' compensation alleviates the suffering of families with injured or ill bread winners, but does little to prevent the injuries or illnesses in the first place. At first it was believed that employers would do what was right in order to avoid workers' compensation costs. But, for the most part that never happened. In the 1950s and 1960s the call for safety in U.S. workplaces became a roar that was eventually heard on Capitol Hill. Until 1970, no uniform and comprehensive legal provisions existed for the protection of workers against workplace hazards.

In 1970, the U.S. Congress considered the fact that job-related accidents accounted for more than 14,000 deaths annually. Nearly 2.5 million

American workers were disabled at the time. Ten times as many person-days were lost from job-related disabilities as from strikes, yet employers were doing little to avoid these losses. An estimate of new cases of occupational diseases in 1970 totaled three hundred thousand!

The cost was so staggering that you would think employers would have taken action based on the loss of production, medical expenses, disability compensation, and the burden on the nation's commerce. Yet too often, the top managers were heard to say, "To hell with them! If they won't work around [you fill in the hazard], let'em work some place else. Fire'em!" Congress, though, recognized that the human cost was staggering and passed Public Law (PL) 91-596 called the Occupational Safety and Health Act (OSHAct) of 1970 (Title 29 U.S. Code Section 651) on November 5, 1970. Notably, OSHAct was passed by a bipartisan Congress and signed into law by one of the more conservative presidents, Richard Nixon. Congress stated as its goal in the Act: "*. . . to assure so far as possible every working man and woman in the Nation safe and healthful working conditions and to preserve our human resources.*"

OSHAct has been amended once by PL 101-552, November 5, 1990.

Scope of the OSHAct

In general, OSHAct covers all employers and their employees in the fifty states, the District of Columbia, Puerto Rico, and all other territories under federal government jurisdiction. Coverage is provided either directly by federal OSHA or through an OSHA-approved state program.

An *employer* per OSHAct is any person engaged in a business affecting commerce who has employees, not including the United States federal government or any state government or political subdivision of a state, such as a local government. The Act applies only to private employers and employees engaged in the various fields of business, such as manufacturing, construction, longshoring, agriculture, law, medicine, charity, disaster relief, organized labor, and private education. The Act even covers religious groups to the extent that they employ workers for secular purposes.

Some people, in addition to the governments mentioned above, are explicitly not covered by the Act. Self-employed persons are not covered, but their employees are. Family farms in which immediate members of the family are the only employees are not covered.

OSHA does not regulate working conditions that are regulated by other federal agencies. However, even when the other agency is authorized to regulate safety and health working conditions in a particular industry (such as the Nuclear Regulatory Commission in the nuclear industry), OSHA standards apply in those specific areas not covered by the other agency. As OSHA develops standards of its own, standards issued under the Walsh-Healey Act, the Services Contract Act, the Construction Safety Act, the Arts and Humanities Act, and the Longshoremen's and Harbor Workers' Compensation Act are superseded.

Federal agency heads are responsible for providing safe and healthful work conditions for their employees. An Executive Order (an order from the President) requires agencies to comply with OSHA standards consistent with safety and health issues of private employers. OSHA conducts federal workplace inspections only in response to employees' reports of hazards and as part of a special program that identifies federal workplaces with higher than average rates of injuries and illnesses. Federal agency heads are required to operate comprehensive occupational safety and health programs that include recordkeeping, analysis of injury/illness data, training at all levels, and self-inspections to ensure compliance. OSHA conducts thorough evaluations of these programs to assess their effectiveness.

The federal sector authority of OSHA differs from its private sector authority. Most significant is the fact that OSHA cannot impose monetary penalties against another federal agency for failure to comply with OSHA standards. Compliance issues that cannot be resolved at the local level are raised to the next higher organizational level. This elevation of the discussion is repeated until an agreement is reached that resolves the issue. The highest level of discussion includes either the Director of OSHA or the Secretary of Labor and the other federal agency head, depending on that agency's level in the federal government. Also, OSHA does not have the authority to protect a federal employee who reports uncontrolled hazards in the workplace—a *whistleblower.* OSHA protects private sector employees by authority of the OSHAct. The Whistleblower Protection Act of 1989, however, affords present and former federal employees an opportunity to file their reports of reprisal with the Office of Special Counsel, U.S. Merit Systems Protection Board. This law does not cover postal workers or employees of certain intelligence-gathering agencies.

OSHA provisions also do not apply to state and local governments in their role as employers. However, any state that wants OSHA approval for its private sector occupational safety and health program must provide a program that covers its state and local government employees that is at least as effective as its program for private sector employees. States may also have programs that cover only public sector employees, their private sector employees being covered by federal OSHA by default.

Occupational Safety and Health Administration

Under OSHAct, the Occupational Safety and Health Administration (OSHA) was created under the U.S. Department of Labor. The primary purpose of OSHA is to encourage employers and employees to reduce workplace hazards and to implement new or improved occupational safety and health programs in keeping with the goal of Congress in passing the law. OSHA also is tasked with providing for research in occupational safety and health. The objective of the research is to develop innovative ways of dealing with occupational safety and health problems.

To reduce accidents, OSHA is supposed to establish separate but dependent responsibilities and rights for employers and employees. Congress recognized that employees couldn't be held accountable to the extent that employers can be for the achievement of better safety and health conditions in the workplace. Therefore, OSHA maintains a reporting and recordkeeping system in order to monitor job-related injuries and illnesses. Also, OSHA was tasked to establish training programs to increase the number and competence of occupational safety and health personnel, not only for their own staff but for staffing industry as well.

In order to force employers to take responsibility for improving conditions in the workplace, OSHA is required by law to develop mandatory job safety and health standards and to enforce them effectively. For those states that have their own occupational safety and health programs, OSHA must provide development, analysis, evaluation, and approval of the programs. OSHA continually reviews and redefines specific standards and practices, true to its basic purposes and initial mission.

Though not generally popular, OSHA has done a good job over the years. OSHA has been too strict for some, too lenient for others. In my opinion, OSHA has always implemented its mandates fully, firmly, and fairly to all parties. In all its procedures, from standard development through implementation and enforcement, OSHA guarantees employers and employees the right to be fully informed, to participate actively, and to appeal its actions. In other words, if you do not know what is going on, it is your own fault.

Authority to Promulgate Standards

OSHA can begin standards-setting procedures on its own initiative. The agency may also respond to petitions from other parties. Among these are the Secretary of Health and Human Services (HHS); the National Institute for Occupational Safety and Health (NIOSH); state and local governments; nationally-recognized standards-producing organizations; nationally-recognized employer representative organizations or representatives; labor representatives; or any other interested person.

OSHA publishes its intentions to propose, amend, or revoke a standard in the *Federal Register* as a Notice of Proposed Rulemaking (NPR). Previous to this, OSHA will be likely to have published an Advanced Notice of Proposed Rulemaking (ANPR) as a heads-up notice to concerned parties. The ANPR can be used to solicit information used in drafting a proposal. The NPR includes the terms of the new or modified rule and provides a specific time for public response. At least thirty days from the time of publishing the NPR must be allowed for public response, but usually sixty or more days are given.

Interested parties may then submit written arguments and pertinent evidence. Persons who submit information may request a public hearing on the proposal, even if none has been announced in the NPR. OSHA will schedule the hearing and publish an announcement of time and place for it in the *Federal Register.* You see then how important the *Federal Register* is for a full understanding of compliance issues.

After the close of the comment period and public hearing, OSHA must publish in the *Federal Register* the full, final text of any standard amended or adopted and the date it becomes effective. This notice includes a *preamble,* which explains the standard and the reasons for implementing it. If you have a particular standard that is difficult to understand, get a copy of the preamble and study it. That may clarify the issue for you. OSHA may also publish notice of its judgment that no standard or amendment needs to be issued in a matter.

Advisory Committees

If OSHA determines that a specific standard is needed, any of several advisory committees may be called upon to develop specific recommendations. OSHA has two standing advisory committees, but *ad hoc* committees may be appointed to examine special areas of concern. All advisory committees must have members representing management, labor, state agencies, and HHS. The occupational safety and health professions and the general public may also be represented on the committees.

The two standing advisory committees are the National Advisory Committee on Occupational Safety and Health (NACOSH) and the Advisory Committee on Construction Safety and Health. NACOSH advises, consults with, and makes recommendations to the Secretary of HHS and the Secretary of Labor on matters regarding administration of OSHAct. The Construction Advisory Committee advises the Secretary of Labor on formulation of construction safety and health standards and other regulations.

NIOSH Recommendations

OSHAct establishes the National Institute of Occupational Safety and Health (NIOSH) as an agency of the Department of HHS. The agency conducts research on various safety and health problems, provides technical assistance to OSHA, and recommends standards for adoption by OSHA. While conducting its research, NIOSH may make workplace investigations and gather testimony from employers and employees. For research purposes, NIOSH may require that employers measure and report employee exposure to potentially hazardous materials. NIOSH may even require as part of their research that employers provide medical examinations for employees and have them tested to determine the incidence of occupational illness among them. Rather than the employer, NIOSH may pay for research examinations and tests.

Variances

Employers may request a *variance* from a standard if they cannot fully comply by the effective date due to shortages of materials, equipment, or professional or technical personnel. A variance may also be requested if the employer can prove that his facilities or methods of operation provide employees with protection against the regulated hazard that is at least as effective as facilities or methods of operation required by OSHA.

Employers located in states with OSHA-approved state plans should apply to the state for a variance. Companies that operate facilities under federal OSHA jurisdiction and also in state plan states may apply directly to OSHA for a single variance that will be effective in all of their locations. OSHA will work with the states involved in such a request for variance to determine if a variance can be granted that satisfies state as well as federal requirements.

Variances are not retroactive. An employer who has been cited for a standards violation may not seek relief from that citation by applying for a variance. Even if a citation is outstanding, though, an employer may file a valid variance application. The original violation will stand regardless of the outcome of the application.

Temporary Variance. A *temporary variance* may be granted when the employer cannot comply by its effective date due to unavailability of professional or technical personnel, material, equipment, or because the necessary construction or alteration of facilities cannot be completed in time. The variance must be applied for within a reasonable amount of time prior to the effective date of the standard. The application must demonstrate to OSHA's satisfaction that the requester is making every effort to safeguard employees in the meantime. The requester must have put in force an effective program for coming into compliance with the standard as quickly as possible.

A temporary variance is granted for the period needed to achieve compliance, but for no more than one year. Two renewals of the variance may be granted for up to six months each. Two years from the effective date of a standard, or from the startup of a new facility, the facility must be in full compliance.

The application must identify the standard or portion of a standard from which the variance is requested. The reasons why the variance is requested must be carefully and completely explained. The requester must document any measures already taken and those measures that are planned. A compliance schedule must be included.

Variance requests may not be submitted without the full knowledge of employees. Workers must be informed of the application for a variance. A copy of the variance application must be given to the employees' authorized representative. A summary of the application must be posted wherever notices are normally posted. The employer must certify in the application that all of the above actions have been taken. Employees must also be informed that they have a right to request a hearing on the application.

OSHA does not grant temporary variances simply because a company cannot afford to pay for the alterations, equipment, or personnel necessary to comply. Neither will a variance be granted to employers who make no attempt to comply with the requirements within a reasonable amount of time prior to the effective date of the standard at issue.

Permanent Variance. A *permanent variance* grants the requester an alternative to a particular requirement or standard once and for all. Such a variance may be granted to employers who prove their condition, practices, means, methods, operations, or processes provide a safe and healthful workplace as effectively as would compliance with the standard, or part of it, at issue. OSHA weighs the requester's evidence and arranges a variance inspection and hearing, as appropriate. If the request is found to be valid, a permanent variance is prescribed that details specific exceptions from the standard and the employer's responsibilities under the ruling. A variance has the force of law.

When applying for a permanent variance, you must inform employees of the action and of their right to request a hearing. Anytime after six months from the issuance of a permanent variance, the employer or employees may petition OSHA to modify or revoke it. OSHA may also modify or revoke the variance on its own accord.

Interim Order. In order for employers to continue operating under existing conditions pending a variance decision, an *interim order* may be applied for. This application may be made at the same time as, or after, the application is filed for a variance. Reasons why the order should be granted must be included in the interim order application. If OSHA denies the request, it will notify the requester of the reason for denial.

If the interim order is granted, the employer and other concerned parties are informed of the order and its terms are published in—you guessed it—the *Federal Register*. The employer must inform employees of the order. A copy must be given to the employees' authorized representative. Also, a copy must be posted wherever notices are normally posted.

Experimental Variance. An employer may participate in an experiment to demonstrate or validate new job safety and health techniques. If either the Secretary of Labor or the Secretary of HHS has approved that experiment,

OSHA may grant an *experimental variance* to permit the experiment. Unsanctioned experiments are not recommended and are to be discouraged.

Other Variance. The Secretary of Labor may also find that a variance is justified when the national defense is impaired. For information and assistance in requesting such a variance, contact the nearest OSHA office. Employees and the general public may also petition OSHA for modification or revocation of standards.

Authority to Adopt Regulations

Regulations are OSHA requirements that do not specify means, methods, practices, equipment, or personnel. For instance, recordkeeping rules are regulations, as opposed to standards. Everything said about promulgation procedures and granting of variances for standards apply to regulations as well.

National Institute for Occupational Safety and Health

The OSHAct established the National Institute for Occupational Safety and Health (NIOSH) under the Department of Health and Human Services (formerly Health, Education and Welfare). A specific assignment for NIOSH was to establish a hazard evaluation program for safety and health issues.

Employer Duties under the OSHAct

Under OSHAct, employers are assigned very specific duties. Administrative and recordkeeping duties are discussed below. Abatement duties for specific hazards are discussed in the chapters that follow.

Specific Standards

In carrying out its duties under OSHAct, OSHA must disseminate legally enforceable standards. By authority of OSHAct, OSHA standards may require those practices, means, methods, or processes that are reasonably necessary and appropriate to protect workers on the job.

Although it may be hard to believe, most OSHA standards were written by industrial experts as consensus standards and OSHA adopted them *in toto* as OSHA standards. In other words, if you are confused by an OSHA standard, it is probably because you need a professional engineer to read it for you. Contrary to popular opinion, not everyone is a safety engineer. As an employer, though, it is your responsibility to become familiar with standards applicable to your establishment. The employees are responsible only for complying with all rules and regulations that are applicable to their own actions and conduct.

States with OSHA-approved occupational safety and health programs must set standards that are at least as effective as the federal standards. Many state plans adopt standards identical to the federal ones—a testimony to their effectiveness if not their clarity.

Definition of Standard

A standard is a regulation that sets forth specific methodology for maintaining a safe and healthful workplace with respect to the hazard it is designed to abate. As mentioned above, in order to get a running start in the 1970s, OSHA took existing engineering consensus standards that dealt with safety issues and made them the law of the land by adopting them as OSHA standards. No longer voluntary, the new standards did not fit every situation in every workplace. Therefore, a lot of discontent has been leveled at OSHA from the start. Hard-nosed OSHA compliance officers who insisted that the standards be followed to the letter of the law exacerbated the situation. Recently, OSHA has adopted a softer approach and the compliance officers I have dealt with in recent years have been more than reasonable in interpreting how a standard matches the workplace.

Types of Standards

Permanent OSHA standards fall into four major categories: General Industry, Maritime, Construction, and Agriculture. Some confusion reigns about which standard applies to a facility. Many people think that only general industry standards apply to their manufacturing plant, for example. However, if your maintenance crew, or other employees, are involved in tearing down a section of offices to make a new office or work area, they are performing construction work and come under the Construction Standards for that work. If they paint the building, they are under Construction Standards. If your plant is located on a waterway and receives materials and transports product by ship, it is conceivable that General Industry, Construction, and Maritime standards all apply to the facility for various tasks.

Under certain limited conditions, OSHA is authorized to set emergency temporary standards that take effect immediately and remain in effect until superseded by a permanent standard. The emergency standard must be necessary to protect the endangered workers' exposure to toxic substances or agents or physically harmful agents or conditions. OSHA publishes the temporary emergency standard in the *Federal Register.* This announcement also serves as a proposed permanent standard subject to the usual procedure for adopting a permanent standard. A *final rule* should be made within six months however. The validity of an emergency temporary standard may be challenged in a federal Court of Appeals.

Duty to Comply

Employers have a duty under the law to comply with the intentions of Congress in OSHAct. If nothing else, the General Duty Clause, discussed below, may have been violated.

Section 5(a)(2) Violations

This clause requires each employer to comply with occupational safety and health standards promulgated by OSHA.

Section 5(b) Violations

This part of the OSHAct is rarely, if ever, used by OSHA in enforcement. This clause states

> each employee shall comply with occupational safety and health standards and all rules, regulations, and orders issued pursuant to this Act which are applicable to his own actions and conduct.

General Duty Clause

Perhaps the best known, but least understood, OSHA citation is the General Duty Clause. No need to search in the CFR for it. The General Duty Clause is Section 5(a)(1) of the OSHAct of 1970. Where OSHA has not issued a specific standard, employers are responsible for following OSHAct's general duty clause.

Scope

The general duty clause applies to employers.

Clause Text

The general duty clause of OSHAct states that each employer

> shall furnish . . . a place of employment which is free from recognized hazards that are causing or are likely to cause death or serious physical harm to his employees.

Section 5(a)(1) Preempted by Standards

Section 5(a)(2) duties, where appropriate, preempt the duties of 5(a)(1). Specific standards always take precedence over the general duty. However, when a specific standard does not exist to cover a situation that OSHA deems in violation of the intent of the law, the General Duty Clause can and will be used.

Elements of Violation

Does a specific standard apply to a workplace? Yes? Does a requirement of the standard apply to a specific situation in the workplace? Yes? Is the workplace situation in compliance with the requirements of the standard? No? Then a violation exists. Does a hazardous situation exist that is not covered by a specific standard? Yes? Then the General Duty Clause has been violated.

Four elements make up a General Duty Clause violation. First, the employer failed to keep the workplace free of a hazard to which employees of that employer were exposed. Secondly, the hazard was recognized as such. Third, the hazard was causing or was likely to cause death or serious physical harm. Finally, a feasible and useful method to correct the hazard must have existed. An unabated hazard cited under the general duty clause must be reasonably foreseeable. All the factors which could cause a hazard need not be present in the same place at the same time in order to prove foreseeability of the hazard. For instance, an explosion need not be imminent.

Recordkeeping and Reporting

Prior to OSHAct, no centralized, systematic method existed for monitoring occupational safety and health problems. Statistics on job injuries and illnesses were collected by a few private organizations for specific objectives. National statistics were based on unreliable projections made from these private statistics. OSHAct gave OSHA the basis for consistent nationwide statistics gathering procedures.

Basic Recordkeeping

Employers of twenty or more employees (eleven or more in the construction industry) must maintain records of occupational injuries and illnesses as they occur. Employers with nineteen or fewer employees (ten or fewer in construction) are exempt from keeping records unless they have been selected by the Bureau of Labor Statistics (BLS) to participate in the Annual Survey of Occupational Injuries and Illnesses. All employers regardless of size are required to maintain records on fatalities and multiple hospitalizations. The number of employees pertains to the entire firm, not the number at one establishment.

The purpose of recordkeeping is to permit BLS to compile statistics and identify high hazard industries. Another purpose is to inform employees of the status of their employer's safety and health record. OSHA also uses the records to assist in safety and health inspections. Employers in state plan states are required to keep the same records as everyone else.

Besides the two forms discussed below and the BLS statistical report, many specific OSHA standards have additional recordkeeping and reporting requirements.

Scope

OSHA recordkeeping is not required for all employers in retail trade, finance, insurance, real estate, and service industries. These are businesses in Standard Industrial Classification (SIC) 52-89 (except building materials and garden supplies, SIC 52; general merchandise and food stores, SIC 52 and 54; hotels and other lodging places, SIC 70; repair services, SIC 75 and 76; amusement and recreation services, SIC 79; and health services, SIC 80). A few regularly exempt employers (exempt from recordkeeping only; no private employer is exempt from OSHA standards) have to maintain records as explained above when BLS designates them for its annual survey. They will be notified in advance and provided with the necessary forms and instructions. Exempt employers must, however, comply with the OSHA poster and report to OSHA any accident that results in one or more deaths or the hospitalization of three or more employees.

Mechanics of Recordkeeping

An *occupational injury* is any injury, such as a cut, fracture, sprain, or amputation, that results from a work-related accident or from exposure involving a single incident in the work environment. An *occupational illness* is any abnormal condition or disorder, other than one resulting from an occupational injury, caused by exposure to environmental factors associated with employment. Occupational illness includes acute and chronic diseases that may be caused by inhalation, absorption, ingestion, or direct contact with toxic substances or harmful agents (such as sound or vibration).

All occupational illnesses must be reported regardless of severity. Illnesses are probably the most underreported of workplace safety and health incidences.

All occupational injuries must be reported if they result in one of the conditions listed in Figure 11-1.

Employers must maintain injury and illness records for each *establishment,* defined as a single physical location where business is conducted or where services are performed. If your employees work in diverse locations, you must keep records in the place where employees report for work. An example would be a power company or communications company with line workers or security system installation technicians who service various locations from a truck. If employees do not report to work at the same location each day, for instance a sales representative who works out of her home and

Figure 11-1
Recordable Occupational Injuries

- Death (regardless of the length of time between the injury and death)
- One or more lost work days
- Restriction of work or motion
- Loss of consciousness
- Transfer to another job
- Medical treatment (other than first aid)

car, the records must be kept at the place where employees are paid or at the base from which they operate.

Recordkeeping Forms. OSHA recordkeeping forms are maintained on a calendar year basis. You do not send the forms to OSHA or to any other agency, but they must be available for inspection by representatives of OSHA, HHS, BLS, or the designated state agency. Records must be maintained for five years at the establishment. Only two forms are required for recordkeeping purposes: the OSHA 300 Log and the OSHA Form 301 or its state equivalent.

OSHA 300 Log. OSHA Form 300 (formerly OSHA 200) is the *Log and Summary of Occupational Injuries and Illnesses* and is the chief instrument in the OSHA recordkeeping system. Each recordable occupational injury and illness must be logged on this form within six working days from the time the employer learns of it. If the log is prepared at a central location by automatic data processing equipment, a copy current to within forty-five calendar days must be present at all times in the establishment. Substitute forms are acceptable if as detailed, easy to read, and understandable as the OSHA 300 Log.

OSHA Form 301. The OSHA Form 301 (formerly OSHA 101) is the *Injury and Illness Report.* Form 301 contains much more detail about each injury or illness as a unique event. This form must also be completed within six working days from the time the employer learns of the work-related injury or illness. You may not be familiar with this instrument if your state has an equivalent form. In many places, an insurance or workers' compensation form is used, as it contains all the required information.

BLS Annual Survey. Employers selected to participate in the annual statistical survey receive OSHA Form 300S in the mail soon after the close of the year. Use your OSHA 300 Log as the source of information to complete and return the 300S to BLS. Small business employers with ten or fewer employees are notified at the beginning of the year and supplied with a 300 Log so that the appropriate information may be collected in order to complete the 300S Form at the end of the year.

Posting Requirements. A copy of the totals and information following the *fold line* of the last page of the 300 Log for the year must be posted at each establishment wherever notices to employees are customarily posted. This copy must be posted no later than February 1, and kept in place at least until March 1, of the year following the year in which the injuries and illnesses occurred. For example, the 300 Log for 1998 must be posted from February 1 to March 1, 1999. Even though no injuries or illnesses occurred in a given year, zero must be entered on the totals line, and the form must be posted the following February 1.

Fatalities/Multiple Hospitalizations

If an on-the-job accident results in at least one death or three or more victims being hospitalized, the employer must report the accident in detail to the nearest OSHA office within twenty-four hours. A telephone call may be made, followed by a written report in a timely fashion. All employers, regardless of the number of employees, must file the report. In states with state plans, employers make the report to the state agency responsible for safety and health programs.

Access to Records

OSHA estimates that more than thirty-two million employees are potentially exposed to toxic substances and harmful physical agents, such as noise and vibration, that may severely impair health. Regardless, employees are poorly informed about the toxic exposures they face and the potential health effects. In 1980, OSHA issued a standard requiring employers to provide employees with information to assist them in the management in their own safety and health. OSHA revised the standard in 1988 to eliminate certain recordkeeping and protect trade secrets.

Nevertheless, employees have a right to access employer-maintained exposure and medical records. The standard limits access only to those employees or former employees who have been or will be exposed to toxic substances or harmful physical agents. This knowledge, theoretically at least, allows employees to determine patterns of health impairment and disease and to establish causal relationships between disease and exposure to particular hazards. Easy access to these records was intended by OSHA to result in the decreased incidence of occupational exposure and aid in designing and implementing new control measures.

For the purposes of the standard, *Access to Employee Exposure and Medical Records* (29 CFR 1910.2120), *access* means the right and opportunity to examine and copy records. Access must be provided in a reasonable manner and place. If access cannot be provided within fifteen days of a proper request, the employer must state the reason for the delay and specify the earliest date when the records will be made available. Responses to initial requests or new information that is sought following the initial request have to be provided without cost to the employee or his or her representative.

If no exposure records exist on the requesting employee, the employer must provide records of other employees with similar job duties to those of the employee making the request. Access to these records does not require the written consent of the other employees. Exposure records must reasonably indicate the identity, amount, and nature of the toxic substances or physical agents to which the employee has been exposed.

Access to Records of Another Employee. Access to the medical records of another employee, not the requesting employee, may only be provided with the specific written consent of that employee. Figure 11-2 gives a sample nonmandatory authorization letter for the release of employee medical record information to a designated representative. See the discussion below regarding disclosure of terminal illness or psychiatric conditions. A physician, nurse, or other responsible health care person who maintains medical records may delete from requested medical records the names of persons who provided confidential information concerning the employee's health status.

An employee or his or her designated representative (or OSHA) can have access to any analyses that were developed using information from exposure or medical records about the employee's working conditions or workplaces. Personal identities, such as names, addresses, social security numbers, payroll numbers, age, race, and sex, must be removed from the data analyses prior to access.

Employer Disclosures. An employer may withhold trade secret information, but must provide information needed to protect employee health. In that case, the employer must release the trade secret information, but may condition access on a written agreement not to abuse the trade secret or to disclose the chemical's identity (popularly called a *secrecy agreement* or sometimes a *confidentiality agreement).*

An employer may delete from records any trade secret that discloses manufacturing processes or the percentage of a chemical substance in a mixture. However, the employer has to identify that such deletions have been made. If the deletion impairs the evaluation of where or when expo-

Figure 11-2
Sample Authorization Letter for the Release of Employee Medical Record Information to a Designated Representative (Nonmandatory)

I, (full name of worker/patient), hereby authorize (individual or organization holding the medical records) to release to (individual or organization authorized to receive the medical information), the following medical information from my personal medical records: (describe generally the information desired to be released). I give permission for this medical information to be used for the following purpose: (describe), but I do not give permission for any other use or redisclosure of this information. (The employee may also wish to (1) specify a particular expiration date for this letter; (2) describe medical information to be created in the future that the employee intends to be covered by this authorization letter; or (3) describe portions of the medical information in his or her records that he or she does not intend to be released as a result of this letter.

____
Full name of employee or legal representative

____
Signature of employee or legal representative

____
Date of signature

sure occurs, the employer must provide alternative information that is sufficient to permit the requester to make such evaluations.

The employer may withhold a specific chemical identity when it is a demonstrable trade secret, it is so stated to the requester, and all other information on the properties and effects of the toxic substance is disclosed. Nevertheless, the specific chemical identity must be disclosed to a treating physician or nurse when that physician or nurse declares that an emergency exists and the identity is necessary for treatment. When the emergency is over, the employer may require the physician or nurse to sign a confidentiality agreement.

The employer must provide access to a specific chemical identity in *nonemergency* situations to an employee, an employee's designated representa-

Figure 11-3
Activities Requiring Non-emergency Disclosure of Specific Chemical Identity

- Exposure assessment (for chemicals to which the employees are or will be exposed)
- Air monitoring (in the workplace to determine employee exposure levels)
- Pre-assignment or periodical medical surveillance (of exposed employees)
- Medical treatment (to exposed employees)
- Selection or assessment of personal protective equipment (for exposed employees)
- Design or assessment of engineering controls or other protective measures (for exposed employees)
- Conduct health effects studies (due to exposure)

tive, or a health care professional, if it will be used for one or more of the activities listed in Figure 11-3.

The employer may require a written statement of need for the non-emergency disclosure of information. Also the employer may inquire into the reasons why alternative information will not suffice. Also, the requester may be required to sign a confidentiality agreement and is not allowed to use it for any purpose other than the health need stated or to release it under any circumstances, except to OSHA.

If the employer elects not to disclose the specific identity of a chemical requested by a health professional, employee, or designated representative a specific procedure must be followed. Figure 11-4 summarizes these steps.

An employee, designated representative, or health professional may refer such a denial to OSHA for review and comment.

When an employer ceases business, all employee medical and exposure records must be provided to the successor employer. When no successor exists, the employer must inform the current affected employees of the access rights at least three months prior to cessation of business and must notify the

Figure 11-4
Procedure for Denial of Information

If the identity of a specific chemical is not disclosed, the employer must

- Provide a written denial.
- Provide the denial within thirty days of the request.
- Provide evidence that the chemical identity is a trade secret.
- Explain why alternative information is adequate.
- Give specific reasons for the denial.

Director of the National Institute for Occupational Safety and Health (NIOSH) in writing at least three months prior to the disposal of records.

Each employer must preserve and maintain accurate medical and exposure records for each employee. The access standard imposes no obligation on the employer to create records, but does apply to any medical or exposure records created by the employer in compliance with other OSHA standards or at his or her own volition.

Exposure records and data analyses based on them must be kept for thirty years. Medical records must be kept for the duration of employment plus thirty years. Background data for exposure records, such as laboratory reports and worksheets, need be kept only for one year. Records of employees who have worked for less than one year need not be retained after employment, but the employer must provide these records to the employee upon termination of employment. First aid records of one-time treatment need not be retained for any specified period of time.

OSHA does not mandate the form, manner, or process by which an employer preserves a record, except that chest X-ray films must be preserved in their original state. Three months before disposing of these records, employers must inform the director of NIOSH.

OSHA. Medical and exposure records are to be made available, on request, to OSHA representatives to examine and copy.

Employees. The employer may give the employee copies of the requested material. Alternatively, the employer may give the employee the actual records and the use of mechanical copying devices so the employee may copy the records. Finally, in lieu of the first two procedures, the employer may lend the employee the records for copying off site. Prior to employee access to medical records, physicians on behalf of employers are encouraged to discuss with employees the contents of their medical records. Physicians may also recommend ways that employers can disclose medical records other than by direct employee access. Where appropriate, a physician representing the employer can elect to disclose information on specific diagnoses of terminal illness or psychiatric conditions only to an employee's designated representative, and not directly to the employee.

Upon initial employment, and at least annually thereafter, employees must be told of the existence, location, and availability of their medical and exposure records. The employer must also inform each employee of his or her rights under the access standard and make copies of the standard available. Employees must also be told who is responsible for maintaining and providing access to records.

Employee Representatives. Upon request, the employer must provide the records or copies of the records to the employee's designated representative. Union officials are one example of designated representatives. Attorneys are another. Doctors, dentists, and other medical practitioners and organizations may also be the designated representative. When requesting exposure records on individuals without their written authorization, union representatives must indicate an occupational health need for the requested records.

Recordkeeping Variances

Employers wishing to set up a recordkeeping system different from the one required by OSHA regulations may apply for a recordkeeping variance. The *petition* must detail and justify your intended procedures. Submit the petition to the regional commissioner of BLS for the area in which the workplace is located. In state plan states, only BLS can grant a recordkeeping variance. A copy of the petition must be given to the employees' authorized representative. A summary of the petition must be posted wherever notices are normally posted. Employees are given ten days to submit to BLS their own written data, views, or arguments.

Prohibition of Discrimination

Employees have a right under OSHAct to expect safe, healthy workplaces and to not fear punishment for that demand. Employers may not punish or discriminate against workers for exercising their rights summarized in Figure 11-5.

If an employee is exercising these rights, his or her employer may not retaliate in any way. Such retaliation violates the OSHAct. This means that the employee may not be fired, demoted, stripped of seniority, stripped of any earned benefits, transferred to an undesirable job, threatened, or harassed.

Once a worker has exercised a right under OSHAct he or she is protected at least to the extent that an employer needs to be careful about how to proceed with any discipline of the employee. That does not mean that the employee cannot be disciplined at all. It does mean that it had better be pretty clear that the discipline is not in any way related to the exercise of his or her OSHAct rights. For instance, if you have knowingly allowed this worker to leave work early in the past, you may be found guilty of violating OSHAct if you suddenly start holding the employee accountable for this behavior after the employee has protested hazardous conditions. I have even seen two employees file a stack of safety and health complaints (there were

Figure 11-5
Workers' Safety and Health Rights

- To complain to their employer, union, OSHA, or any other government agency about job safety and health
- To file safety or health grievances
- To participate on a workplace safety and health committee or in union activities concerning job safety and health
- To participate in OSHA inspections, conferences, hearings, or other OSHA-related activities

three dozen complaints filed between them) because they suspected they were being laid off at the end of the week and they believed that this action would prevent that. The ironic thing was that the complaints they made were regarding their own unsafe behaviors and work practices!

Similarly, if you know that a number of employees are doing the same thing wrong, you cannot single out the one who filed a safety and health complaint.

Text of Prohibition

The prohibition is found in Section 11(c)(1) of the Act.

> No person shall discharge or in any manner discriminate against any employee because such employee has filed any complaint or instituted or caused to be instituted any proceeding under or related to this Act or has testified or is about to testify in any such proceeding or because of the exercise by such employee on behalf of himself or others of any right afforded by this Act.

Scope

The prohibition applies to employers and any agent of the employer, such as legal counsel, technical consultant, management consultant, or other contracted assistance.

Procedure

If a worker believes that he or she has been punished for exercising safety and health rights, he or she must contact the nearest OSHA office within thirty days of learning about the alleged discrimination. This is called an 11(c) complaint and a union representative may file it on behalf of another employee. The worker does not need to complete any forms. OSHA does that for the worker, asking what happened and who was involved.

The text of 11(c)(2) reads:

> Any employee who believes that he has been discharged or otherwise discriminated against by any person in violation of this subsection may, within thirty days after such violation occurs, file a complaint with the Secretary alleging such discrimination. Upon receipt of such complaint, the Secretary shall cause such investigation to be made as he deems appropriate. If upon such investigation, the Secretary determines that the provisions of this subsection have been violated, he shall bring an action in any appropriate United States district court against such

> person. In any such action the United States district courts shall have jurisdiction, for cause shown to restrain violations of paragraph (1) of this subsection and order all appropriate relief including rehiring or reinstatement of the employee to his former position with back pay.

OSHA investigates 11(c) complaints. If it is found that the employee was illegally punished for exercising OSHAct rights, OSHA asks the employer to restore that worker's job earnings and benefits. If necessary, and OSHA can prove discrimination, OSHA takes the employer to court. The worker does not pay any legal fees because the matter is a case of the U.S. v. the employer.

Employees in state plan states may file their complaint with either OSHA or the state equivalent agency. Employees of truckers and certain other employees in the trucking industry have up to 180 days to file their complaint with OSHA in accordance with their protection under Section 405 of the *Surface Transportation Assistance Act* (STAA). Employees of primary and secondary schools, whether public, private, or Department of Defense schools overseas, who are discriminated against for complaining of exposure to asbestos in their schools, have ninety days to file a complaint with OSHA under Section 211 of the *Asbestos Hazard Emergency Response Act* (AHERA). An employee who reports an unsafe intermodal container or another violation under the *International Safe Container Act (ISCA) of 1977* is protected under Section 7 of ISCA and has sixty days after a discrimination violation to report it to OSHA.

Elements of Violation

Did an employee file a complaint with OSHA about unsafe conditions in the workplace? Yes? Was he or she terminated after the complaint was filed? Or, was the employee given an undesirable job assignment after the complaint was filed? Or, was the employee told in any way that an action he or she deems undesirable is taken as a result of the complaint filed? If the answer to any of these three is yes, then in OSHA's eyes a violation has occurred.

I have seen two cases where desperate employees attempted to use this protection to save their jobs in a pending layoff. Trivial safety complaints were filed with OSHA. The investigating compliance officer, however, recognized the complaints as trivial and, finding no other violations, did not issue any citations. When the layoff came, the employees filed discrimination charges. However, their safety complaints were filed with OSHA after it was announced they would be included in the layoff but before the layoff actually occurred. In this instance, their 11(c) complaint was ruled meritless.

Protection under Other Federal Laws

Other laws also have whistle blower protection clauses. Where these apply, OSHA may not be the investigating agency for the discrimination complaint.

Protected Conduct

Section 11(c) of OSHAct is intended to protect bona fide claims of unsafe conditions in the workplace. A bona fide claim does not mean that OSHA finds a violation and issues a citation. Rather it means that the claim was filed with OSHA in good faith. Regardless of whether the claim turns up real violations, OSHA will protect the employee who filed the complaint. However, claims that are not filed in good faith, such as a desperate act to save one's job, are not protected.

Employees' Right to Refuse Dangerous Work

Walking off the job due to potentially unsafe conditions is not ordinarily an employee right. Doing so may result in disciplinary action by the employer. An employee does, however, have the right to refuse to be exposed to an imminent danger. Assuming the refusal was made in good faith, OSHA rules protect the employee if the steps of Figure 11-6 are followed.

Section 13(a) of OSHAct spells out the procedure by which OSHA deals with imminent danger. The United States district courts have jurisdiction. OSHA makes a petition to the court to restrain conditions or practices where

Figure 11-6
Refusing Exposure to Imminent Danger

- Where possible, the employee asked the employer to eliminate the danger, and the employer failed to do so; and
- The danger is so imminent that normal OSHA enforcement procedures would take too long to eliminate the danger; and
- The danger facing the employee is so grave that a reasonable person in the same situation would conclude a real danger of death or serious physical harm exists; and
- The employee has no reasonable alternative to refusing to work under the conditions such as reassignment or PPE.

a danger exists that could reasonably be expected to cause death or serious physical harm immediately or before the imminence of such danger can be eliminated through the normal enforcement procedures. The court order may require such steps to be taken as OSHA considers necessary to avoid, correct, or remove such imminent danger. The order may also prohibit the presence of any individual in locations or under conditions where such imminent danger exists. A logical exception is made in three circumstances. First, an exception is made for those individuals whose presence is necessary to avoid, correct, or remove such imminent danger. Second, those required to maintain the capacity of a continuous process operation to resume normal operations without a complete cessation of operations is permitted. Third, where a cessation of operations is necessary, the necessary personnel to accomplished shutdown in a safe and orderly manner is allowed.

Enforcement

To enforce its standards, OSHA is authorized under the OSHAct to conduct workplace inspections.

Inspections

Every establishment, even those with ten or fewer employees, is subject to inspection by OSHA compliance safety and health officers. These men and women are presumably chosen for their knowledge and experience in the occupational safety and health field. Federal and state compliance officers are "vigorously trained," to use OSHA's own words, in OSHA or state standards and in the recognition of safety and health hazards.

How OSHA Selects a Company for Inspection. OSHA does not inspect without reason, though the reason may not be apparent until OSHA informs you. Reasons for an OSHA inspection are:

- Wall-to-wall program inspection
- Employee complaint
- Follow-up visit to verify abatement
- Investigation of an accident noted on a Workers' Compensation Report
- Investigation of a fatality or hospitalization of three or more workers in the same accident

How does OSHA prioritize its inspections, given that it cannot possibly inspection every one of some six million workplaces every year? Obviously, the squeaking axle gets the grease. OSHA recognizes that the worst situations need attention first. Therefore, OSHA has established a logical system of inspection priorities.

Inspection Priorities. *Imminent Danger.* Imminent danger situations get the top priority. An *imminent danger* condition is one where a reasonable certainty exists that a danger can be expected to cause death or serious physical harm immediately, or before the danger can be eliminated through normal enforcement procedures.

Serious physical harm is any type of harm that could cause permanent or prolonged damage to the body or which, while not damaging the body on a prolonged basis, could cause such temporary disability as to require inpatient treatment in a hospital. *Permanent or prolonged damage* has occurred, for example, when a body part is crushed or severed; an arm, leg, or finger is amputated; or sight in one or both eyes is lost. Such damage also occurs when a part of the body is either rendered functionally useless or substantially reduced in efficiency, both on and off the job. For example, permanent damage has occurred when the bones in a limb have been so severely shattered that mobility or dexterity are permanently reduced. Temporary disability requiring treatment as an inpatient in a hospital includes simple fractures, concussions, burns, or wounds involving substantial loss of blood and requiring extensive suturing or other healing aids.

Injuries or illnesses that are difficult to observe are classified as serious if they inhibit a person in performing natural functions, cause reduction in physical or mental efficiency, or shorten life. Health hazards may constitute imminent danger situations when they present a serious and immediate threat to life or health. For a health hazard to be considered an imminent danger, two factors must be present. First, toxic substances—such as dangerous fumes, dusts, or gases—must be present. Secondly, exposure to these substances must be capable of causing immediate and irreversible harm to such a degree as to shorten life or cause reduction in physical or mental efficiency, even though the resulting harm is not immediately apparent.

Imminent Danger Procedures. Upon inspection, if an imminent danger situation is verified, the compliance officer will ask the employer to voluntarily and immediately abate the hazard and to remove the employees from the exposure. Should the employer fail to comply with this request, OSHA will go to the nearest federal solicitor and obtain from the U.S. District Court appropriate legal action to correct the situation. Before the OSHA compliance officer leaves the workplace, however, he or she will advise all affected employees of the hazard and post an imminent danger notice.

The District Court can produce a restraining order resulting in the immediate shutdown of all operations, or at least of the section of the operation where the danger exists, by federal marshals.

Employees may, with impunity, refuse in good faith to be exposed to imminent danger. While they do not have the right to do this with all hazards, they are protected by law in the case of imminent danger. You may not require them to risk life and limb.

Employee Complaints. The second priority is given to investigate employee complaints of alleged violations of standards or of unsafe or unhealthful working conditions. OSHAct gives every employee the right to request an OSHA inspection when the employee feels he or she is in imminent danger from a hazard or when he or she feels a violation of an OSHA standard exists that threatens physical harm. OSHA will maintain confidentiality if the employee requests not to be identified. The area office will notify the employee of any action it takes regarding the complaint. If requested, OSHA will hold an informal review of any decision not to inspect.

Programmed High-Hazard Inspections. Third priority is given to programmed or planned inspections. Certain industries are selected for inspection on the basis of such factors as the death, injury, and illness incidence rates. In some industries, employees are inherently exposed to toxic substances and these, too, receive programmed inspections. Special emphasis may be regional or national in scope, depending on the distribution of the workplaces involved. States with their own safety and health programs may use somewhat different systems to identify and prioritize high-hazard industries for inspection.

Catastrophes and Fatal Accidents. Assuming that an imminent danger does not continue to exist, in which case these inspections get a much higher priority, the next priority is the investigation of catastrophes involving the hospitalization of three or more employees or the death of one or more. The investigation is made to determine if OSHA standards have been violated and to avoid recurrence of similar accidents. Normally, then, these inspection-investigations are given fourth priority precisely because they are after-the-fact, even though official literature places them at number two priority.

Other Programmed Inspections. Area offices conduct routine inspections of low-hazard workplaces on a random basis. Call the OSHA Office of Management Data Systems for more information (telephone 202-219-8576).

Follow-up Inspections. The lowest priority is given to inspections that are conducted to determine whether previously cited violations have been corrected. If an employer is found to have failed to abate a violation, the compliance officer informs the employer that he or she is subject to receive a Notification of Failure to Abate for the alleged violations. The employer may face additional penalties while such failure to abate violations persists.

How OSHA Conducts an Inspection. Prior to inspection, the compliance officer will familiarize himself with as many relevant factors about the workplace as practical. He or she will take into account the compliance history of the establishment, the nature of the business, and the particular standards that are likely to apply. The compliance officer will also select appropriate equipment for detecting and measuring fumes, gases, toxic substances, noise, and other chemical and physical agents that may exist in the workplace.

Inspector's Credentials. As soon as the compliance officer arrives at the establishment, he or she displays official credentials and asks to see an appropriate management representative. At most facilities that representative will be the safety professional. The receptionist or guard should have a call list with names and internal telephone numbers for potential management representatives to call, in the order of preference. Any one of these representative should have sufficient awareness of operations and the authority to be able to decide whether to grant entry or deny it. Do not deny entry unless you know of the existence of an obvious violation of an OSHA standard that can be fixed quickly. However, if you and your management are committed to occupational safety and health, why is a plan not already in place to fix the problem?

Always scrutinize the compliance officer's credentials to assure validity, unless you are personally familiar with the officer. An OSHA compliance officer should carry a U.S. Department of Labor identification card bearing his or her photograph and a serial number that can be verified by calling the nearest OSHA office. Compliance officers do not mind you placing this call for verification purposes and will not penalize you in any way for making it. Unfortunately, unscrupulous sales persons have been known to pose as OSHA compliance officers in order to get into plants and sell their wares or services. Anyone who promotes the sale of a product or service is not an OSHA compliance officer. In fact, posing as a compliance officer is a violation of the law. Report the suspected impostor at once to local law enforcement agencies.

An OSHA compliance officer is authorized by the OSHAct to enter without delay and at reasonable times, any factory, plant, establishment, construction site or other areas, workplaces, or environments where work is performed by employees. He or she is also authorized to inspect and investigate during regular working hours, and at all other reasonable times, and within reasonable time limits and in a reasonable manner, any such place of employment and all pertinent conditions, structures, machines, apparatus, devices, equipment, and materials therein, and to question privately any such employer, owner, operator, agent, or employee.

The compliance officer will announce the scope and reason for the visit. If she makes this announcement in the lobby, ask him or her to repeat it during the opening conference. If the inspection is the result of a complaint, obtain a copy of the complaint. Reasons for a visit from your local OSHA office are noted in the previous section.

Opening Conference. You should maintain a list of people who attend the OSHA inspection opening conference. The receptionist, security guards, each person on the attendance list, and all human resources and industrial relations offices should have a current copy of the list. The safety professional, plant manager, operations manager, maintenance manager, human resources and/or industrial relations manager, and the employees' authorized representative should also attend, at a minimum. The employees' representative is required to attend the conferences and inspection by law. The senior managers should be present at the conferences—at least to hear first hand what the issues are and how well your facility faired. The senior maintenance person is needed because he or she is in charge of repairs. Sometimes standards violations can be fixed during the inspection, especially if your inspection team includes a friendly person with the ability to get things done.

During the opening conference the compliance officer explains why the facility was selected for an inspection. He or she will again identify himself or herself and indicate whether he or she is a Compliance Safety Officer

(CSO), a Compliance Health Officer (CHO), or a Compliance Safety and Health Officer (CSHO). The officer will next ascertain whether an OSHA-funded consultation program is in progress or whether the facility is pursuing or has received an inspection exemption. Normally he or she will already know this, but will ask to verify his or her information. If so, the inspection is usually terminated.

Assuming the inspection continues, the CSHO will explain the purpose of the inspection, its scope, and the standards that apply. The employer will be given copies of applicable safety and health standards as well as a copy of any employee complaint that may be involved. If the employee has requested confidentiality, the name will be expunged from the complaint form and will not be revealed.

The employer is then asked to select an employer representative to accompany the CSHO during the inspection—usually you, the safety manager. You have every right to choose the maintenance representative to also accompany anyone you wish. Ensure that the CSHO is not intimidated in any way and keep in mind that too many cooks can spoil the pot. Otherwise, it is not the concern of the CSHO how many persons accompany the inspection party.

An authorized employee representative also is given the opportunity to accompany the CSHO. He or she will not proceed until this representative arrives, so you should allow that person to attend the opening conference. If, for some reason, the employee representative declines to accompany the CSHO the inspection will proceed anyway. However, if you present an obstacle to the employee representative, the CSHO will wait until you relent or will leave to get a warrant (and a federal marshal).

Where a bargaining unit exists within the plant, employees are typically represented by one of the union officials. Otherwise, one or more member of the plant safety committee may accompany the inspection. If the plant does not have a safety committee the CSHO will determine if any employee presented as the employees' representative can represent the interests of the other employees. Under no circumstances may the employer select the employee representative for the walkaround the employer can only suggest possible personnel.

When no employee representative is forthcoming or if the CSHO determines the representative may have been improperly selected by management, the CSHO will consult with a reasonable number of employees concerning safety and health matters in the workplace. Such consultations may be held privately and may widen the scope of the inspection. Feel free to sit in on the consultations, but if the CSHO or the employee asks you to leave, you must. The best method is to just sit down as the employee does and wait quietly to be told to leave or not.

Before conducting the walkaround, the CSHO typically examines the OSHA 300 Log to see if it is maintained currently and to calculate the plant's incident rate (IR). The IR will be compared to the average for the SIC code number for your industry. If your IR is half or less of the average, the original inspection scope will stand. If, on the other hand, your IR is greater than half the SIC average IR, the CSHO will call his or her supervisor, who may order that the scope be expanded to a wall-to-wall compliance inspection.

Inspection Tour. Escort the CSHO directly to the site for any type of limited inspection. If you note signs that the CSHO may broaden the scope of the inspection, notify your corporate legal counsel at once.

While talking with employees in the workplace where the alleged hazard is located, the CSHO will make every effort to minimize work interruptions. The CSHO observes conditions, consults with employees, takes photographs, takes instrument readings, and examines records.

Regarding the use of photography, discuss the use of cameras with corporate legal counsel and get a written policy before an inspector arrives at your door. The employer is not allowed to prohibit the taking of photographs, even if a trade secret is involved. However, you may certainly challenge the taking of photographs and it is a good practice to request that the compliance officer make sketches in lieu of taking photographs. If he or she insists on the camera, do not interfere.

If a trade secret is involved and the officer insists on photographs, do not hesitate to ask him or her to sign a confidentiality agreement and to preserve the confidentiality of information regarding the trade secret. If the compliance officer appears to be taking misleading photographs of non-trade secret processes, take photographs yourself. It is always a good practice to duplicate everything the inspector does. Experienced CSHOs expect this behavior on your part and may even be more comfortable with you if you demonstrate that you know what you are doing. Every little edge in your favor, no matter how small, may be needed.

Trade secrets observed by the CSHO must be kept confidential. An inspector who releases confidential information without your authorization is subject to a $1,000 fine and/or one year in jail. The employer may also require that the employee representative have a security clearance for any area in question.

Employees are consulted by the CSHO during the inspection tour. The CSHO may stop and question employees in private about safety and health conditions and practices in their workplaces. Employees who cooperate with OSHA are protected under OSHAct from discrimination for exercising safety and health rights. However, you are not required to interrupt their production.

If a machine or process is down before the inspection you do not have to start it up, even if the inspector asks. A CSHO cannot cite you for a missing or inoperative safety device or warning sign if a machine or process is not operating. He can only cite violations observed on operating equipment.

Posting and recordkeeping are checked. The CSHO will inspect records of deaths, injuries, and illnesses, which the employer is required to keep. He or she will check to see if a copy of the totals from the last page of the OSHA 300 Log has been posted. He or she will also look for the OSHA workplace poster (OSHA 2203) to ensure that it is prominently displayed. It used to be that failure to post the OSHA 2203 was cited as an alleged violation of OSHA regulations and a penalty was assessed. Now the OSHA CSHO brings a copy of the poster to the inspection and watches as the employer corrects the violation on the spot. A citation is still issued for no poster, but the "no penalty" is assessed if the poster is put up while the inspector watches. Where records of employee exposure to toxic substances and harmful physical agents have been required, the CSHO will examine them.

During the inspection the CSHO will point out to the employer representative any unsafe or unhealthful working conditions observed, even if not within the scope of the inspection. At the same time, the CSHO will gladly discuss with you potential corrective measures, if you desire.

Some apparent violations can be corrected immediately, and should be. That is why you need to bring the maintenance manager along. When they are corrected on the spot, the CSHO records such actions to help in judging the employer's good faith in compliance. Even though corrected, the apparent violations may still serve as the basis for a citation and/or a notice of proposed penalty.

Closing Conference. Limit attendance at the closing conference. Of course the CSHO will be there and so will the designated employee representative. Beyond that, yourself, the plant manager, the maintenance supervisor, and perhaps the manager of human resources/industrial relations should be in attendance.

The compliance officer will review all possible citations with you. Now is the time for free discussion of problems and needs. Frank questions and answers are the order of the day. At all times, however, maintain cool heads.

For each possible citation, request the CSHO to tell you the hazard created, the job functions exposed to the hazard, the standard alleged to be violated, and the suggested means of abatement. Ask him or her to give you a copy of the notes and take thorough notes yourself. It is a good idea to have a stenographer attend the closing conference to do this for you. Do not admit to any violations except for any obvious violations corrected during the walk around. This is not the time to admit violations. Also, during the closing conference express your desire to request as much time as reasonably needed for the abatement of these citations.

The employer is told of appeal rights and the informal conference. Procedures for contesting citations should be explained by the CSHO at the closing conference. The CSHO will not, however, indicate any proposed penalties at this time. Only the OSHA area director has that authority and those penalties are determined only after the completion of a full report by the CSHO.

During the closing conference you may produce records to show compliance efforts and to provide information that may help OSHA determine how much time may be needed to abate an alleged violation. Sometimes more than one closing conference is needed. When health hazards are being evaluated, for instance, the CSHO may have to wait on laboratory results. A second closing conference may be needed when these results are reported.

Warrants

If you wish, it is your right, based on a 1978 Supreme Court ruling *(Marshall v. Barlow's Inc.)*, to demand a warrant when the compliance officer arrives. This is not usually recommended, but it is your right nevertheless. After all, an inspection is a purposeful search for evidence of wrong-doing or at least evidence of negligence. Your constitutional right, in case you do not remember, is covered under Amendment IV:

> The right of the people to be secure in their persons, houses, papers, and effects, against unreasonable searches and seizures, shall not be violated, and no warrants shall issue, but upon probable cause, supported by oath or affirmation, and particularly describing the place to be searched, and the persons or things to be seized.

Constitutional and Statutory Provisions. Your constitutional right to demand a warrant is one thing, but keep in mind that OSHA is not without authority. Under the OSHAct, OSHA is given authority to inspect to verify the allegations of an employee complaint. Also, OSHA has statutory authority to investigate fatalities and the hospitalization of three or more employees who were injured in the same incident. Nevertheless, if you think it is in your best interest to do so, then by all means ask for a warrant. Since I am not a lawyer, I recommend you discuss this thoroughly with your legal counsel—do that now, before an inspector arrives at your door.

Requirements for a Warrant. In order to get a warrant, the compliance officer must approach the federal magistrate with reasonable cause. This cause can be as simple as "It is your turn to be inspected on a random basis." Or it can be based on the fact that your incident rate is higher than the national average for your SIC code. To get the warrant the compliance officer must specify the scope of the intended inspection. You can bet the scope will be expanded from the original intended inspection. Typically, the compliance officer returns with the warrant and a team of inspectors. One person who will return with the compliance officer in this case is the federal marshal. Hopefully, by the time the marshal hands you a copy of the warrant, you will have agreed upon a strategy with your corporate legal counsel.

Challenging a Warrant. When the OSHA compliance officer returns, the accompanying federal marshal will have the warrant in hand. You still have a right to have a copy of it and to review it before the inspection begins with the opening conference. While people are gathering for the opening conference, call your corporate legal counsel and comply as advised. When the federal marshal reads the warrant during the opening conference, inform the marshal and compliance officer how the plant will cooperate. Avoid being antagonistic, but be firm in your requests.

Types of Inspections *vis a vis* Warrants. When a warrant is issued, the inspection must proceed exactly as outlined in the warrant.

Advance Notice. Inspections are conducted without notice. Alerting an employer in advance of an OSHA inspection can bring a criminal fine of up to $1,000 and/or six months in jail. Under special circumstances OSHA may give less than twenty-four hours notice to an employer. Figure 11-7 lists these circumstances.

Figure 11-7
Circumstance under which Advance Notice Is Permissible before an OSHA Inspection

- An imminent danger requires immediate correction
- Inspection must take place after regular business hours
- Inspection takes special preparation
- Notice is required to assure the appropriate parties are present
- Cases where the OSHA area director determines that advance notice would produce a more thorough or effective inspection

Consent or Warrant to Inspect Workplace. You may give your consent to the inspection or exercise your right to demand a warrant for the search. Normally, as said before, let OSHA inspect your workplace without demanding a warrant.

Consent, Subpoena, or Warrant to Inspect Records and Files. Whether you explicitly say, "Go ahead and search my records and files," or merely show the inspector where they are and by your silence give consent, the inspector will halt the inspection before taking them without your consent. However, if he or she holds a subpoena or a warrant, your consent is no longer required.

What Is in Plain View? Anything in plain view is open game for the inspector. If, for instance, he or she is present to inspect your hazard communication program due to an employee complaint, but while walking to the place where the inspection will take place sees that a machine guard is missing, he or she may cite that as a violation. Or if walking down the sidewalk on the way into the plant, he or she sees a violation, that may also be cited, even if you subsequently demand a warrant. Anything the inspector can see without effort, he or she may cite, with or without a warrant.

Planning the Walkaround

Assure that one person in your facility has been assigned primary responsibility as the OSHA contact. Also, name two alternates. Reception personnel must be trained to give prompt notice of the compliance officer's arrival. Some recommend contacting your corporate legal counsel before agreeing to the scope of the inspection.

Take detailed notes of the compliance officer's activities. Include a description of all areas inspected. Write down the rationale for the existence of these violations. Summarize the highlights of employee interviews. List all requests for information other than the OSHA-300 log or the OSHA-301. Itemize any corrective actions taken during the inspection. Write down the details of exposure monitoring. Forward a copy of your inspection notes to the corporate legal counsel.

A maintenance supervisor should be part of the walk around team and he or she should have attended the opening conference. As the walk around proceeds, apparent violations may be corrected. For instance, say you encounter a blocked aisle as you enter the manufacturing plant. Turn to your maintenance supervisor and in no uncertain terms ask him or her to take care of it—now! Another common example of an obvious violation is the blocked access to emergency eyewash. Unsafe walking or working surfaces are another set of easy-to-fix-now obvious violations often spotted on a walkaround. No excuses—have them fixed before the closing conference. The compliance officer knows that on any given day he or she will discover some violation in your plant, but show OSHA you mean business. Correcting apparent violations helps demonstrate good faith compliance. Even if corrected immediately, however, a citation can be issued for violations that are fixed before the end of the inspection. These days OSHA is likely not to cite you, but that could change with a changing political climate. Do not immediately correct circumstances, however, that are not obvious violations of a standard.

Document Control

Understand and follow guidance for responding to OSHA requests for information. The compliance officer has a right to see several documents.

The first priority is the current OSHA 300 log, plus the logs of the previous five years. The compliance officer will ask for the man-hours worked in the last two years. This information allows him or her to calculate your incident rate and lost workdays rate. If your incident rate is more than your SIC code average, the officer calls his or her supervisor and gets the scope of the inspection broadened to a wall-to-wall, if not that broad already.

The first report of injury forms or the OSHA 301 form for the current and previous years will probably be requested. Any OSHA 300 form is subject to review. Any records of exposure to any substance may be needed, especially those that are required to be maintained by regulation. Training records, medical surveillance records, and exposure assessment records are all fair game.

One document the compliance officer will want to see for sure is the Hazard Communication Program. He or she will also want to examine Material Safety Data Sheets (MSDSs) for substances encountered during the inspection. Some other possibilities for requested documents are:

- Respiratory protection training
- Fit test records
- Noise measurements
- Hearing conservation training
- Hazardous materials emergency response training
- Process safety management training
- Process Safety Management Program

This list is not all inclusive.

Do not allow the compliance officer to begin industrial hygiene monitoring until the following information is provided:

- Sampling method
- Sampling time
- Job functions to be monitored
- Specific regulatory exposure level addressed

If this information cannot be supplied before the monitoring proceeds, contact your legal counsel. Request the compliance officer to give you time to prepare for side-by-side monitoring.

Preparing Employees to Be Interviewed

Insist on being present during salaried employee interviews. Hourly employees, on the other hand, may meet privately with the compliance officer. But remember to remain seated in the interview room until the compliance officer tells you to leave.

Employee Participation in the Inspection

Hourly employees have a right to confer privately with the compliance officer. However, nothing precludes you from asking if an employer representative can be present. The worst thing the compliance officer can do is say "No." A good practice, mentioned elsewhere, is to remain seated in the interview room when the employee enters. If the compliance officer does not tell you to leave, you can stay. Remember, the employer representative may also be present at all salaried employee interviews, and cannot be required to leave.

Limitations on Employee Rights. Employees are not entitled to a group of participants, only one. They are not entitled to leave their work stations to approach the compliance officer at will during the inspection.

Representatives of Employees. The OSHAct requires that an employee representative be part of the inspection team. Selection of the representative is made by the employees, not management. If your facility has a bargaining unit, an officer of the union generally acts as the employee representative. Otherwise, gather your employees and let them choose one of their own. It is best to have done this beforehand. The employees get only one representative. You are not required to let them have more than one representative. The representative should be in attendance at the opening conference.

OSHA Accident Investigations

OSHA investigates all workplace fatalities and also any accidents that hospitalize three or more workers. You are required to report catastrophic accidents to OSHA within eight hours of learning about them. Regardless of the inspection prioritization method used by the area office, the incident will be investigated.

Be aware that if OSHA finds, as a result of its investigation, a willful violation of an OSHA standard or the general duty clause of OSHAct was the cause of the catastrophe, someone in your facility may face federal criminal charges. Generally, but not always, the plant manager is implicated because of his or her overall responsibility for the safety and health of all employees. Typically, the immediate line supervisor is also implicated because of his or her proximity to the incident. What I am saying is that usually two people—the immediate line supervisor and the plant manager—would be the ones led away in handcuffs by the federal marshals. Right after you call OSHA to report such an unfortunate incident, call your corporate attorneys.

Citations

The silly thing about getting an OSHA citation is that you know in advance what they are going to cite. All you have to do is get a copy of the top ten or twenty OSHA citations for a given year and there you have it! For instance, in 1994 the number one violation was failure to have a written Hazard Communication Program. Figure 11-8 lists the top ten violations for 1997.

An interesting aside: when I started drafting this chapter the latest version of this list I had close at hand was 1994. When I updated it for the latest fiscal year figures, I had to make only minor changes. If OSHA were not around to hammer us, would we ever get the point on our own? I wonder.

In addition to the frequent citations in Table 11.7, OSHA reports that they frequently cite companies for failure to maintain an OSHA 300 Log and Summary of Occupational Injuries and Illnesses (1904.2(a)). Too many companies also fail to post the OSHA poster required by 1903.2(a)(1). OSHA now delivers the poster to you when they come to inspect. Post it before they leave, unless you have an up-to-date poster on the bulletin boards already.

OSHA Instruction CPL 2.111 supports the Paperwork Reduction Act effective October 2, 1995. The impact of CPL 2.111 will be less violations, but not necessarily smaller ones in terms of penalties.

The area director determines what citations, if any, will be issued to the company inspected, and what penalty, if any, will be proposed for each citation. Citations inform the employer and employees of the regula-

Figure 11-8
Ten Most Frequently Violated Standards - 1997 General Industry (29 CFR 1910.X unless noted)

1200(e)(1)	Hazard Communication - written plan
1200(h)	Hazard Communication - employee information and training
212(a)(1)	Machine Guarding - point of operation
100(a)	Head Protection - protective helmets
1904.2(a)	Recordkeeping - OSHA 300 Log
215(b)(9)	Abrasive Wheel Machinery - guard adjustment
151(c)	Medical Services/First Aid - eye wash/safety shower
147(c)(1)	Lockout/Tagout - energy control program
219(d)(1)	Mechanical Power Transmission - pulley guards
OSHAct 5(a)(1)	General Duty Clause - safety and healthful workplace

tions and standards alleged to have been violated and of the proposed length of time set for their abatement. You will receive citations and notices of proposed penalties by certified mail. You must post a copy of each citation at or near the place a violation has occurred for three days, or until it is abated, whichever is longer. You must comply with these posting requirements even if you contest the citation.

Legal Counsel. Do not take action in response to a citation without consulting with corporate legal counsel. Immediately forward a copy of your citations to legal counsel when you receive them. Before responding to the citations, secure an action plan from your legal counsel, at least by telephone.

If you are not contesting the violation, promptly notify the OSHA area director by letter when you have complied with abatement. When abating the violations cited, follow all instructions in the accompanying letter from the OSHA area office. Do this within the time set forth in the citation. Pay any penalties itemized in the citation. The notification you send to the area director is called an *Abatement Certification.* A simple, signed letter will do for other-than-serious violations. More detailed proof of compliance is required for serious, willful, repeat, or failure-to-abate violations. Consult with the area director about certifying abatement in these cases.

If your legal counsel directs, proceed to the informal conference or contest the citation as warranted. Ask for the informal conference before deciding to contest the citation. However, the informal conference is always to be recommended 1) because it is your right and 2) you may use this opportunity to contest citations and negotiate penalties downward. Figure 11-9 summarizes the potential benefits of the informal conference.

You are especially encourage to take advantage of the informal conference if you foresee any difficulties complying with any part of the citation. The informal conference cannot extend the fifteen working day period you

Figure 11-9
Potential Benefits of the Informal Conference

- A better understanding of the violations cited
- A more complete understanding of the applicable standards
- Negotiate an Informal Settlement Agreement
- Learn alternative ways to correct violations
- Negotiate more favorable abatement dates
- Present problems concerning employee safety practices as mitigating circumstances
- Obtain answers to questions you may have
- Resolution of disputed citations and penalties in a friendly legal setting
- Avoid more formal proceedings in a potentially hostile legal setting

had to file a Notice of Intent to Contest. The informal conference does not take the place of the written notice if you plan to contest.

You may not leave the employees out of the process. Authorized representatives of the employees have the right to participate in any informal conference or in any negotiations between the OSHA area director or the Regional Administrator and the employer. Also, you may not limit employee participation at an informal conference or negotiation meeting.

Let me make one thing clear about OSHA citations. No company has any reason to have more than a handful of administrative or non-serious violations as the result of any compliance inspection. To have more than a handful of serious violations is a loud and clear message not only to OSHA but to the whole world: you do not care in the least about your most precious resource—your employees. Tear down all your slogan posters to the contrary—they are meaningless to you.

Any plant that not only has a written health and safety program, but also has gone the next step to implement systematic identification, evaluation, and control of hazards will not fear OSHA inspections. Any program that goes beyond mere compliance with the OSHA standards to really address hazards in a timely manner may have a few administrative chinks in its armor, but nothing serious will be found. You know what is needed: management commitment, employee involvement, continuous improvement, work site analysis, hazard prevention and control, and effective training.

Citation and penalty procedures may differ somewhat in states with their own occupational safety and health programs.

Elements of a Citation

If a violation exists, OSHA will issue you a Citation and Notification of Penalty detailing the exact nature of the violation and any associated penalties. This document is called the "citation" for short.

Features. The citation does three things. First, it informs you of the alleged violations by citing chapter and verse from the OSHA standards, regulations, or general duty clause that you violated and stating how you failed to comply. Secondly, the citation sets a proposed time period within which you are expected to correct *(abate)* the violation. Finally, the citation proposes a monetary penalty as a civil punishment for violation of OSHAct. The law states that your misdeeds cannot benefit you while a compliant employer has to pay for compliance. Therefore, you are expected to give up some of you supposedly ill-gotten profits if you are found to be in violation of the Act.

Consequences. Keep in mind that normally these are administrative law procedures. Therefore you are within your rights to insist on representation or counsel. Not only may you have a lawyer represent you, but you may also bring along a consulting engineer, physician, industrial psychologist, or other professional who can look after your best interests and speak expertly for you. The penalties, too, are administrative matters, although any resulting publicity may be a *de facto* punishment. The exception is that the proceedings come under federal criminal law and will be tried in a federal court if a death or catastrophe occurred.

Procedure. If an inspection was initiated by an employee complaint and no citation is issued, the complainant or the employees' authorized representative may request an informal review of the decision. Employees may not contest citations, amendments to citations, penalties, or lack of penalties. They may, however, contest the time allowed for abatement in the citation. They may also contest their employer's Petition for Modification of Abatement (PMA) if it requests an extension of the abatement period.

Employees may contest the PMA within ten working days of its posting or within ten working days after an employees' authorized representative has received a copy. Within fifteen days of the employer's receipt of the citation, an employee may submit a written objection to OSHA. The area director forwards the objection to the Occupational Safety and Health Review Commission (OSHRC), which operates independently of OSHA. Finally, employees may request an informal conference with OSHA to discuss any issues raised by an inspection, citation, notice of proposed penalty, or employer's notice of intention to contest.

When issued a citation or notice of a proposed penalty, an employer may request an informal meeting with the OSHA area director to discuss the case. The area director is authorized to negotiate and enter into settlement agreements that revise citations and penalties to avoid prolonged legal disputes.

Contesting a Citation. Upon receiving a citation, the employer must correct the cited hazard by the prescribed date, unless he or she contests the citation or abatement date. Factors reasonably beyond the employer's control may prevent the completion of abatement by the prescribed date. The employer who has made a good faith effort to comply but cannot may file a PMA for an extended date.

The written PMA should specify all steps that have been taken to achieve compliance. The additional time needed to achieve complete compliance must be specified. The reasons such additional time is needed must be thoroughly explained. All temporary measures to safeguard employees against the cited hazard must be listed. Certify that a copy of the PMA is being posted at or near each place where the violation occurred. Also, certify that you have given the employees' authorized representative a copy of the PMA.

Timeline of Contest. If you decide to contest either the citation, the prescribed time for abatement, or the proposed penalty, you have fifteen working days from the time the citation and proposed penalty are received to notify the area director in writing. A verbal disagreement does not suffice.

This written notification is a Notice of Contest. It has no specific format requirements, but it must clearly identify your basis for contesting the citation, notice of proposed penalty, abatement period, or notification of failure to abate violation. A copy of the Notice to Contest must be given to the employees' authorized representative. If the affected employees are not represented by a collective bargaining unit, a copy of the notice must be posted in a prominent place, or else served personally upon each unrepresented employee.

If the written Notice of Contest is filed within the required fifteen working days, the area director will forward the case to the OSHRC, an independent agency that assigns the case to an administrative law judge.

Types of Violations. Violations come in several types, which are discussed below.

Willful. A *willful violation* exists under OSHAct where evidence shows that the employer committed an intentional and knowing violation of the Act and the employer is also conscious of the fact that what he is doing constitutes a violation of the Act. If the employer was not consciously violating the Act but was aware that a hazardous condition existed yet made no reasonable effort to eliminate the condition—that, too, is a willful violation.

Repeated. If, upon reinspection, an employer is found in violation of a section of the standard, regulations, rule, or order violating the General Duty Clause, which has been previously cited, a *repeated violation* exists. Do not confuse this type of violation with a failure to abate situation. To be the basis of a repeat citation, the original citation must be final. If the citation is under contest, the citation may not be the basis for a repeat violation. A citation for a repeated violation bears considerably heavier penalties than the original citation.

Now hear this! When the compliance officer returns to the office with his or her list of violations, a file review is made for repeat violations of the facilities and any sister facilities within the same OSHA jurisdiction. A violation can be cited, as repeated, if the employer has been cited for a substantially similar violation anywhere in the nation within the past three years! Therefore, it behooves you to share information with your sister facilities, lest one of you earns a repeat violation.

Serious. If a substantial probability of death or serious physical harm could result from the hazardous condition which exists, the violation is considered *serious.* The compliance officer need only prove that an accident that might result from the violation of a regulation would have substantial probability of resulting in death or serious physical harm. The compliance officer does not have to prove that such a substantial probability of the accident itself exists.

Non-serious or *Other Than Serious.* This type of citation is issued in situations where an accident or an occupational illness resulting from a violation of a standard would probably not cause death or serious physical harm, but which would have a direct or immediate relationship to the safety and health of employees. The lack of guardrails at a height from which a fall would probably result in no injury or in mild sprains or cuts and abrasions at the worst is an example of a non-serious violation.

De Minimus. When a violation of an OSHA standard is not immediately or directly related to a safety or health citation, it is considered a *de minimus citation.* The de minimus citation involves no fine. Similarly, the citation carries no requirement for abatement. OSHA will take no further enforcement action, even if the employer does not correct the de minimus condition. If, for instance, the compliance officer were to write a citation for the height of letters on an exit sign—which are not in strict conformance with the size requirements of a standard—that would be a de minimus violation.

Failure to Abate. In this situation, OSHA determines on reinspection that the employer has failed to correct the previously cited conditions. As with repeated violations, failure to abate citations are assessed with heavy penalties. The difference is that in the failure to abate citation the original condition was never abated, while in the repeated citation the original condition was abated but another just like it has occurred without being timely abated.

Imminent Danger. In the case of imminent danger, OSHA can take immediate action, including asking you to shut a process down until the hazard is abated. Imminent danger situations mean that someone's life or health is in danger if something is not done right away.

Civil Penalties

A proposed penalty of up to $7,000 for each other-than-serious or serious violation cited is possible for civil penalties. For willful or repeated violations the ante is upped to $70,000. However, if a death has occurred due to a willful violation the matter becomes criminal. Failure to abate prior violations may bring a civil penalty on a daily basis.

Penalty Amounts. A discretionary penalty for other-than-serious violation may be adjusted downward from by as much as 80 percent depending on the employer's good faith demonstrations to comply with OSHAct, history of violations, and size of business. When the calculated penalty amounts to less than $60, no penalty is proposed.

A mandatory penalty of up to $7,000 for each serious violation is proposed. Penalties for serious violations may be adjusted downward, based on the employer's good faith, history, the gravity of the alleged violation, and size of business.

Willful violations bring proposed penalties of up to $70,000, as mentioned above, but in no case may the penalty be reduced to less than $5,000 for each violation. A proposed penalty for a willful violation may be adjusted downward for size of business and history. Credit for good faith is not typically given, however, as bad faith is implied in the case of a willful violation.

For a repeated violation the proposed penalty (maximum $70,000) is not adjusted downward. Failure to correct a prior violation may bring a penalty of up to $7,000 per day the violation continues to exist beyond the prescribed abatement date.

Violations of the posting requirements can bring a civil penalty of up to $7,000.

Calculation of Penalties. The dollar value of penalties is based on a certain minimum times a gravity factor, the size of the business, the good faith of the employer, and the employer's history of previous violations.

Gravity of Violation. The gravity of the violation is the primary consideration in determining penalty amounts. It is the basis for calculating the basic penalty for both serious and other violations. To determine the gravity of a violation two assessments are made. First, the severity of the injury or illness that could result from the alleged violation is considered. Second, the probability that an injury or illness could occur as a result of the alleged violation is considered.

The classification of an alleged violation as *serious or other-than-serious* is based on the severity of the injury or illness that could result from the violation. This classification constitutes the first step in determining the gravity of the violation. A severity assessment is assigned to a hazard according to the most serious injury or illness that could reasonably be expected to result from an employee's exposure. *High severity* is assigned if death from injury or illness could occur. Injuries involving permanent disability or chronic, irreversible illnesses are also assigned high severity. *Medium severity* is assigned where injuries or temporary, reversible illnesses resulting in hospitalization or a variable but limited period of disability could occur. A *low severity* is assigned if the potential injuries or temporary, reversible illnesses would not result in hospitalization or would require only minor supportive treatment. Finally, the category of *minimal severity* is reserved for other-than-serious violations. Although such violations reflect conditions that have a direct and immediate relationship to the safety and health of employees, the injury or illness most likely to result would probably not cause death or serious physical harm.

Probability Assessments. After the severity is determined, the *probability assessment* is made and factored into the penalty. The probability that an injury or illness will result from a hazard has no role in determining the classification of a violation but does affect the amount of the penalty to be proposed. Probability is categorized as either greater or lesser probability. *Greater probability* results when the likelihood that an injury or illness will occur is judged to be relatively high. *Lesser probability* results when the likelihood that an injury or illness will occur is judged to be relatively low.

The following circumstances are normally considered by OSHA when violations likely to result in injury or illness are involved.

- Number of workers exposed
- Frequency of exposure or duration of employee overexposure to contaminants
- Employee proximity to the hazardous conditions
- Use of appropriate personal protective equipment (PPE)
- Existence of a medical surveillance program

All of these factors are considered together in arriving at a final probability assessment. Any other pertinent working conditions the compliance officer thinks appropriate are also factored into the penalty calculation. When strict adherence to the probability assessment procedures result in an unreasonably high or low gravity, the probability is adjusted as appropriate, based on the professional judgment of the compliance officer and approved by the area director. Such decisions are documented in the case file.

Penalty Reductions. The GBP may be reduced by as much as 95 percent depending upon the employer's good faith, size of business, and history of previous violations. Up to 60 percent reduction is permitted for size. Up to 25 percent reduction is given for good faith. As much as 10 percent can be knocked off for history of previous violations.

Since these adjustment factors are based on the general character of a business and its safety and health performance, the factors are generally calculated only once for each employer. After the classification and probability ratings have been determined for each violation, the adjustment factors are applied, subject to limitations. Penalties assessed for violations that are classified as high severity and greater probability are adjusted only for size and history. Penalties assessed for violations that are classified as repeated can be adjusted only for size. Penalties assessed for regulatory violations, which are classified as willful, can be adjusted for size. Penalties assessed for serious violations, which are classified as willful, can be adjusted for size and history. If any violation is classified as willful, no reduction for good faith can be applied to any of the violations found during the same inspection. The theory is that an employer cannot be willfully in violation of the Act and, at the same time, be acting in good faith.

A maximum penalty reduction of 60 percent is permitted for small businesses. Size of business is measured on the basis of the maximum number of employees of an employer at all workplaces at any one time during the previous twelve months. The rates of reduction to be applied are as follows:

Employees	Percent reduction
1-25	60
26-100	40
101-250	20
251 or more	0

When a small business (1-25 employees) has one or more serious violations of high gravity or a number of serious violations of moderate gravity, the inspecting officer may recommend to the area director that only a partial reduction in penalty be given for size of business because the indication is that there is a lack of concern for employee safety and health.

The 25 percent credit for good faith normally requires a written safety and health program. In exceptional cases, the compliance officer may recommend the full reduction for a smaller employer (1-25 employees) who has implemented an efficient safety and health program, but who has not committed it to writing. An effective program provides for appropriate management commitment and employee involvement; worksite analysis for the purpose of hazard identification; hazard prevention and control measures; and safety and health training. A reduction of 15 percent is normally given if the employer has a documentable and effective safety and health program, but with more than only incidental deficiencies. No reduction is given to an employer who has no safety and health program or where a willful violation is found. Only these percentages (15% or 25%) are used to reduce penalties due to the employer's good faith. No intermediate percentages can be used.

A reduction of 10 percent can be given to employers who have not been cited by OSHA for any serious, willful, or repeated violations in the past three years.

Assessment of Penalties. The gravity-based penalty (GBP) is an unadjusted penalty and is calculated in accordance with strict procedures. The GBP for each violation is first determined based on judgment, combining the severity assessment and the final probability assessment.

For serious violations, the GBP is assigned on the basis of the following scale:

Severity	Probability	GBP	Gravity
High	Greater	$5,000	High
Medium	Greater	$3,500	Moderate
Low	Greater	$2,500	Moderate
High	Lesser	$2,500	Moderate
Medium	Lesser	$2,000	Moderate
Low	Lesser	$1,500	Low

The gravity of a violation is defined by the GBP. A high gravity violation is one with a GBP of $5,000 or greater. A moderate gravity violation is one with GBP of $2,000 to $3,500. A low gravity violation is one with a GBP of $1,500. The highest gravity classification (high severity and greater probability) is normally reserved for the most serious conditions. Such conditions would include those involving danger of death or extremely serious injury or illness. If the area director desires to achieve a deterrent effect, a GBP of $7,000 may be proposed. The reason for this determination is documented in the case file.

For other-than-serious safety and health violations, no severity assessment is made. The area director may authorize a penalty between $1,000 and $7,000 for an other-than-serious violation when it is determined to be appropriate to achieve the necessary deterrent effect. The reason for such a determination is documented in the case file.

Probability	GBP
Greater	$1,000 - $7,000
Lesser	$0

OSHA has the liberty to assign GBP in some cases without using the severity and the probability assessment procedures outlined in this section when these procedures cannot appropriately be used.

Grouped Violations. Combined or grouped violations are normally considered as one violation and assessed one GBP. The severity and the probability assessments for combined violations are based on the instance with the highest gravity. OSHA does not necessarily complete the penalty calculations for each instance of a combined or grouped violation if it is clear which instance will have the highest gravity.

For grouped violations, two considerations are kept in mind when calculating the severity. The severity assigned to the grouped violation is not less than the severity of the most serious reasonably predictable injury or illness that could result from the violation of any single item. If a more serious injury or illness is reasonably predictable from the grouped items than from any single violation item, the more serious injury or illness serves as the basis for the calculation of the severity factor. Two considerations are also kept in mind when calculating the probability of grouped violations. The probability assigned to the grouped violation is never less than the probability of the item that is most likely to result in an injury or illness. If the overall probability of injury or illness is greater with the grouped violation than with any single violation item, the greater probability of injury or illness serves as the basis for the calculation of the probability assessment.

Penalty Enhancement. The area director can order that the calculated penalty be increased for certain reasons. The two factors considered are *egregious acts or repetition of the violation.*

Egregious policy. In egregious cases, an additional factor of up to the number of violation instances may be applied by OSHA in calculating penalties. Penalties calculated with this additional factor proposed with the concurrence of the Assistant Secretary.

Repeated violations. Section 17(a) of the Act provides that an employer who repeatedly violates the Act may be assessed a civil penalty of not more than $70,000 for each violation. Each repeated violation is classified as serious or other-than-serious. A GBP is then calculated for repeated violations based on facts noted during the current inspection. Only the adjustment factor for size, appropriate to the facts at the time of the reinspection, can be applied.

For employers with two hundred fifty or fewer employees, the GBP is doubled for the first repeated violation and quintupled if the violation has been cited twice before. If the Area Director determines that it is appropriate to achieve the necessary deterrent effect, the GBP may be multiplied by ten.

For employers with more than two hundred fifty employees, the GBP is multiplied by five for the first repeated violation and multiplied by ten for the second repeated violation.

For other-than-serious violations where no initial penalty was assessed, a GBP penalty of $200 is assessed for the first repeated violation, $500 if the violation has been cited twice before, and $1,000 for a third repetition.

Section 17(a) of the Act provides that an employer who willfully violates the Act may be assessed a civil penalty of not more than $70,000 but not less than $5,000 for each violation. OSHA may adjust the penalty for size of employer using one half the credit listed above. A 10 percent credit may be given for a good history in the past three years. No credit may be given for good faith, which is inconsistent with a willful violation.

	Penalties to be assessed				
Total percent reduction	0%	10%	20%	30%	40%
High Gravity	$70,000	$63,000	$56,000	$49,000	$42,000
Moderate Gravity	$55,000	$49,500	$44,000	$38,500	$33,000
Low Gravity	$40,000	$36,000	$32,000	$28,000	$25,000

Criminal Penalties

Criminal penalties are levied for specific violations of OSHAct and also for general offenses. Typically, a criminal offense has an element of intent, which must be proved by the prosecution. However, I do not wish to give you legal advice, so consult with an attorney if even a hint of criminal charges is made.

Specific OSHAct Offenses. If a willful violation of a standard has resulted in the death of an employee and the employer is convicted, the criminal offense is punishable by a court-imposed fine of up to $250,000 for an individual and $500,000 for a corporation. Also, the individual responsible and the plant manager (who is *always* responsible) may be imprisoned for up to six months. Both imprisonment and a fine may be imposed simultaneously.

General Federal Offenses. Assaulting a compliance officer, or otherwise resisting, opposing, intimidating, or interfering with a compliance officer in the performance of his or her duties is a criminal offense. If convicted, the accused is subject to a fine of not more than $5,000 and imprisonment for not more than three years.

Falsifying records, reports, or applications, upon conviction, can bring a fine of $10,000 or up to six months in jail, or both.

Occupational Safety and Health Review Commission

The Occupational Safety and Health Review Commission (OSHRC) is an independent federal agency not associated with OSHA or the Department of Labor.

Independent Adjudicatory Agency

The Occupational Safety and Health Review Commission is established by Section 12(a) of OSHAct. The principal office of the Commission presides in the District of Columbia. However, the Commission holds hearings and conducts other proceedings throughout the country.

Organization

The Commission is composed of three members who are appointed by the President, by and with the advice and consent of the Senate, from among persons who by reason of training, education, or experience are qualified to carry out the functions of the Commission. The President designates one of the members of the Commission to serve as Chairman.

The term of a member of the Commission is six years, except that when a vacancy is caused by the death, resignation, or removal of a member prior to the expiration of the appointed term that position can be filled only for the remainder of the unexpired term. A member of the Commission may be removed by the President for inefficiency, neglect of duty, or malfeasance in office.

Pre-Hearing Procedures

The Chairman appoints administrative law judges and other employees as deemed necessary to assist in the performance of the Commission's functions.

Documentation Discovery

Any person may be compelled to produce books, papers, or documents to the Commission.

Depositions

The Commission may order testimony to be taken by deposition in any proceedings pending before it at any state of such proceeding. Any person may be compelled to appear and depose, and produce documentary evidence before the Commission. Witnesses whose depositions are taken, and the persons taking such depositions, are entitled to the same fees as are paid for like services in the courts of the United States.

Commission Hearings

The OSHRC receives Notices of Contest forwarded by OSHA area directors and assigns them to administrative law judges. If the judge schedules a hearing, both the employer and employees have the right to attend the hearing and participate in it. The OSHRC does not require that they be represented by attorneys, but by attorneys who are familiar with labor laws and OSHAct are recommended.

An administrative law judge may hear and make a determination upon any proceeding instituted before the Commission and any motion in connection therewith. He makes a report of any determination that constitutes a final disposition of the proceedings. The report of the administrative law judge becomes the final order of the

Commission within thirty days after such report by the administrative law judge, unless within such period any Commission member directs that such report be reviewed by the Commission.

Administrative Law Judge

The judge may disallow the contest if it is found to be legally invalid. Otherwise, a hearing may be scheduled for a public place near the employer's workplace.

Evidence

The evidence is limited to the OSHA case file, the employer's rebuttal at earlier hearings, and evidence taken during discovery and depositions.

Post-Hearing Procedures

Once the administrative law judge has ruled, any party to the case may request a further review by OSHRC. Any of the three OSHRC commissioners may bring a motion to have the Commission review a case. Commission rulings may be appealed to the U.S. Court of Appeals having jurisdiction.

Appeals in states with state plans are generally heard by a state review board or equivalent authority.

Defenses

An affirmative defense is any matter which, if established by the employer, will excuse the employer from a violation that has otherwise been proved by the compliance officer. The burden of proof is on the employer. Although affirmative defenses must be proved by the employer, OSHA is prepared to respond whenever the employer is likely to raise or actually does raise an argument supporting such a defense. The compliance officer considers the potential affirmative defenses that the employer may make and attempts to gather contrary evidence when a statement made during the inspection fairly raises a defense.

Invalidity of the Standard

The employer must prove that the standard cited is invalid. In other words, it is somehow defective as promulgated and cannot be applied. Despite what you may think about OSHA standards, I wish you luck if you try this defense.

Preemption of Cited Standard

The employer must prove that the cited standard is preempted by another standard, which he or she is in compliance with.

Preemption of General Duty Clause

The defendant must prove that a specific standard, which is complied with, preempts the citation of the General Duty Clause in the alleged violation. In other words, a violation does not really exist, because the employer is complying with some requirement that supersedes the General Duty Clause.

Preemption of the Act

In this instance, the employer must show that some other federal law preempts the coverage of the OSHAct. This defense is only successful where the other law is strictly complied with.

Multi-Employer Workplace

In a multi-employer workplace, if the cited employer had no control over the alleged violation, this defense can be used.

Infeasibility

This defense is used where compliance with the requirements of a standard is either functionally impossible or would prevent performances of required work. It can only be used successfully if no alternative means of employee protection is available.

Greater Hazard Caused by Strict Compliance

If compliance with a standard would result in greater hazards to employees than noncompliance, the employer has an affirmative defense. However, two conditions are placed on this. First, no alternative means of employee protection can be available. If any are available, the question will be, "Why did you not use them in a good faith attempt to comply?" Secondly, an application of a variance would have to be inappropriate for the same reason of good faith effort before the citation.

Unforeseeable Employee Misconduct

If unforeseeable or unpreventable employee misconduct is claimed, the violative condition must be unknown to the employer and in violation of an adequate work rule, which was effectively communicated and uniformly enforced.

Reasonable Promptness

Section 9(a) of the Act requires OSHA to issue a citation to the employer with reasonable promptness. To facilitate the prompt issuance of citations, the area director may issue citations that are unrelated to health inspection air sampling, prior to receipt of sampling results.

Section 9(c) states that no citation may be issued after six months following the occurrence of a violation. Accordingly, a citation can not be issued where any violation alleged therein last occurred six months or more prior to the date on which the citation is actually signed and dated. However, where the actions or omissions of the employer concealed the existence of the violation, the time limitation is suspended until such time that OSHA learns or could have learned of the violation.

Lack of Particularity

Section 9(a) of OSHAct requires that each citation be in writing. It must, with particularity, describe the nature of the violation, including a reference to the provision of the Act, standard, rule, regulation, or order alleged to have been violated. OSHA cannot simply cite you by saying, "We think that some hazard exists here, but we're not sure just what it is." OSHA must quote chapter and verse and show how you did not comply with it.

Statute of Limitations

In criminal cases, the usual statute of limitations applies. This time starts with the discovery of the crime, as I understand it. You may wish to consult with your attorney.

Federal Courts and OSHAct

For criminal proceedings and the issuance of warrants and subpoenas, OSHA turns to the federal courts.

U.S. District Courts

OSHA may require the attendance and testimony of witnesses and the production of evidence under oath in the U.S. District Court. In case of a contumacy, failure, or refusal of any person to obey such an order, the district court may issue an order requiring the person to appear to produce evidence and give testimony. Any failure to obey the court order may be punished as contempt of court.

The U.S. District Court is also where discrimination hearings are held.

Upon the filing of imminent danger petitions, the district court has jurisdiction to grant such injunctive relief or temporary restraining order pending the outcome of an enforcement proceeding. The proceeding is held in accordance with federal civil procedures. However, no temporary restraining order issued without notice shall be effective for a period longer than five days.

Civil penalties owed to OSHA are paid to the Secretary for deposit into the Treasury of the United States and accrued to the United States. Unpaid penalties may be recovered in a civil action in the name of the United States brought in the U.S. district court for the district where the violation is alleged to have occurred or where the employer has his principal office.

U.S. Courts of Appeal

No decision on a permanent OSHA standard is reached without due consideration of the arguments and information received from the public in written submissions and hearings. Any person who may be adversely affected by a final or emergency standard may, nevertheless, file a petition for judicial review of the standard with the U.S. Court of Appeals for the circuit in which the objector lives or has his or her principle place of business. This petition must be filed within sixty days of the rule being promulgated. Filing an appeal does not delay the enforcement of a standard unless the Court of Appeals having jurisdiction specifically orders a delay, in which case the order holds for all jurisdictions pending the appellate ruling on the matter.

Anyone adversely affected or aggrieved by an order of the OSH Review Commission may obtain a review of such order in any U.S. court of appeals for the circuit in which the violation is alleged to have occurred or where the employer's principal office is located. Alternatively, the appeal may be filed in the Court of Appeals for the District of Columbia Circuit. The filing must occur within sixty days following the issuance of the Commission order. The appeal is in the form of a written petition praying that the order be modified or set aside. A copy of the petition is transmitted by the clerk of the court to the Commission and to the other involved parties. Thereupon the Commission files in the court the record in the proceeding. Upon such filing, the court has jurisdiction of the proceeding and of the question determined therein. It also has the power to grant such temporary relief or restraining order as it deems just and proper. It may make and enter upon the pleadings, testimony, and proceedings set forth in the record a decree affirming, modifying, or setting aside in whole or in part, the order of the Commission and enforcing the same to the extent that such order is affirmed or modified.

The commencement of appeal proceedings, unless ordered by the court, does not operate as a stay of the order of the Commission. No objection that has not been urged before the Commission can be considered by the court, unless the failure or neglect to urge such objection is excused because of extraordinary circumstances. The findings of the Commission with respect to questions of fact, if supported by substantial evidence on the record considered as a whole, is held conclusive.

If any party applies to the court for leave to adduce additional evidence and shows to the satisfaction of the court that such additional evidence is material and that reasonable grounds exist for the failure to adduce such evidence in the hearing before the Commission, the court may order such additional evidence to be taken before the Commission and to be made a part of the record.

The Commission may modify its findings as to the facts, or make new findings, by reason of any additional evidence. The Commission then files modified or new findings, with respect to questions of fact. If the new findings are supported by substantial evidence now on the record and are conclusive, its recommendations for the modification or setting aside of its original order are considered by the court. Upon the filing of the record with it, the jurisdiction of the court is exclusive. Its judgment and decree is final, except that it is subject to review by the Supreme Court of the United States.

Conclusion

Unfortunately, the strategy of making people do the right thing by instilling fear of sanctions is chock full of limitations. Eckenfelder gives the litany. OSHA has a had a string of frustrated administrators who have been hampered in achieving laudable goals. Nevertheless, the system is based on big government, as Eckenfelder points out, and government has rarely demonstrated the efficiency and effectiveness of the private sector in solving problems. Besides, rightly or wrongly, Americans distrust government. The nation was born that way and apparently does not like government any more today in the waning twilight of the twentieth century than it did in the eighteenth century.

The OSHA compliance hammer has little impact on workplace culture, which motivates management and labor equally to either cooperate or fight each other. The regulatory solution uses the wrong driver—compliance statistics. The whole foundation of safety through regulatory compliance is based on fear.

The foregoing discussion notwithstanding, now is not the time to dismantle OSHA. Although I agree with Eckenfelder and others that OSHA's track record at accident rate reduction is dismal, I do not agree that it is inconclusive or negative. Without OSHA, I feel confident that the rate of workplace injuries, illnesses, and fatalities would be the same or worse than it was in 1970 on the eve of the OSHA era.

In my three decades of experience, which began at the birth of OSHA, I have been exposed to too many superiors who wanted to evade compliance by all sorts of criminal schemes. How can we preach cooperation, culture sensitivity, and trust when so many of these type managers are still around? The sad truth is, we cannot.

References

Access to Medical and Exposure Records. U.S. Department of Labor. OSHA Publication 3110. U.S. Government Printing Office, 1989.

All about OSHA. U.S. Department of Labor. OSHA Publication 2056. U.S. Government Printing Office, 1995.

Employer Rights and Responsibilities Following an OSHA Inspection. OSHA Publication 3000. U.S. Government Printing Office, 1996.

Porpora, Charles. "What to Do When OSHA Knocks at Your Door." *Engineer's Digest.* January 1998, pp. 68-72.

Roughton, James E. "OSHA Recordkeeping Revision: Is It a Better Method?" *Professional Safety.* May 1997, pp. 38-43.

12

Hazardous Chemical Exposure Management

The management of exposure to hazardous chemicals has become foremost in the minds of many people in the twilight decade of the twentieth century. Perhaps this may be interpreted as a natural result of the chemophobia written and talked about so much in the 1970s and 1980s. However, as a practicing industrial hygienist, let me assure you that chemical exposure is as old as recorded history. Suffice it to say that without the impetus of OSHA regulations, chemicals would less often be handled safely in the workplace than they are. Today much more emphasis is placed on indoor air quality and quality of work life than even fifteen years ago.

Chemical Hazard Exposure

Chemical contamination of our immediate environment—the workplace—happens in many and varied ways. Liquids evaporate at finite rates. Gases escape through valves and fittings. Solid particles divide and are thrown into the air by rotating parts of machines.

In one case where chemophobia gripped the workplace, a newly assigned safety professional set up a picture-based sign system to simplify identification. Unfortunately, management of chemical safety is not that simple, nor does it have to be complex and frustrating, as is more often the case.

One exposure that is often overlooked in industrial hygiene investigations is work surfaces. Burke mentions doorknobs, table tops, and toilet seats. Control panels and workbench surfaces are also potentially contaminated surfaces. These are called *secondary contamination sources* because they are spread from primary sources by poor hygiene and housekeeping practices. Detection of surface contamination relies on wipe samples.

The best way to handle hazardous chemicals is to substitute less hazardous materials where feasible. The best way to do this is to stop the hazardous stuff right at the door! In other words, don't let it into your plant in the first place. I know this is not always feasible and thus the remainder of this chapter will address other solutions. But, the point needs to be made that preventing a hazardous material from ever coming aboard is far superior to even the best-engineered systems and the best-written administrative procedures.

Certain chemical acceptance issues are tougher to deal with than others. Figure 12-1 illustrates some of the tougher issues.

We've taken a little aside here. Now let's examine chemicals and how they affect us, starting with the most basic effect.

Chemical Burn Management

Most people readily claim to have a healthy respect for corrosive chemicals, yet will turn around and mishandle a lead-acid battery. Corrosives, irritants, sensitizers, and other skin damagers are probably more common in the workplace than inhalation hazards.

Dermatitis, while less dramatic than an acid burn, is a serious and debilitating disease. In some cases, chemical dermatitis can be a permanent effect. Typical symptoms include cracked, swollen, and bleeding skin. A bad case of dermatitis is not unlike severe dishpan hands. When the chemical defats the skin, cracking and bleeding occur. Two types of chemical dermatitis are contact irritant dermatitis and allergic dermatitis.

Irritants are similar to corrosive chemicals except that their effect on the skin is less dramatic. A chemical could be both corrosive and an irritant—strong and dilute acid, for instance. Contact irritant dermatitis typically follows skin exposure to solvents and cutting fluids. Figure 12-2 lists some skin irritants.

Allergic dermatitis stems from second or subsequent contact with a chemical that triggered an allergic reaction on first contact.

A skin notation with its PEL is the only warning OSHA gives of a chemical that attacks or absorbs through the skin. Skin notations are rare, though, and do not suggest exposure limits for skin attacking chemicals. For corrosives the problem is simple: no exposure. For irritants and absorbent toxic chemicals, the problem is undefined at present.

The problem with potential dermal exposures is that exposure assessment is limited by a lack of accurate sampling methods. Surrogate skin sampling is performed by patch testing or analyzing contamination on garments. Removing chemicals from the skin by washing or wiping has also

Figure 12-1
Tough Chemical Issues

- Fear of cancer
- Protective measures slow work process, but production goals remain the same
- Statements in a MSDS or on a label seem to call for more sophisticated protection than management seems willing to provide
- Management has never monitored the air for exposure level or has not communicated the results effectively

Figure 12-2
Common Irritants

Acrolein
Allyl
Alcohol
Allyl Glycidyl Ether
Ammonia
Ammonium Chloride
n-Butyl Acetate
Caprolactam
Chlorine
Chloroacetyl Chloride
o-Chlorobenzylidene Malononitrile
Diethylamine
Diethylenetriamine
Ethyl Benzene
Glutaraldehyde
2-Hydroxypropyl Acrylate
Methyl-2-cyanoacrylate
Phosphoric Acid

Figure 12-2 *(continued)*

Potassium hydroxide
Sodium Bisulfite
Sodium Hydroxide
Thioglycolic Acid
1,2,4-Trichlorobenzene
Triethylamine
Tetrasodium Borate Salts

After Ness

been practiced. The accuracy of these methods is limited, however, and does not give real time results.

In areas where corrosive or irritant chemicals are handled, check work surfaces, tools, process equipment, and storage and transfer areas for the presence of contamination on surfaces that might be transferred to skin. The usual way of making this check is to conduct a wipe test and having the wipe patch analyzed for the contaminant of concern. No magic limit exists, so mere presence of the contaminant indicates potential for skin exposure.

If potential for exposure is verified in the work area, spread the investigation out to see how far the contaminant has been transported by feet, clothing, and portable containers. Decontaminate surfaces as feasible. Then place warning signs at the sources to indicate the potential for contamination upon contact and the need for personal protective equipment.

Train workers on how to don and doff protective clothing and equipment without contaminating themselves. Remember, no glove or protective suit is impervious. Some amount of chemical will eventually permeate any artificial barrier. Manufacturers supply breakthrough tables showing the time it takes specific chemicals to permeate their gloves or suits. Once the protective articles have been in contact with a contaminant for the length of its breakthrough time, a new glove or suit is needed. Select personal protective equipment that has a longer breakthrough time than your worker safely needs or else provide plenty of protective clothing changes.

Understanding pH

Acidity is measured as the concentration of hydrogen ions [H^+] in an aqueous solution. Since the measurement of [H^+] in terms of molar concentration is a cumbersome process, scientists express it as its negative logarithm, or pH.

$$pH = -\log[H^+] \qquad [12.1]$$

which may also be written as

$$pH = \log\frac{1}{[H^+]} \qquad [12.2]$$

The potential of any aqueous solution to be corrosive depends on its pH value. Normally a range represents the pH scale from 0 to 14. Although true pH can range outside these values, we need not concern ourselves with that technicality here. We will adopt the same definition that the U.S. EPA has for a corrosive material: if the pH is less than 2.5 or greater than 12.5, the material is corrosive. Figure 12-3 lists the pH of some common liquids.

Acid Burns

If strong enough, acids are capable of destroying living tissue. Most inorganic and organic acids that have a pH greater than about 2.5 are primary irritants. These irritants cause irritant contact dermatitis.

Figure 12-3
pH of Some Common Liquids

Lemon Juice	2.4
Orange Juice	3.5
Pure Water	7.0
Blood, Tears	7.4
Milk of Magnesia	10.6
Ammonia	11.5

Acid burns do their damage immediately. After you stop the assault of the acid on the flesh, you can see the damage it has done and that is the extent of it. As acid attacks the flesh, a protein is precipitated that acts as a barrier to further damage.

Alkaline Burns

Strong alkalis, or basic liquids, dissolve fat. Thus, other chemical agents are able to more easily penetrate the protective barrier of the skin. The following inorganic alkalis are skin irritants: ammonium hydroxide, calcium chloride (ice salt), sodium carbonate (soda ash), and sodium hydroxide (caustic soda). The organic alkalis, especially the amines, are irritants. So are metallic salts. The more severely irritating of the latter are arsenicals, chromates, mercurials, nickel sulfate, and zinc chloride.

Actually caustics are more injurious than acids. What may not look too bad after one day may turn into a nasty condition later. The alkali keeps on destroying tissue even after you think you have stopped its progress.

Neutralization

When a strong acid or base contacts a person's skin, do not apply the opposite strong base or acid to neutralize the corrosive material. That is the way you neutralize these materials in laboratory glassware or even in process vessels, but not on human skin. I heard of an overly excited supervisor who once saw an employee doused in sulfuric acid. He ran to a nearby caustic soda tank and grabbed a hose used to fill small containers with 50% caustic soda. He used the hose to douse the poor victim with caustic in an attempt to neutralize the acid. The resulting neutralization caused thermal burns in addition to the acid burns and the caustic stripped the victim of underlying fatty tissue. He would have fared better had the supervisor done nothing to help.

The correct response is to douse the person with fresh water—lots of fresh water. At sulfuric acid plants you will find the typical eye wash stations and emergency showers. More to the point of a sulfuric acid dousing, you will find pools of water in little sheds every few hundred feet. With a strong acid eating away at flesh, you do not want to use a neutralization reaction with a strong caustic going on at the same time.

Emergency Eyewashes, Showers, Baths

Emergency deluge showers capable of thirty gallons per minute or more or immersion baths should be located no more than ten to fifteen feet from strong acids or bases. (See Chapter 14 on eyewashes.) In no case should the shower or bath be further than one hundred feet from the potentially corrosive, irritant, or skin absorbing chemical.

In applying skin ointments, barrier creams, and soaps, whether for an emergency dermal contact or routine hygiene, keep in mind the three primary reasons for these skin care products. The first reason for applying a skin care product should be for cleanliness, whether you must remove dirt, microorganisms, or toxic chemicals. Harsh chemicals such as turpentine or acetone or alcohol should not be used to cleanse the skin. For best results, skin care experts recommend washing with specially designed cleansers and lukewarm water as soon as possible after the exposure. The second reason you may wish to use a skin care product, probably not the same one you cleaned with, is to restore the natural barrier that is the skin system by moisturizing the skin. Typically, soaps and detergents dry the skin out, remove natural oils, and, if strong enough, defat the skin. Good skin restoring products contain emollients, aloe, glycerin, and the gentlest cleansers feasible. The products are best used before and after the exposure. Damaged skin makes it easier for microorganisms and chemicals to enter through the skin barrier and get into the blood stream and other bodily systems. Finally, to help with this latter problem, products are available that serve themselves as a barrier against microorganisms or hazardous chemicals. Engineering controls, gloves and other protective clothing, and safe work practices should be insisted on to prevent dermal exposure, but barrier creams are a safety backup. In some cases—for instance where skin damage is already in progress or still healing—barrier creams are definitely advised.

PPE for Burn Protection

Gloves for chemical burn protection are typically made of rubber or plastic. Figure 12-4 gives some glove recommendations. Always check the manufacturer's literature for compatibility tables. While providing protection from acids, alkalis, organics solvents, and other fluids that may burn the skin, gloves are not totally impervious to these materials. Therefore, it is

Figure 12-4
Gloves for Chemical Burn/Irritation Protection

Natural Rubber for alcohols, dilute aqueous solutions, and bases.
Neoprene for oxidizing acids.
Polyvinyl chloride (PVC) for strong acids and bases, alcohols, less dilute aqueous solutions, and salts of acids and bases.

essential to choose the right type of glove for the task at hand. Manufacturers of gloves generally provide matrices showing suitable applications for each type of glove they offer. The manufacturer's literature may also list the breakthrough time for each chemical substance. This time is important. For instance, if the breakthrough is listed as two hours, then every two hours your employees need to dispose of their gloves and wear new ones. Usually, however, breakthrough will allow a pair of gloves to be worn for one or more shifts before replacement becomes necessary. Employees should always check gloves for cracks and holes before donning them. If damage is found, the gloves should be disposed of, even if they are brand new.

An apron should be worn to protect the body from splashes of chemicals. Regular safety glasses are not suitable for handling hazardous chemicals—therefore, goggles are required to protect the eyes. A face shield is required when transferring most corrosive chemicals. It is even a good practice to wear the goggles under the face shield. A chemical-resistant hood may be worn over the head and shoulders in case of extreme danger from a chemical. For toxic chemicals, a chemical splash suit with self-contained breathing apparatus gives complete protection against spills, splashes, and vapors.

A key factor in PPE selection is permeation. Use the manufacturer's published permeation times, but in some cases have a laboratory conduct permeation tests for you. Situations requiring this are the use of a chemical the manufacturer has not tested or breakthroughs have been experienced at far less than the stated time of permeation. The key questions to ask are:

- How long will the employee be exposed to the chemical?
- How much skin contact can be tolerated?

Protective clothing does not permit evaporation of sweat. Therefore, body heat builds up while wearing protective clothing. Longer and more frequent rest breaks may be required in order to prevent heat exhaustion. Give the worker plenty of electrolyte-enhanced fluids to drink.

In Figure 12-5, Johnson gives us some questions to consider when selecting protective clothing for burn prevention or exposure to toxic substances. Figure 12-6 lists the levels of protection.

HAZARD COMMUNICATION

Theoretically, the OSHA Hazard Communication Standard (29 CFR 1910.1200) helps reduce the incidence of illnesses and injuries caused by

Figure 12-5
Assessing Need for Protective Apparel

1. What hazardous materials or conditions might the employees be exposed to while performing their usual duties?
2. What hazardous materials exposure could reasonably be expected on site?
3. What information is available about the components or ingredients in the hazardous substances?
4. How quickly can information be communicated to exposed personnel?
5. What concentrations of pollutants can be expected on site?
6. What level of danger is anticipated for the exposed workers?
7. What level of protection is available to workers?
8. What level of protection is necessary?
9. Is the protective apparel compatible with the materials being used?
10. How many hours at a time will the workers wear their protective clothing?

Figure 12-6
Levels of Protection

Level A provides the highest level of protection and includes positive-pressure self-contained breathing apparatus (SCBA), chemical vapor protective clothing, chemical resistant inner gloves, chemical resistant outer gloves, chemical resistant boots with steel toe and shank, two-way radio communications, optional hard hat under suit, optional long underwear, and optional coveralls all under the protective suit.

Level B provides the next level of protection where the highest level of respiratory protection is needed but less skin protection is required. An SCBA, chemical splash clothing (coveralls, overalls, and long-sleeved jacket or disposable coveralls), chemical resistant inner gloves, chemical resistant outer gloves, chemical resistant outer boots with steel toe and shank, two-way radio communications. Optional coveralls under the suit and hard hat.

Level C provides minimum respiratory protection and skin protection. Used where skin absorption hazards do not exist and includes use of a full-face, air-purifying respirator, optional escape mask, chemical splash protective clothing such as a one-piece coverall or hooded two-piece suit, chemical resistant outer gloves, optional inner chemical resistant gloves, chemical resistant boots with steel toe and shank, optional cloth coveralls inside chemical resistant protective clothing, optional two-way communications, optional hard hat.

Level D is common work clothes. No specific respiratory or skin protection. Should be worn only when no respiratory or skin hazards are present.

chemical hazards in the workplace by informing employees of the nature and effect of the hazardous materials they work with. In the author's experience, however, employees haven't a clue as to what the information means and, consequently, they largely ignore it. In the twelve years that the standard has been in effect, and at a plant with eight hundred employees, fewer than ten persons have ever requested a Material Safety Data Sheet (MSDS) from me outside of training sessions. That amounts to less than one request per year. Many attorneys geared up for mass litigation in 1986 when the standard took effect, yet where are all the law suits? They never happened. We'll discuss this apparent employee apathy and its causes later.

At any rate, the standard requires the development of Material Safety Data Sheets (MSDSs) and their communication to all employees exposed to chemical hazards. An MSDS describes the physical and chemical properties of products, health hazards, routes of exposure, precautions for safe handling and use, emergency and first aid procedures, reactivity data, and control measures. Information on an MSDS aids in the selection of safe products and their safe handling and use, and helps employees to respond effectively to emergency situations.

At least that is the theory. It does not happen as often in practice. MSDSs are written so vaguely—partly because the manufacturer of the product cannot possibly know of all the ways in which the product will be used or abused—that the information therein is of little practical use to the average employee. Also, manufacturers play games with the wording, partly out of fear of litigation. For instance, a common claim found in the section where ingredients should be listed is "This product contains no hazardous chemicals as defined by OSHA." First of all, the OSHA definition of a hazardous chemical practically includes water. Secondly, such MSDSs frequently contain a whole page of information in the Emergency/First Aid section and Health Effects section. Another example is use of the phrase "use adequate ventilation" in the Use and Handling section. These examples are merely the tip of the iceberg. Do you wonder why employees largely ignore MSDSs? Finally, what does the average employee understand about the significance of vapor pressure, boiling point, and evaporation rate? How many understand the effect of partial pressure of each ingredient on the evaporation of the whole mixture? How many understand diffusion or other dispersion mechanisms? How many understand that cancer is not a single disease but a whole slew of them? In my experience, very few indeed—and good luck trying to teach them these physical, chemical, and medical principles!

Labels are an important factor in hazard communication. Though mostly forgotten, they are equally important with MSDSs in effective hazard communication. What OSHA, EPA, and you and I call a "label," DOT calls a "marking." Usually it is a piece of paper that is pasted onto the container and which contains this information:

- Name, address, and telephone number of either the manufacturer or distributor of the product contained.
- The chemical names of specific hazardous chemicals contained and their specific dangers.

- Work practices and PPE required to safely handle the material contained.
- Recommended first aid procedures for acute exposure.
- Contingency recommendations for spills and other releases to the environment.

DOT labels are the diamond-shaped stickers placed on containers that match the placards on the truck or rail car. Neither marking nor label should be removed or defaced or damaged until the container has been:

- Cleaned of residue
- Purged of vapors
- Potential hazards have been removed if the first two actions are inadequate

Toxic Chemical Management

Handling of toxic chemicals means some level of exposure will occur. The best we can do with a toxic chemical in the workplace is to follow certain safe practices in order to minimize exposure. After that (assuming risk is still too high), ventilation and personal protective equipment integrated with air monitoring and medical surveillance programs are called for. But there is one thing, if practical, that we may be able to do at the outset that has more value than all these measures—and that is substituting safer materials. By all means, elimination of the toxic species by substituting a safer material is always to be preferred over all other options.

Candidates for substitution include surface primers, surface coatings, adhesives, solvents, and cleaning compounds such as degreasers and parts cleaners. When selecting replacement materials or evaluating new materials for use in your plant, consider the impact the materials will have on:

- Employee health
- Training
- Slips and falls
- Fire and explosion
- Air emissions
- Wastewater
- Hazardous waste

Examine all these potential impacts or you may find yourself in a pickle. When investigating the potential impact on employee health, do not rely solely on the Material Safety Data Sheet (MSDS). One plant did so and bragged to its sister plants that it had replaced a degreasing compound that is a nephrotoxin (kidney-attacker) and hepatotoxin (liver-attacker) with a safer compound. However, an in-depth toxicological study on the new compound revealed that it was a mutagen—a substance that may damage genetic codes in employees of child-bearing age. Which would you prefer?

Table 12-1 lists some chemicals that cause occupational liver disease.

Table 12-1
Some Causes of Occupational Liver Disease

Disease	General Agent	Specific Example
ACUTE HEPATITIS		
Acute Toxic Hepatitis	Chlorinated Hydrocarbons	Carbon tetrachloride Chloroform Dinitrophenol
	Nitroaromatics	Dinitrobenzene Dioxin
	Ether	Polychlorinated
	Halogenated aromatics	biphenyl DDT Chlordecone Chloribenzenes Halothane
Acute Cholestatic Hepatitis	Epoxy resin	Methylenedianiline
	Inorganic element	Yellow phosphorus
Subacute Hepatic Necrosis	Nitroaromatic	TNT
CHRONIC LIVER DISEASE		
Fibrosis/Cirrhosis	Alcohol Inorganic element Haloalkene	Ethyl alcohol Arsenic Vinyl chloride
Angiosarcoma	Haloalkene	Vinyl chloride (rubber worker disease)
Biliary Tree Carcinoma	Unknown agents	

Table 12-2
Some Occupational Allergens

General Agent	Specific Examples
Anhydrides	Hexahydrophthalic Anhydride (HHPA) Himic Anhydride (HA) Phthalic Anhydride (PA) Tetradichlorophthalic Anhydride (TCPA) Trimellitic Anhydride (TMA)
Isocyanates	Hexamethylene Diisocyanate (HDI) Methylene Diphenyl Diisocyanate (MDI) Paratolyl Diisocyanate (PTI) Toluene Diisocyanate (TDI)
Metals	Chromium Nickel Platinum Vanadium
Miscellaneous	Azodicarbonamide Sulfonechloramide

Table 12-2 lists some occupational allergens.

When selecting new materials you must abide by some ground rules as well as making the determination discussed above. Alternative materials must perform the task at least as well as the existing material. Given that safe materials are generally not as widely available, as affordable, or as useful as their marketing representatives tout, the alternative material must often be able to outperform the existing material. Normally, safer materials cost more than the materials they are meant to replace. That puts even more demand on the alternative material's performance in order to save enough money to pay for itself. No matter how hard you work, some applications may demand materials that have no practical substitute. After nine years of testing alternative materials, one aerospace company has been able to find substitute materials for 65 percent of its parts cleaning applications, but a hazardous vapor degreaser remains the material of choice for the other 35

Figure 12-7
Known Reproductive Hazards

Benzene—chromosome damage, fetotoxicity, growth retaerdation.
1,2,-Dibromo-3-Chloropane (DBCP)—male sterility.
2-Ethoxyethanol—fetotoxicity, congenital malformations, male sterility.
2-Ethoxyethanol Acetate—spermatoxicity, teratogenesis.
2-Methoxyethanol—teratogenesis, testicular atrophy, female infertility.
2-Methoxyethanol Acetate—testicular atrophy, hypospadia.

percent of cleaning tasks. Unfortunately, this hazardous solvent replaced a safer one that was a stratospheric ozone-depleting chemical.

Not much has been done to study the effects of chemicals on fertility, menstruation, or the reproductive process. The little data that is available has come from animal testing, which makes it difficult to apply to humans due to metabolic and pharmacokinetic differences.

Female exposure to reproductive hazards is time dependent, since the egg is carried from birth. The only critical male exposure is prior to conception, since sperm is produced periodically. Figure 12-7 lists some of the known reproductive hazards.

A mutagenic effect occurs when the parent genetic material is altered. When this occurs, typically, a permanent heritable change occurs, which shows up in future offspring. Children conceived before the damaging exposure are unaffected. Mutagens alter either somatic or germinal cells. Mutations in the somatic cells are not inherited, but may be implicated in the etiology of cancer. Mutations in the germ cells can be inherited. Hall refers to experimental evidence that suggests that most chemical carcinogens are mutagens and that many mutagens are carcinogens.

While most of the discussion below pertains to airborne contaminants, some chemicals, which are not irritants or corrosives *per se,* are toxic when absorbed through the skin. Figure 12-8 lists some chemicals that are extremely hazardous when absorbed through the skin.

Figure 12-8
Skin Caution: Dangerous to Touch!

Aniline
o-Cresol
Dimethyl Sulfide
Formaldehyde
Hydrogen Peroxide (undiluted)
Lindane
Mercuric Oxide
Methyl Vinyl Ketone
Nicotine
Phenol
Toluene 2,4-Diisocyanate
Trichloroethylsilane

Nicotine? Yes, nicotine from green tobacco leaves causes serious illness when absorbed through the skin, often requiring hospitalization. This disease is called Green Tobacco Sickness. For industries where toxic chemical absorption through the skin is a reality, a skin care program is recommended. Figure 12-9 is a recommended program.

Indoor Air Quality

Indoor air quality has long been the domain of industrial hygienists and safety engineers. Today a lot is heard about sick building syndrome and building-related illnesses, which are discussed below. Indoor air quality problems can be investigated from the perspective of sources, contaminant of concern, or signs and symptoms. Table 12-3 lists typical IAQ complaints in order of frequency.

Comfort Zone = 68-76°F (20-24.5°C); 40-60% RH;
no odors; no drafts; no stuffiness

Figure 12-9
Skin Care Program

- Survey and evaluate existing processes for skin contamination.
- Determine the types of soils in each process area.
- Identify skin cleansing products required in each area.
- Provide instructions on how to wash properly at every wash basin.
- Identify skin medicating products that are needed and supply them in first aid or medical office.
- Identify skin protective products that are needed in each area and stock them where employees can have access.
- Promote skin care with periodic training sessions and awareness briefings.

Table 12-3
Frequency of IAQ Complaints

Complaint	Frequency of Complaint, %
Eye irritation	81
Dry throat	71
Headache	67
Fatigue	53
Sinus congestion	51
Skin irritation	38
Shortness of breath	33
Cough	24
Dizziness	22
Nausea	15

Complaints other than those listed in Table 12-3 suggest that something else is the problem.

Rosalind Anderson reports a link between respiratory depression and offgassing products of wallcovering. In experiments with mice, she notes a depressed respiratory rate after only ten minutes exposure to these offgases. The decrease in respiratory rate is proportional to the linear feet of wallcovering. The rate is also proportional to the number of products in the offgas, indicating that their effect is additive.

Figure 12-10 lists typical air contaminants found indoors. Biological contaminants will be discussed in Chapter 18.

Toxicity of Chemical Agents

Polynuclear aromatic hydrocarbons (PAHs) are both carcinogenic and mutagenic. They occur naturally as part of combustion but are mostly intro-

Figure 12-10
Typical Indoor Air Contaminants

Fibers		
Acetate	Flocking	Rayon
Asbestos	Hemp	Rock Wool
Cotton	Linen	Silk
Dacron	Nylon	Wool
Fiberglass	Orlon	
Fumes		
Metals	Smoke	Welding
Irritant Solvents		
Benzene	Naphthalene	m-Xylene
Cumene	Tetrachloroethylene	o-Xylene
o-Dichlorobenzene	Toluene	p-Xylene
Ethylene Dichloride	1,1,2-Trichloroethane	
Monochlorobenzene	Trichloroethylene	
Particulate Matter		
Arsenic	Gypsum	Powders
Asphalt	Hops	Polyvinyl Chloride
Carbon	Lead	Salts
Cement	Metals	Sand
Coal	Minerals	Sulfate
Cotton	Ores	Stone
Fertilizer	Oxides	Wheat

Figure 12-10 *(continued)*

Vapors		
Chlordane	Herbicides	Radon (gas)
Formaldehyde	Pesticides	

duced by industrial combustion processes. Figure 12-11 lists some common PAHs.

Both UV and fluorescence detectors can be used for the determination of PAHs. UV detectors are used at 254 nm with no wavelength program, an adequate sensitivity for EPA Method 610. Fluorescence detectors with wavelength programming give more sensitivity and selectivity. A fluorescence detector in series with a UV detector provides the best sensitivity testing.

Table 12-4 lists the content of trace elements in human blood and hair. Toxicity lies somewhere beyond these levels.

Figure 12-12 lists the target organs for various metals.

Table 12-5 lists absorption by different routes of exposure for several metals.

In Figure 12-13, the route of elimination for various metals is summarized.

Alkyl and aryl mercury compounds are more toxic than inorganic forms of mercury.

Figure 12-11
Common PAHs

Acenaphthene	Dibenzo(a,h)anthracene
Acenaphthylene	Fluoranthene
Anthracene	Fluorene
Benzo(a)anthracene	Indeno(1,2,3-cd)pyrene
Benzo(a)pyrene	Naphthalene
Benzo(b)fluoranthene	Perylene
Benzo(g,h,i)perylene	Phenanthene
Benzo(k)fluoranthene	Pyrene
Chrysene	

Table 12-4
Trace Elements in Blood and Hair

Metal	Blood, ppm	Hair, ppm
Antimony	0.005	0.2
Arsenic	0.7	0.2
Boron	0.09	—
Cadmium	0.009	1.0
Chromium	0.003	1.0
Cobalt	0.0005	0.004
Copper	1.5	15.0
Lead	0.04	4.0
Mercury	0.005	1.5
Nickel	0.03	3.0
Selenium	0.2	0.8
Tin	0.015	1.0
Vanadium	0.02	0.03

Figure 12-12
Target Organs of Toxic Metals

Metal	Target Organs
Aluminum	Lungs, liver, thyroid, bone, brain
Cadmium	Liver, kidneys, pancreas, joints, arteries, salivary glands, periosteum
Copper	Ceruloplasmin, albumin, liver, brain, heart, kidneys
Iron	Hemoglobin, myoglobin, iron-containing enzymes (catalase, peroxidase), liver, spleen, heart, brain, pancreas, joints skeleton, liver
Lead	Kidney, brain, thyroid, pituitary glands
Mercury	

Table 12-5
Absorption of Metals by Route of Exposure

Metal	Inhalation	Skin	Ingestion
Aluminum	incomplete		incomplete
Beryllium	nearly complete		
Cadmium	nearly complete		nearly complete
Trivalent Chromium	incomplete		incomplete
Hexavalent Chromium	nearly complete		nearly complete
Iron	nearly complete		nearly complete
Lead	nearly complete		nearly complete
Manganese	nearly complete		negligible
Elemental Mercury	nearly complete		
Inorganic Mercury			nearly complete
Organic Mercury	nearly complete	nearly complete	nearly complete
Nickel	fairly complete	fairly complete	fairly complete
Selenium	nearly complete		nearly complete
Thallium	nearly complete	nearly complete	nearly complete
Inorganic Tin			incomplete
Organotin			fairly complete
Alkyl Tin		fairly complete	
Vanadium	nearly complete		negligible

The extreme measure of toxicity is the term *immediately dangerous to life and health* (IDLH). A material concentration at IDLH or greater can cause serious harm to health, including death, irreversible health damage, or impediments to the ability to escape from an exposure to the material. IDLH exposure is not associated with a duration. The only response is immediate evacuation. Since part of the definition of IDLH is the potential impairment of ability to escape, it is recommended that a detection device with warning alarm be set at some lower concentration, at most one-half the IDLH.

TLVS

The American Conference of Governmental Industrial Hygienists (ACGIH) established threshold limit values (TLV) for various chemicals that represent the upper limit of the time-weighted average (TWA) airborne exposure during a 40 hour work week. TLVs are concentrations usually

Figure 12-13
Predominant Routes of Elimination, Metals

Metal	Route of Elimination
Aluminum	urine
Antimony	urine
Arsenic	urine, hair
Beryllium	urine
Cadmium	urine, hair (but mostly absorbed)
Chromium	urine
Cobalt	urine
Copper	bile
Iron	urine
Lead	urine, hair
Manganese	bile
Inorganic Mercury	urine, feces
Short-chain Alkyl	feces
Mercury	feces
Aryl Mercury	urine
Alkoxyalkyl Mercury	urine
Nickel	urine (24-hr composite)
Selenium	exhaled (as metabolite dimethylselenide)
Selenium	urine
Thallium	urine
Inorganic Tin	urine
Organotin	bile
Vanadium	urine
Zinc	feces

expressed in terms of volume of the contaminant per volume of air. These concentrations are determined on the basis of published toxicology studies on animals, epidemiology studies of human cohorts, clinical case studies, and analogies to other materials. Concentrations greater than the TLV-TWA are risky with respect to health effects. However, TLV conditions are safe concentrations for continual exposure of nearly all workers (but not all) without adverse health effects. TLVs are designed to protect healthy workers only. Because of biological and physiological variability, TLVs might not be appropriate protection for a small fraction of workers.

A TLV skin notation means that the material can be absorbed through the skin. Care must be taken to wear personal protective clothing with these chemicals. Among hazardous chemicals having skin notations are benzidine, epichlorohydrin, and xylene.

Short-term exposure limits (TLV-STEL) are concentrations that should never be exceeded. The STEL concentration is permitted for a 15 minute period.

A permissible exposure limit (PEL) is a concentration established by OSHA in 29 CFR 1910.1000 that must not be exceeded in the workplace without immediate use of personal protection equipment and eventual implementation of engineering and administrative controls to abate exposure levels. The PEL for each chemical represents the 1968 ACGIH TLV for that chemical. OSHA attempted to update PELs in the early 1990s by was blocked by court action. The current TLVs are better informed exposure limits to use while OSHA irons out its with the legislative and judicial branches of federal government.

EPA also regulates air concentrations of some of the same materials that have been assigned TLVs and PELs. The primary difference between EPA air emission limitations and PELs and TLVs is that the PELs and TLVs are based on an 8 hour workplace exposure to healthy workers, whereas the EPA limit is based on a 24 hour exposure of the most vulnerable members of the community: children, the elderly, and health-impaired individuals.

Industrial Hygiene Surveys

When and where should industrial hygiene surveys be undertaken for chemical exposure? Jobs that take place near combustion processes are candidates for survey, carbon monoxide being the chief contaminant of concern (COC). Jobs that expose employees to infectious pathogens or potential host fluids want to be surveyed. An example is the handling of bloodborne pathogens by hospital personnel or industrial employees who provide first aid treatment. Another set of COCs are lubricants, chips, dust, and mists associated with metal parts fabrication. Any dusty work environment is a candidate to be surveyed. Also, mixing, blending, and open transfer of, as well as dipping into, liquids may need to be surveyed. Degreasing with solvents is another candidate because several COCs may be present around coating operations.

When determining the concentration of gaseous mixtures for comparison with TLVs, the gases must have similar toxic effects.

Table 12-6
Industrial Hygiene Chemical Monitoring Techniques

Direct Reading			Laboratory Analysis			
Electronic		Colori-metric Indicators	Grab	Full Shift		
General	Specific	Tubes		Personal		Area
Explosive atmosphere Total organic vapor Dust/fibers	O_2 H_2S CO CO_2 SO_2 Organic vapors	Hundreds available Diffusion tubes Personal samplers	Evacuated container Tedlar/Mylar bag	BREATHING ZONE Filters Cyclones Sorbents Impingers Dosimeters Passive	BIOLOGICAL Breath Blood Urine Hair	Filters Cyclones Sorbents Impingers Dosimeters Passive

Table 12-6 summarizes basic industrial hygiene monitoring techniques, adapted from Anderson.

Physical Chemistry of Gases and Vapors. Gases take the volume and shape of their container. A liquid absorbs heat to evaporate, forming vapor, which expands to the size and shape of its container. The amount of liquid that evaporates, q (acfm), can be calculated

$$q = \frac{19.6 \times 10^{-3} \bullet V^{0.5} \bullet A \bullet P^{vap} \bullet \left(M^{-1} + 0.035\right)^{0.25}}{M^{0.167} \bullet d \bullet \left(T + 273\right)^{0.05} \bullet \left(L \bullet P^{atm}\right)^{0.5}} \qquad [12.3]$$

where

V = air speed over pool, fpm
A = surface area of pool, ft^2
P^{vap} = vapor pressure of liquid, mmHg
M = molecular weight of liquid
d = air density correction factor
T = temperature of pool surface, °C
L = length of pool in direction of air flow, ft
P^{atm} = barometric pressure, mmHg

Let's consider a spill of a paint thinner composed of xylene at a paint booth. The ambient temperature, T, is 24°C. We'll assume the air is flowing, V, at 100 fpm, P^{atm} = 760 mmHg. The xylene pool measures one foot by forty inches, A = 3.5 ft². Air flow is across the short axis of the pool, L = 1 ft.

The concentration, C_a, of a vapor in air can be calculated

$$C_a\left(ppm\right) = \frac{q \times 10^6 \times K}{Q} \qquad [12.4]$$

where

K = mixing factor, 0.1 to 1 (perfect)
Q = dilution air volume flow rate, acfm

$$Q = V \bullet A \qquad [12.5]$$

If the numerator of equation 12.2 is generation rate, G,

$$C = G/Q \qquad [12.6]$$

If the room is divided into a near and far field, the general Equation 12.4 becomes,

$$C_{N,eq} = \frac{G}{Q} + \frac{G}{\beta} = \frac{G}{\left(\frac{\beta}{\beta + Q}\right)Q} \qquad [12.7]$$

β is the interfield flow rate. A good estimate to use is 175 cubic feet per minute.

The effects of dispersion can be estimated by the following equation for concentration:

$$C = \frac{G}{2\pi Dr} \qquad [12.8]$$

where

D = diffusivity coefficient (from handbooks)
r = distance from source, feet

Particle Behavior. With a dust concentration of 15 grams/m^3, you will be able to see practically nothing. Visibility will be close to zero. Any concentration less than that and you need to conduct air monitoring to determine exposure concentration. Guess what. OSHA PELs for PM are less than 15 g/m^3—way less. Take samples and measure the PM in your workplace.

How particles behave physiologically depends on particle size. Large PM, greater than ten microns in size, tends to settle out of the air quickly. If such PM reaches the nose, it is filtered out by nose hair. PM smaller than ten microns can be inhaled and where it ends up in the respiratory system again depends on particle size as well as water solubility. Coarser, more soluble particles impact near the nasal passages or in the back of the throat. Finer, less soluble particles go deeper and enter the bronchial passage and upper lungs. The finest, least soluble particles may end up deep in the lungs in the tiny air sacs called alveoli where oxygen is exchanged with carbon dioxide in the blood.

Asbestos is considered PM but is measured in fibers per cubic centimeter rather than in the traditional concentration terms. The PEL for asbestos is 0.2 fibers/cc.

Crystalline silica is a natural component of the earth's crust. Also known as quartz, crystalline silica is found in sand and granite. Breathing air that is contaminated with crystalline silica particles leads to a lung condition known as silicosis. Crystalline silica creates fibrosis or scar tissue formation in the lungs, leading to reduce ability to extract oxygen from the air. Silicosis has no cure. Figure 12-14 lists some steps to minimize exposure to crystalline silica.

The PEL for crystalline silica in general industry is:

$$PEL = \frac{10 mg/m^3}{(\%Quartz) + 2} \qquad [12.9]$$

Physical Chemistry of Respirable Air. With respect to particulate matter (PM) such as dust, metal fume, mist, and fibers suspended in air as aerosols, the particle size is the important factor. Size is determined by measuring the

Figure 12-14
Minimizing Crystalline Silica Exposure

- Provide appropriate engineering controls.
- Make employees aware of the health effects of crystalline silica and how smoking increases the damage.
- Place warning signs at locations where crystalline silica is present in potentially hazardous concentrations.
- Conduct air monitoring.
- Provide type CE positive pressure abrasive blasting respirators for sandblasting.
- Provide respirators approved for crystalline silica dust for other processes.
- Prohibit the alteration of respirators by employees.
- Prohibit facial hair on employees who must wear tight-fitting respirators.
- Have employees who work in dusty areas change into disposable or washable work clothing at the worksite.
- Have employees shower and change into clean clothing before leaving the worksite.
- Prohibit eating, drinking, use of tobacco, or application of cosmetics in crystalline silica dust areas.
- Require employees to wash hands and face before eating, drinking, smoking, or applying cosmetics outside the exposure area.

diameter of the particle and expressed in *microns.* Concentration of particles in air is expressed in *milligrams per cubic meter* (mg/m^3) of air sampled or *millions of particles per standard cubic feet* (mppscf) of air sampled. The concentration of fibers is reported as *fibers per cubic centimeter* (fcc) of air sampled.

Typically, PM as dust is generated from solid material by mechanical means: grinding or crushing. *Dust particles* vary from 0.1 to over 100 microns in diameter. *Mist* is suspended droplets of liquid that have either condensed from the vapor state or generated by spraying or atomizing. *Fibers* are particles whose length-to-width ratio are 3-to-1 or greater. *Fumes* are created when a volatilized material that was formerly solid condenses in cool air. The smallest of all PM generally, fumes are one micron and less in diameter. The best example of a fume is welding fumes.

Wilsey, *et al,* give the inhalability I of an aerosol as

$$I = 0.5\{1 + \exp(-0.06d_{ac})\} \qquad [12.10]$$

for d_{ac}, particle aerodynamic diameter, 100 microns and less.

Fluids take the shape of their container and have no form on their own. A gas is a fluid that will expand to occupy its container. A liquid is a fluid that does not expand to occupy its container. Vapors are the volatile versions of liquids and solids heated to boiling point or sublimation temperature, respectively. Vapors will condense as the air temperature cools the material.

Calculating Concentrations. Gases and vapors are measured and reported either in terms of volumetric concentration, ppm, or mass concentration, mg/m^3. The two units are related, if you know the molecular weight of the species measured, and in many cases you do. An assumption can be made about molecular weight if not.

The volumetric concentration is calculated from measured data.

$$C_V = \frac{m \bullet 10^3}{V} \bullet \frac{24.46}{M} \qquad [12.11]$$

where

C_V = volumetric concentration, ppm
m = mass of substance, mg
V = air volume, L, and
M = molecular weight, g/mole.

The constant 24.46 represents the volume of one mole of the gas or vapor in liters at standard temperature and pressure. If STP other than 25°C and 760 mmHg is used, or if the conditions at the time of measurement are not STP, the constant must be adjusted accordingly. Mass concentration is calculated as

$$C = \frac{m \bullet 10^3}{V} \qquad [12.12]$$

where

C is the concentration in mg/m^3

The mass concentration of aerosols is determined in the same manner.

Markiewicz reminds us that in screening chemicals we can use the vapor pressure to calculate a saturated concentration of the vapor of that liquid in still air.

$$ppm = \frac{P^{vap}}{760} \times 10^6 \qquad [12.13]$$

Critiquing Exposure Assessment Data. Crystalline silica in respirable dust is limited by formula instead of a fixed number. The OSHA PEL for crystalline silica, which depends on the amount of crystalline silica in the dust, is

$$PEL = \frac{10\,mg/m^3}{\%SiO_2 + 2} \qquad [12.14]$$

As an example, consider a dust that contains 28% crystalline silica. The PEL is this case would be 0.333 mg/m^3 or 333 $\mu g/m^3$. What if the dust also contains 20 percent amorphous silica? The PEL for crystalline silica is still 333 $\mu g/m^3$.

NIOSH has recommended an REL for crystalline silica set at 50 $\mu g/m^3$. Since REL is time weighted over ten hours, the appropriate PEL, if OSHA were to adopt it, would be 62.5 $\mu g/m^3$ assuming no exposure for the next two hours after the eight hour shift. The ACGIH TLV®-TWA is 100 $\mu g/m^3$ for respirable quartz, which is one form of crystalline silica, and 50 $\mu g/m^3$ for respirable cristobalite and tridymite, which are the other two forms.

One method of exposure assessment we have not discussed is biomonitoring. This consists of measuring biomarkers in urine, blood, hair, and other parts of the body or its fluids.

Table 12-7 gives half-lives of a few chemicals. An extensive review of literature is required when using the biomonitoring technique.

Respirators and Other PPE

Depending on the specific chemical, airborne exposure to chemical

Table 12-7
Chemical Half-lives in the Body

Chemical species/body part	Half-life (avg ± S.D.)
4,4′-MDA/urine	68.75 ± 10.78 hr
4,4′-MDA/blood plasma	15.50 ± 5.00 da

MDA= methylenedianiline

vapors and particulate matter leads to a variety of health effects, such as skin irritation, loss of vision, damage to the respiratory system, cancer, heart disease, and other chronic illnesses. Respirators and other protective clothing can help reduce the exposures to a level that does not exhibit health effects. If engineering and administrative controls are inadequate protection, or in the interim while such controls are waiting to be implemented, or during emergency situations, the use of respirators is indicated and the key is proper selection.

For chemicals that may be absorbed through the skin, Figure 12-15 lists some glove recommendations.

For dimethyl mercury and other highly toxic compounds that can dissolve straight through these gloves, wear highly resistant laminated gloves under a pair of long-cuffed, unsupported heavy-duty glove such as neoprene or nitrile.

Warning!
Do not use latex or PVC gloves for protection against highly toxic chemicals that absorb through the skin.

Respirator Selection. When determining which of many respirator types to use, keep in mind that hazardous atmospheres come in three types: 1) oxygen-

Figure 12-15
Gloves to Prevent Skin Absorption

Butyl for esters, glycol ethers, and ketones. Poor protection provided for chlorinated solvents and hydrocarbons.

Fluoroelastomer for aromatics, chlorinated solvents, esters, ethers, and ketones (except acetone). Poor protection for amines, some esters, and some ketones. Check compatibility matrix carefully.

Natural rubber provides fair protection for aldehydes and ketones. Poor protection is provided for grease, oil, and other organics that may dissolve the rubber.

Neoprene for aniline, glycol ethers, and phenol.

Nitrile for aliphatics, greases, oils, perchloroethylene, trichloroethane, and xylene. Fair protection provided for toluene. Poor protection for benzene, many ketones, methylene chloride, and trichloroethylene.

Polyvinyl alcohol for aliphatics, aromatics, chlorinated solvents, esters, ethers, and ketones (except acetone). Poor protection provided for acetone and light alcohols.

deficient atmospheres, 2) chemical-contaminated atmospheres, and 3) atmospheres that are both oxygen-deficient and chemically contaminated.

An oxygen-deficient atmosphere, by OSHA definition, is one that has less than 19.5 percent oxygen by volume at sea level. Below this level, the oxygen does not have enough partial pressure to migrate across the membranes of the alveoli in the lungs and the body begins to starve for oxygen. Close to 19.5 percent the problem is minor. The lower the oxygen level gets, the more pronounced the problem. Eventually the exposed person passes out and ultimately he or she dies of suffocation. Oxygen-deficient spaces often include silos, boilers, tanks, ship holds and voids, sewers, and basements or other below ground spaces where organic material decomposes.

Various chemicals may contaminate an atmosphere. Some people have a limited idea of what a chemical is, thinking that a chemical is a liquid other than water. However, gases and solids may also contaminate the atmosphere. When a liquid evaporates, the gas it forms is referred to as a vapor.

Dust is solid particles, called particulate matter. Sprays are aerosols or fine mists of liquid droplets. Fumes are solid particles that sublime from metals. Many people mistakenly refer to vapors as fumes. Smoke is particulate matter from combustion processes that contain carbon as soot.

Air Respirators. To protect persons from such atmospheres as discussed above two basic types of respirators are available with many variations: air-purifying and supplied-air. The three basic types of air-purifying respirators are particulate removing, vapor and gas removing, and combination respirators. The two types of supplied-air respirators are: air-line respirators and self-contained breathing apparatus (SCBA).

Air-purifying respirators have limited service applicability. They are only used to remove contaminants from the air and must never be worn in oxygen deficient atmospheres because, they are simply filters and do not supply any air or oxygen. The air flow through the filters is powered by human lungs. If those lungs are starved of oxygen, chances are they will not have the strength to pull air through the filters and it would not do any good even if they could. Figure 12-16 lists the necessary conditions for using air-supplied respirators.

Where multiple chemical hazards are present, respirator wearers are faced with stopping work periodically to monitor the atmosphere—if they select cartridges that deal with only one or two of the hazards. Assuming oxygen deficiency is not one of the hazards, new cartridges have been designed that can remove up to ten chemical species at once. These combination cartridges can remove ammonia, chlorine, chlorine dioxide, formaldehyde, hydrogen chloride, hydrogen fluoride, methylamine, organic vapors, and sul-

Figure 12-16
Use Air-Purifying Respirators When:

- The ambient oxygen level will sustain life.
- The identity of the contaminant is known.
- The concentration of the contaminant is known.
- The concentration of the contaminant does not exceed the limitations of the facepiece or the filter cartridge.

Figure 12-17
Protection Factors of Air-Purifying Respirators

Half-face respirator	PF = 10
Full-face respirator	PF = 50

fur dioxide. A paint/pesticide prefilter may be added for protection from dusts, enamels, fumes, lacquers, mists, paints, and pesticides.

Different types of air-purifying respirators afford different degrees of protection, as shown in Figure 12-17. Protection factor (PF) is the ratio of the concentration of a contaminant outside the respirator to the concentration of the contaminant inside the respirator. For instance, with PF = 10, the concentration of the contaminant inside the respirator is one-tenth the outside concentration.

An example of how to use the PF is as follows. The PEL for carbon monoxide is 35 ppm. Therefore if exposed to CO, the concentration inside a respirator must not exceed 33 ppm. A sample of air is measured in a forklift maintenance room and CO is found to be 975 ppm. The workers are wearing half-face respirators.

$$\text{Half-face} = 975\text{-}(10)(35)\text{= } 625 \text{ ppm}$$

In this situation, an overexposure situation exists even with the respirators. Would full-face respirators be acceptable?

$$\text{Full-face} = 975\text{-}(50)(35)\text{= } <0$$

In this case, a full-face respirator would provide acceptable protection.

Another important selection to make is the appropriate filter cartridge. A filter filled with activated carbon is used for organic vapors. A material called hopcalite is used as a catalyst for protection against carbon monoxide.

Supplied air respirators provide breathing air in oxygen deficient areas from an outside source by air hose to the mask or hood of the respirator. Air-line respirators are tight-fitting. A pressure demand nozzle allows the user to control the airflow to the mask. Grade D breathing air is used. The air supply is virtually unlimited. However, the hose length is limited to three hundred feet or less due to friction.

Dust respirators may be used as a stop-gap measure until adequate ventilation is installed.

A frequent concern about respirator use is the potential effects of fatigue and work load while wearing them. In a study conducted by the University of Maryland for the U.S. Army Edgewood Research Development and Engineering Center, subjects were tested while performing cognitive, psychomotor, and motor tasks with and without respirators (Johnson, *et al*). Subjects were also deprived of food intake during half the test events while performing tasks with respirators. Performance data, psychological responses, and physiological parameters observed during the study show that wearing respirators has little effect on task performance, even with food deprivation. Keep in mind that the test subjects were young people of military age, implying excellent physical condition. In the workplace, on the other hand, we deal with people of all ages, up to and including retirement and all levels of fitness and general health. Many employees are not in the best physical condition, in fact can be in very poor physical condition. Also, in industry, we have many more people with debilitating medical problems than we used to have. Therefore, a physical examination focusing on respiratory capacity, while not expressly required by OSHA, is highly recommended before having anyone routinely wear a respirator. This examination is especially critical in the case of employees who may have medical conditions, such as asthma, that may predispose them to have problems with respirators.

Respirator Maintenance. Maintenance-free respirators are pre-assembled and packaged at the factory. They have no spare parts and are not designed to be disassembled. These respirators are simply discarded when the filter media becomes contaminated or the assembly is damaged. Maintenance-free respirators are assigned to only one employee.

Low-maintenance respirators have a low-cost facepiece with replaceable filter cartridges, but no other spare parts. More than one employee can use this type, with appropriate cleaning between use by different employees.

Conventional air-purifying respirators are designed for heavy-duty multiple reuse and maintenance. Replaceable components are provided.

Respirator Fit Testing. You cannot simply give a person a respirator and tell him or her to put it on and wear it. Certainly you need to teach each respirator user how to don and doff the device and to let him or her practice these activities under your supervision. Part of this training is checking to see whether or not the respirator fits tightly enough to do the wearer any good. Such a check is called a *fit test.*

Not everyone can wear a respirator. Some have facial features that do not permit a good seal between the facepiece and the skin of the face. Typically, beards, long mustaches, and long side burns interfere with this seal. People with dentures may have trouble getting a good seal on their facepiece. The strap tension also affects the facepiece seal. Some people are apparently claustrophobic and cannot calmly keep anything on their faces, even when it is for their own good. A certain percentage of persons are going to have cardiac problems or problems with the respiratory system that will be further stressed by the wearing of respirators.

Except for the latter case, which ought to be identified by having a physician examine all potential routine respirator users, the fit test is used to determine whether an individual can wear a respirator with a good seal. The wearer, with respirator donned, is exposed to a challenge agent while performing OSHA-prescribed exercises. The challenge agent may be irritant smoke (stannic chloride), banana oil (isoamyl acetate), or saccharin. The irritant smoke, if it gets past the seal, produces an involuntary response by the wearer. He or she will rip the respirator off their face. Hence, you know that a good seal was not obtained.

The banana oil test relies on the ability to detect odors. However, as a check, OSHA requires the wearer to read aloud the so-called "Rainbow Passage" since the agent, isoamyl acetate, will cause slurring of speech and the passage is nonsensical, giving opportunity for the effect of the agent to reveal itself. Saccharin, the artificial sweetener, is a taste test. Regardless of the agent used, the wearer must perform certain exercises while wearing the respirator in the presence of the challenge agent. These exercises include normal and deep breathing, moving the head from side to side, moving the head up and down, and talking. If these are performed without the wearer detecting the presence of the agent, then the fit is good and the respirator is sealed to his or her face.

Hazard Communication

The key to hazard communication is:

> **P** – Program
> **I** – Inventory
> **T** – Training
> **L** – Labeling
> **S** – Sheets (as in MSDS)

The whole premise of the Hazard Communication Standard is that if you tell employees about the hazards involved with the chemicals they work with daily, they will take responsibility for protecting themselves. Despite two serious flaws in the program, it is the best thing we have and it does provide a lot of information—albeit inadequate, for the adept. I have beaten the program to death elsewhere but briefly, the information on MSDSs is largely junk, and employees either do not use MSDSs at all or they unjustly use them.

You must devise a written hazard communication program containing the information of Figure 12-18.

Ventilation

In selecting a system to control a particular problem, you need a three-pronged strategy: measure the problem, design the control equipment to abate the problem, and verify that the problem has been solved, making adjustments as necessary.

The control strategy may be expanded to provide details. First, identify and classify each operation. Are its emissions particulate matter? Gas or vapor? Toxic? Asphyxiating? Some combination of these? Classify the operation by its contaminants of concern. Then evaluate the concentration of each contaminant of concern. This requires workplace measurements. No one expects you to measure all suspected contaminants at once. Devise a plan with a schedule. Obvious problems and situations where workers exhibit symptoms of exposure should be measured first. Time, cost, and feasibility constraints are taken into account next.

Figure 12-18
Hazard Communication Program

1. **Written Program**
    - Who is in charge?
    - Who will maintain MSDS?
    - Where will MSDS be located?
    - Who will give you a copy of an MSDS if you ask for one?
    - Who can you ask about an MSDS (explain it to you)?
    - Who can you ask about labeling?
    - What type of labeling will be used?
    - Who will notify outside contractors?
    - Who will collect MSDS from outside contractors?
2. **Labeling**
    - Contents.
    - Type of labeling system.
    - Hazard statement.
    - Hazard warning.
3. **Material Safety Data Sheets**
    - Make them easily accessible for employees.
    - Provide a maintenance system to incorporate new and revised MSDS.
4. **Inventory**
    - Include all chemicals on premises.
    - List area where used.
    - Include hazard statement.
5. **Training**
    - Make employees aware of their rights.
    - How to read and understand MSDS.
    - How to read and understand label.
    - Location of MSDS and written program.
    - Written program contents.
    - Location of hazardous materials.
    - Emergency first aid, if exposed.
    - Exposure signs and symptoms.
    - Personal protective equipment.

Figure 12-18 *(continued)*

6. When to conduct training

- Upon initial assignment.
- When a new chemical is introduced.
- When transferred to a new department using different chemicals.
- When employee does not show understanding of training received.

After measurement, before selecting a design for control and abatement where necessary, first consider control by preventive measures. Can you substitute materials? Can you do the process in a safer way? Is new equipment available that eliminates the need for controls? Are less dusty processes available?

Principles of AirFlow

Unrestricted flow, Q, of air in a duct is calculated as follows:

$$Q = 4005 \bullet A \bullet \sqrt{VP/d} \qquad [12.15]$$

where

A = duct area
VP = velocity pressure
d = density correction factor

Air Flow Measurements

In calculating air flow, the diameter of the duct is one of the important factors. However, many, if not most, ducts are built in a rectangular shape. Burton gives us a formula to use called Wright's Approximation:

$$D_C = 1.3 \bullet (a \bullet b)^{0.625} / (a + b)^{0.25} \qquad [12.16]$$

where

D_C = diameter of equivalent circular duct, inches
a = length of one side of rectangular duct, inches
b = length of adjacent side, inches

The equation produces erratic results if the aspect ratio, which is the ratio of sides (a/b or b/a), is greater than 11:1). Otherwise, if you have a duct that measures eighteen inches by three feet, for example (an aspect ratio of 2:1), the equivalent circular diameter, D_C, is 27.42 inches.

Basic System Design

Take advantage of the fact that hot air rises and expands. A chimney, for instance, uses the natural draft created by expanding gas and temperature and pressure difference at the entrance and exit to the stack. The required static pressure, P_D, at the foot of a chimney to create flow is:

$$P_D = P_S + P_{V,exit} + F \quad [12.17]$$

where

P_S = static pressure created by the exhaust fan
P_V = velocity pressure (exit denotes discharge)
F = friction losses in the stack

Fan Selection

Several kinds of fans are used, depending on the system needs. Fans are either radial or axial. Centrifugal fans are radial and come in three groups:

Low pressure centrifugal fans:	$P < 4$ in. H_2O
Medium pressure centrifugal fans:	$4 < P < 12$ in. H_2O
High-pressure centrifugal fans:	$12 < P < 40$ in. H_2O

Impeller configurations for centrifugal fans are straight, forward-curved, and backward-curved blades. The straight blades are also called paddlewheel impellers. Forward-curved blades are also known as radial wheel impellers. For exhausting large volumes of dust-laden air, a heavy industrial fan with radial flow blades are needed. Figure 12-19 provides guidance on selection of centrifugal fan types, after Chicago Blower.

Figure 12-20 lists applications for axial fans.

Air Cleaning

Particulate matter and gas and vapor contamination must be removed from air streams, at least to the point where OSHA and EPA air quality is satisfied. Particulates are removed by gravity or inertia filtration, electrical charge, or wet scrubbing. Gases and vapors are removed through absorption, adsorption, condensation, or combustion.

Figure 12-19
Selection of Centrifugal Fan Types

Type Blading	Applications
Forward curved	Low pressure HVAC: domestic furnaces, central station units, packaged air conditioning equipment
Radial blade	Material handling; high pressure industrial requirements
Backward inclined	General HVAC in industry where light dust may cause erosion
Airfoil blade	Used in larger HVAC and clean air industrial applications; special construction available for dust
Radial tip	Material handling or dirty or erosive applications; more efficient than radial blade

Figure 12-20
Selection of Axial Fan Types

Type Blading	Applications
Propeller	High volume circulation of air in a space; ventilation through wall without duct work
Tubeaxial	Ducted HVAC applications where air distribution downstream is not critical. Also for drying ovens, paint spray booths, and local exhaust hoods.
Vaneaxial	General HVAC with straight through flow. Compact. Many industrial applications.
Inline centrifugal	Low-pressure return air HVAC. Straight through flow.

The effectiveness of different particulate removal control equipment for different particle sizes is shown by size efficiency curves. Air cleaner efficiency is the ratio of the amount of dust collected by the air cleaner to the amount of dust entering the device. For gases and vapors, the removal efficiency is the ratio of the amount of contaminant removed to the amount of contaminant entering.

Electrostatic precipitators (ESPs) consist of a discharge electrode, a collection plate, and a rapper. The dust collects on an electrode called the collection plate, which has a charge opposite from the particles. Particles are charged by a corona produced by the discharge electrode. Particles collected are removed from the collection electrode by rapping the electrode with a mechanical or electrical device. A major advantage of using an electrostatic precipitator is that the force for collection is applied only to the particle, allowing low pressure drop through the collector. Electrostatic precipitators typically have the lowest pressure drop of any type of dust collector. ESPs are used to collect fine particles, especially when the particles are valuable. They are the most efficient means of collecting particles that are smaller than five microns in diameter and are the best devices for removing smoke from an air stream. ESPs can also be used when the gas stream temperature is elevated and/or the humidity is high. An increase in the collecting area of an ESP will increase the efficiency of removal. Collection efficiency is determined by size of particles being collected, physical state of the particles, and charge on the particles. A disadvantage of ESPs is that they are difficult to maintain at peak efficiency.

Mechanical separators are grouped in three categories: gravity chambers, cyclones, and impingement separators. These separators are efficient in removing particles ranging in size from 15 to 40 microns.

Gravity chambers are simple and not very effective except for removing the largest of particles. Typically these devices are used as pre-cleaners for other devices downstream.

The simplest form of impingement separator is a baffled chamber. Impingement separators such as fabric dust collectors provide high removal efficiency and are the most used industrial air cleaning devices. As the fabric, or bag, becomes clogged with filtered particles, the efficiency of the filter increases. The fractional efficiency for five microns and smaller particles is highest in a baghouse over an ESP, spray tower, or even a multiclone. Another advantage is that they require very little space. Bag filters are not sensitive to wide process variations. They are simple to operate and cheap to install. However, they can be expensive maintenance items and use the most electrical power.

Example: A fabric dust collector handles 300 cfm with a dust load of 10 grains per cubic foot. The initial resistance of the fabric is 1″ of water. After six hours of operation, the pressure drop across the fabric reaches a maximum permissible 5″ of water. How soon would 5″ i.w.g. be reached for a dust load of 20 grains per cubic foot and a flow rate of 300 cfm?

Answer: Doubling the dust load, with all other factors being the same, halves the cycle time. In this example, the answer is three hours. If flow is doubled with the same grain loading, resistance increases proportional to cycle time.

A centrifugal collector is a dust collector that depends on the difference in the density of dust and air for separation. The most common of these devices is the cyclone. The average inlet velocity of cyclone separators is 50 fps. The pressure drop across cyclones increases with the square of the inlet velocity. The efficiency of cyclones increase as the dust load increases. The most important factor in cyclone design is the radius. Rotary centrifugal separators are also called dynamic separators. Multicyclone removal efficiency for the collection of particles greater than ten microns in diameter can be as high as 90 percent. The principle advantages of the cyclone dust collector are low cost, low maintenance, and low pressure drop.

In addition to these ultimate control devices, a general control strategy must be adopted in dusty situations. Figure 12-21 is a summary of control strategies suggested in the Smandych, Thomson, and Goodfellow article.

Capture Hoods. Generically, capture hoods come in four varieties: enclosures, booths, capture hoods proper, and receptor hoods. Enclosures surround a process entirely. Booths are enclosures that have one wall missing to give access to the process. The most common example of this is the spray paint booth. The receptor hood is built near the persons or equipment you want to protect. With high velocity dust coming from a source towards the receptor, this hood also captures the dust by an exhaust fan.

Specifically, capture hoods have four subcategories: exterior hoods, assisted exterior hoods, high velocity/low volume exhaust hoods, and downdraft hoods. Exterior hoods induce airflow towards the suction opening and are built as side draft hoods, slot hoods, pull hoods, or lateral suction hoods. Assisted exterior hoods enhance dust capture with air jets that direct the dust stream to the hood. Called push/pull systems, these hoods have small jets of compressed air directed across the source to the hood.

Scrubbers. A wet scrubber works by forming condensation on the particles. The larger, heavier particles slow down and gravity takes over. The mechanism is impingement of liquid droplets on the sides of the collector. The performance of a wet scrubber is determined by emission quality, liquid-gas ratio, and liquid temperature. Wet scrubbers include spray chambers, mechanical scrubbers, and Venturi scrubbers. While most wet scrubbers require little energy and are therefore inexpensive to operate, the Venturi scrubber has high energy requirements.

Mechanical scrubbers include fog filters, mist eliminators, and packed towers. The waterfall curtain in a spray paint booth is another common

Figure 12-21
Dust Control Strategies

Application	Control Strategy
Conveying	
Belt Conveyor	Enclosure Stabilization
Bucket Elevator	Enclosure
Transfer between Conveyors	Enclosure Capture Hood Wet Dust Suppression
Feeding	
Loading/Unloading	Capture Hood Physical Stabilization Wet Dust Suppression
Feeders	Enclosure Wet Dust Suppression
Chutes	Enclosure Wet Dust Suppression
Packaging	
Bags	Enclosure
Drums and Barrels	Enclosure Capture Hood
General Filling	Capture Hood
Processing	
Dryers	Exhaust Filter
Screens	Enclosure
Crusher	Capture Hood Enclosure
Grinder	Capture Hood
Mixers	Enclosure
Transfer between Process Equipment	Enclosure Wet Dust Suppression
Storing	
Stockpiles	Physical Stabilization Vegetative Stabilization Chemical Stabilization

Figure 12-21*(continued)*

Bins	Enclosure
Sand Bins	Enclosure
Transfer to and from Storage	Enclosure Wet Dust Suppression

type of mechanical filter. Wet scrubbers are recommended for removing particulate matter from combustible gases.

Venturis are the most common type of generic orifice-type collector. The particle collection efficiency of a Venturi scrubber is very high due the fact that the high velocity in the throat of the Venturi atomizes water and provides good contaminant-liquid contact. To remove particles of one micron size with efficiencies exceeding 95 percent, Venturi scrubbers typically operate with a pressure drop in the range of six to eight i.w.g.

Recirculation of ventilated air that has been passed through a scrubber is not advisable because improper operation of the unit could result in return of contaminated air to the workplace. Nevertheless, recirculation of air that has been cleansed of nuisance dust is prevalent.

In a packed tower, flooding can be prevented by increasing the surface area. The disadvantage of a packed tower is that dust will plug the packing. This requires an excessive amount of maintenance to keep the packing material cleaning and the air stream flowing with optimal pressure drop. Another disadvantage of packed towers is that efficiency of removal is sensitive to changes in volumetric flow rate. The towers are packed with materials of complex geometry, such as Pall rings or Raschig rings or Berl saddles, to provide high surface area for liquid films. Typically packed towers are more compact in size than other towers and can work at higher temperatures with a constant pressure drop. Packed towers are subject to freezing, however.

Gas Cleaning. Factors of importance in gas cleaning are concentration gradients, large interfacial surface area, and large mass diffusion coefficients. Assuming Gaussian dispersion from a single source, the ground level concentration of a contaminant at any point downwind from the source is inversely proportional to wind speed.

Combustion is either direct flame or catalytic. A complete combustion process yields carbon dioxide and water. Carbon monoxide is one gas that must be burned.

Direct flame combustion, or thermal incineration, is used when the concentration of combustible contaminants is low. Typically, the temperature range of an incinerator is from 1,000 to 1,500°F. The typical residence time of gas in the fire box is from 0.2 to 0.8 seconds. Insurance regulations require that waste gas transported to a direct flame incinerator be maintained at a concentration of 25 percent or less of the lower explosive limit.

During catalytic combustion, fine particles cause the catalyst operating time to decrease. Typically, the after burner operating temperature of a catalytic converter ranges from 600 to 1,000°F.

Gas adsorbers are either fixed or floating bed types. Industrial adsorbent systems are usually regenerative. During adsorption, the amount of contaminant being adsorbed decreases as the system temperature increases. Activated adsorbents have a larger surface area than the same material that has not undergone the process. Gas adsorbers are more effective than other types of gas cleaners for controlling odorous, radioactive, or toxic gases. In fact, adsorption is the most satisfactory method of removing odors from organic gases or vapors.

Testing and Troubleshooting

To troubleshoot a system, you need to understand the basic fan laws, summarized in Figure 12-22.

Figure 12-22
Basic Fan Laws

Volume varies directly with speed (Eqn. 12.18).

Pressure varies directly with density (Eqn. 12.19) and with the square of speed (Eqn. 12.20).

Horsepower varies directly with density (Eqn. 12.21) and with the cube of speed (Eqn. 12.22).

$$CFM_2 = CFM_1\left(\frac{RPM_2}{RPM_1}\right) \quad [12.18]$$

$$P_2 = P_1\left(\frac{D_2}{D_1}\right) \quad [12.19]$$

$$P_2 = P_1\left(\frac{RPM_2}{RPM_1}\right)^2 \quad [12.20]$$

$$HP_2 = HP_1\left(\frac{D_2}{D_1}\right) \quad [12.21]$$

$$HP_2 = HP_1\left(\frac{RPM_2}{RPM_1}\right)^3 \quad [12.22]$$

To test a local exhaust hood, determine the static pressure of the system. A U-tube manometer is usually used to measure static pressure. Take measurements at points where airflow is nearly parallel to the wall of the duct and well downstream of any obstructions. For a hood take the measurements three diameters of exhaust duct downstream. The airflow, Q in cfm, is

$$Q = 4005C_eA\sqrt{SP_h} \quad [12.23]$$

where

C = coefficient of entry
A^e= cross-sectional area of duct, ft[2]
SP_h = static pressure at hood, iwg

The velocity pressure, VP, is measured with a pitot tube. A pitot traverse of the duct is conducted in order to determine the air velocity. Velocity and velocity pressure is calculated as

$$v = 4005\sqrt{VP} \quad [12.24]$$

If only the centerline velocity, v_{cl}, is known

$$v = 0.9v_{cl} \quad [12.25]$$

Troubleshooting Fans with Low System Air Flow

- Check direction of rotation
- Measure rpm
- Check for clogging or corrosion of fan
- Check duct for clogging
- Check for closed dampers
- Check blast-gate settings
- Check for clogged air cleaners
- Check outlets
- Check design of duct and branches
- Check make-up supply

Troubleshooting Hoods with Poor Contamination Capture

- Look for cross-drafts in front of the hood
- Check distance of operation from hood
- Verify that no changes have been made to hood

System Balancing

Balancing a complex air handling system can be a chore. Start with the branch of greatest resistance and determine the flow rate and transport velocity. Calculate all losses: hood losses; acceleration losses; elbows; and branch connections. Total the losses and calculate the static pressure of the branch. Do the same for all branches. Compare the static pressure of the branches.

If $SP_{greater} / SP_{lesser} \leq 1.05$ system balanced.

If the ratio of the greater static pressure to the lesser is less than 1.20, recalculate Q for the branch of least resistance:

$$Q = Q_{lesser}\, SP \sqrt{SP_{greater} / SP_{lesser}} \qquad [12.26]$$

Use this value as the balanced flow rate.

If, however, the ratio of static pressures is greater than 1.20, the branch with the least resistance must be reduced in size to obtain a greater resistance. Begin your calculations all over again. Computer programs are now available to assist you in this process.

Modifying Existing Systems

What happens when you have a ventilation system and exposures are still too high? If that is truly the best you can do, require all exposed employees to wear PPE. However, in many cases the ventilation system was poorly designed and a modification is in order. The first order of business is to evaluate the effectiveness of the existing ventilation system. The required modification may be as simple as conducting some corrective maintenance. The worst case may be to tear it completely out and install a new, more effective system. More often, some situation between these two actions exists.

Energy Considerations

Whenever you make changes to a ventilation system, evaluate the effect on the HVAC systems. Practically speaking, you may require more energy or loss energy efficiency and this adds to the operating cost of the plant.

Building-Related Illness *vs.* Sick Building Syndrome

Indoor air quality problems have, in a few cases, evolved from vague complaints of headaches and sore throats to drawn out legal battles and expensive renovations. While the trend towards litigation is not clear, we can be assured that building occupant complaints are on the rise. For one thing, here at the beginning of the twenty-first century, people's expectations for comfort have risen.

As Odom and Barr point out in a recent article, when an IAQ problem is abated, or otherwise solved, and when the occupants agree that it is solved are two entirely different events. The connecting factor is psychology and that makes it a can of worms for those of us who are not industrial psychologists.

The key is effective communication, which, sad to say, not too many of us are very good at either. Odom and Barr refer to the Sandman equation:

$$Risk = Hazard + Outrage \quad [12.27]$$

Whether you agree with Sandman or not, clearly employees in particular and the public in general do not see risk as we do. Here, outraged individuals have on their mental agendas moral-emotional issues that to safety professionals seem to have little to do with hazard or risk. Outrage is why an employee gets bent out of shape about a strange, but innocuous smell, yet smokes three packs of cigarettes per day. Moral-emotional issues, whether you like the term outrage or not, bias an individual to react with fear or anger instead of reason. The emotional response can, in fact, increase the level of risk that otherwise would exist by altering the person's mental-emotional state and causing irrational behavior.

Figure 12-23 shows the effect of outrage factors as adapted from Odom and Barr.

The progression of an IAQ problem is denial, outrage, and acceptance. When the employees first report symptoms, management typically has the maintenance department adjust climate controls and perhaps change air filters in the HVAC system. The problem as perceived by management is that there is one oversensitive employee or a small group of oversensitive employees. The employee perception is that management denies a problem exists

Figure 12-23
The Effect of Outrage Factors on Risk Perception

Tends to decrease outrage:	Tends to increase outrage:
Natural materials	Manmade materials
Voluntary risk	Involuntary risk
Familiarity with risk	Exotic risk
Controllable	Uncontrollable
Not memorable	Memorable (in the news)
Perceived as fair	Perceived as injustice

and, not uncommonly, management states this opinion unmistakably. Little wonder, the employees' outrage erupts. OSHA may be called in to investigate. If management conducts IAQ monitoring, the employees are not likely to accept the results. The outcome yields either continual grumbling, OSHA action, litigation, or, if management invites the employees to be involved in the investigation and problem correction, final acceptance. Actually, it's a lot more complicated than that, but you get the picture. Act early; involve the employees; communicate well.

Sick Building Syndrome (SBS)—theoretically a building that makes people sick—is pretty much a myth. Buildings are merely containers. Structural factors can not be blamed for sore throats, colds, rashes, fatigue, itchy eyes, coughs, headaches, and any number of other symptoms currently being blamed on buildings themselves. Psychosocial factors such as job dissatisfaction, a poor outlook on life, long commutes to and from work, second-hand smoke exposure, and poor housing have been strongly implicated as the causes for SBS by architectural researchers.

Figure 12-24 summarizes some appropriate questions to ask persons complaining of SBS.

Tact and confidentiality is required if you ask these questions yourself instead of bringing in an industrial psychologist. Take every complaint seriously and if a psychosocial factor is involved perhaps your company needs an Employee Assistance Program as discussed in Chapter 22.

SBS complaints and symptoms are listed in Figure 12-25.

Figure 12-24
Investigation of SBS

1. Have you recently lost much sleep over worry?
2. Have you felt you could not overcome your difficulties?
3. Have you been thinking of yourself as a worthless person?
4. Have you been feeling reasonably happy, all things considered?

Figure 12-25
SBS Complaints and Symptoms

- Nonspecific discomfort
- URT irritation: itchy eyes, coughing, sneezing, sore throat
- Headache, dizziness, nausea
- Fatigue, listlessness, inability to concentrate
- Shortness of breath
- Sensitivity to odors

Thermal discomfort may lead some persons to attribute unrelated symptoms to indoor air quality. Check to see if it's too hot, too stuffy, too cold, too drafty, too dry, or too humid. Eliminate these problems before proceeding with a psychosocial investigation. Also, lighting, noise, and ergonomic risk factors should be eliminated as the potential causes of headaches, fatigue, and eye strain. Nuisance conditions such as visible dust or unusual odors may alarm occupants, leading them to attribute unrelated symptoms to these nuisances. Water-borne or food-borne illnesses may also run their course through the population of a building. Reactions caused by exposure to toxic substances can be identified or eliminated by examining building drawings for connecting routes of exposure and conducting some air monitoring.

Psychogenic sources are individual and organizational stresses that may cause occupants to experience symptoms. *Sociogenic illnesses* are psychogenic symptoms that have been transferred among susceptible occupants. A rare occurrence is *Mass Sociogenic Illness* (MSI) where this phenomenon is pandemic in the workplace. NIOSH did a study on MSI in which it found the characteristics summarized in Figure 12-26.

Sandler gives us an objective mechanism for assessing the potential relationship between work and symptoms reported by an individual. Four criteria must be met in order to say the symptoms are work related: temporality, clear presence of hazard, adequate dose, and elimination of alternative etiologies.

The criterion of *temporality,* according to Sandler, means that the symp-

Figure 12-26
Characteristics of Mass Sociogenic Illness

- Complaints are gender specific
- Verbalized complaints influence others to discover the same complaints
- Labor/management problems exist
- Complaints persist despite removal of supposed causes
- Socially powerful people became affected
- Symptoms and complaints are nonspecific
- Work is boring and repetitive
- Work load is excessive
- Management structure is rigid
- Employees feel they lack control
- A poor work environment exists

toms arise only at work, under the same circumstances, and are consistently reproduced. *Clear presence of hazard* means that we can objectively document a hazard that has the capability of producing the symptoms being complained of. The hazard must be present at a concentration known to produce the symptoms in order to have an *adequate dose.* Finally, by investigating potential nonwork factors, including psychosocial and personal risk factors such as smoking and lifestyle, we meet the criterion of the *elimination of alternative etiologies.*

With these criteria in mind, symptoms fall into three basic causal groups, according to Sandler. Real occupational chemical exposure complaints are easy to understand and deal with. Real complaints that are not occupationally related are also easy to understand, but typically difficult to deal with. The last causal group can be the most difficult to deal with: perceived complaints coupled with psychosocial factors but with no occupational basis. For instance, at one plant where the cancer rate is about one tenth of the local rate for the general public, the employees have not let go of the belief that their low level exposures to chemicals (just barely measurable even) are killing them with cancer. In addition, the few cancers that have occurred in this population have *definitely* been lifestyle types. For instance, smokers whoe develop lung cancer, heavy drinkers who develop liver cancer, or the development of colon cancer, which is diet-related.

Building-Related Illnesses (BRI) is a general class of illnesses, among which are the so-called SBS: sinusitis, rhinitis, asthma, Legionnaires disease, Pontiac fever, humidifier fever, hypersensitivity pneumonitis (HP), flu, colds, and skin irritation.

Particulate Sampling and Control

Table 12-8 summarizes air standards for particulate matter that may cause IAQ problems.

Aerosols are fine particles suspended in air. Some are as fine as 0.001 mm in diameter. Respirable size particulate matter ranges from 1-10 mm diameter. This size aerosol can settle in the nonciliated portion of the lungs and lead to any of several health problems. Dust is mechanically generated aerosol that ranges from 10 to 200 mm in diameter. Allergens are aerosols composed of mold spores, pollen, and animal dander. Pathogen aerosols include bacteria and viruses (see Chapter 18). Radon progeny attached to dust is another hazardous airborne particulate.

Exposure to environmental tobacco smoke (ETS) has led many companies to adopt nonsmoking policies. ETS contributes to the PM load in a work area that is otherwise clean, see Table 12-9.

This is why nonsmokers object so much to the presence of ETS. With tight buildings and lower volumes of fresh air coming in, respirable PM

Table 12-8
IAQ Particulate Standards

Contaminant	Concentration Allowed	Time	Source
Asbestos	0.1 fibers/cc	Action Level	OSHA
Lead	1.5 μg/m³	1 yr	NAAQS
Particles, <10μm	50 μg/m³	1 yr	NAAQS
>10μm	150 μg/m³	24 hr	NAAQS
Radon	4 pCi/L	1 yr	EPA/TSCA

Table 12-9
Particles per Cubic Foot in Normal Work Areas

	0.3 μm	0.5 μm	1.0 μm
Smoking	290,000	220,000	25,000
Nonsmoking	80,000	20,000	1,000

concentration is an order of magnitude greater in smoking environments than in nonsmoking. Combine that with the increased awareness of nonsmoking workers about the hazards of ETS and you understand why the pressure to desmoke work places and public buildings is so great. The current thinking is that smokers do not have a right to do that to unwilling people.

PM is controlled by filters. Arrestance filters are flat, dry, open fiber panels or roll-type. Dust spot resistance filters are constructed of pleated paper, envelopes, or bags made of filter media. High efficiency particulate air (HEPA) filters are built in wooden boxes.

Typical filter efficiencies are described in Table 12-10.

ASHRAE provides a standard for the weight arrestance test and the dust spot test. Military Standard 282 addresses the *dioctyl phthalate* (DOP) test. As you see from the preceding table, the three tests do not produce equivalent numbers, but the three results can be related to one another.

Gas/Vapor Sampling and Control

Air monitoring involves collecting samples of air to be analyzed in a laboratory, unless a direct reading instrument is used. Reasons for air monitoring include emergency situations, cleanup of spills, and ongoing industrial hygiene programs.

Assuming that you know or suspect that a toxic chemical is released in the workplace, the next step is to determine the level of exposure and document the results. Sampling is only required if you estimate that the potential exposure of any employee could be at the Action Level (AL) or greater. If concentrations greater than the AL are anticipated, measure the exposure of the employees at greatest risk. Then identify and measure the exposure of

Table 12-10
Filter Efficiency by Type and Test

Filter Media	Weight Arrestance Test	Dust Spot Test	DOP Test (Mil Std 282)
Fine open foams	70-80	15-20	0
Cellulose mats	80-90	20-35	0
Wool felt	85-90	25-40	5-10
Mats, 5-10µm, ¼″	90-95	40-60	15-20
Mats, 3-5µm, ½″	>95	60-80	30-40
Mats, 1-4µ,fiber	>95	80-90	50-55
Mats, 0.5-2µm, glass	–	90-98	75-90
Wet-laid glass, HEPA	–	–	95+

employees in atmospheres that could possibly be at AL or higher. If these latter exposures turn out to be less than AL, the sampling program may be discontinued. If the exposure exceeds PEL, immediate action is required. If the AL is exceeded, but not PEL, make another measurement within six months of the first.

Table 12-11 lists some standards for gases and vapors that could affect IAQ.

For standards for various air contaminants in general industry, see OSHA 29 CFR 1910 Subpart Z.

Volatile organic compounds (VOCs) irritate mucous membranes, cause headaches and neuropsychological dysfunction. IAQ concentrations of total VOCs 200-500 $\mu g/m^3$ (1-2 ppm) are worthy of further investigation. At 600 $\mu g/m^3$, as many as 20 percent of occupants may complain of URT irritation and headaches. Most people do not smell VOCs until a concentration of 1,000 $\mu g/m^3$ is reached. Sources of these chemicals are stay pressed fabrics, molding and wood adhesives, latex caulks, paint, vinyl rubber moldings, carpet backings, carpet adhesives, neoprene duct liners, office furnishings, foam insulation, HVAC systems, cleaning compounds, bioeffluents from

Table 12-11
IAQ Gas / Vapor Standards

Contaminant	Concentration Allowed	Time	Source
Carbon dioxide	800 ppm	Continuous	OSHA proposed
Carbon monoxide	1,000 ppm	Continuous	ASHRAE
Formaldehyde	9 ppm	8 hr	NAAQS
Nitrogen	0.1 ppm	Continuous	ASHRAE
dioxide	0.05 ppm	1 yr	NAAQS
Ozone	0.05 ppm	Continuous	ASHRAE

humans, and ETS. Many industrial processes emit VOCs.

Acetic acid causes irritation of the eyes, respiratory system, and mucous membranes. Sources are silicone caulking compounds, X-ray development equipment, and photographic development chemicals.

Since it is a natural constituent of the air we breathe in, carbon dioxide (CO_2) is not a contaminant. Our bodies also produce carbon dioxide. However, if the concentration of carbon dioxide exceeds 5,000-10,000 ppm, people start to get drowsy and their respiration rate increases. This may cause irritability and general dissatisfaction without knowledge of the cause.

Carbon monoxide (CO)is a chemical asphyxiant. As little as 10 ppm causes fatigue. Exposure to 25 ppm for 24 hours may impair visual acuity. Only 50 ppm for 24 hours will give a headache, but worse, it leads to irregular heartbeat. One hour of exposure at a level of 500 ppm yields nausea, headache, and mental confusion. Up to 1,500 ppm for one hour may be fatal. Sources of CO are ETS, malfunctioning heating equipment and vents, motor vehicle exhaust including industrial vehicles, internal combustion engines, and steel production.

Formaldehyde causes skin irritation, URT irritation, and eye irritation. As little as 2-10 ppm may cause headache, nausea, vomiting, dizziness, and coughing. Concentrations of formaldehyde as low as 0.10 ppm may cause symptoms in hypersensitive persons and may induce asthma in asthmatics.

Table 12-12 Formaldehyde Emissions from Common Materials	
Material	**µg/m²-day**
Clothing	400-470
Hardwood plywood paneling	1,500-34,000
Medium density fiberboard	17,000-55,000
Paper products	260-680
Particle#board	2,000-25,000
Softwood plywood	240-720
Urea-formaldehyde foam insulation	1,200-19,000

Formaldehyde is a known sensitizer. Stay pressed fabrics, building materials, furnishings, and foam insulation emit formaldehyde. Table 12-12 gives formaldehyde emissions from common materials.

Nitrogen dioxide causes irritation of the URT and eyes. As little as 2-10 ppm may cause headache, nausea, dizziness, vomiting, and coughing. Sources are furnaces, motor vehicles, welding, and combustion equipment.

Ozone is irritating to the eyes, mucous membranes, and URT. Exhaust from internal combustion engines combined with sunlight produce ozone. Electronic equipment and electric motors also produce ozone. Electrostatic precipitators and copy machines also are sources. Lightning is the chief natural source of ozone.

Pesticides such as insecticides, herbicides, fungicides, rodenticides, and others, can cause dizziness, headaches, nausea, and vomiting.

Air Sampling. No matter what the air contaminant, collect samples upwind to establish background levels. Collect samples downwind to measure contaminants leaving a site or work area.

Carry samples to a clean place for cataloging and labeling. Ensure that contamination on sample containers and other equipment is dealt with safely.

Activated carbon is used to collect organic vapors with a boiling point of greater than 0°C. Tenax® or Chromosorb® or some other porous polymer is used to collect high molecular weight hydrocarbons, organo-phosphorus compounds, and certain pesticide vapors. Silica gel or some other polar sorbent is used for aromatic amines. Florisil® is used for polychlorinated biphenyls (PCBs). Inorganic gases can be collected on either silica gel or in impingers filled with specific reagents. Aerosols are often collected in particulate filters.

Diffusion samplers have become an acceptable method of collecting many gases and vapors. Gas molecules will move from higher to lower across a concentration gradient. The diffusion samples provide a space full of stagnant air with low (zero) concentration of the contaminant of interest. Typically, when you buy the badge you also buy the analytical chemistry to read the badge.

Permeation devices rely on natural permeation of a contaminant through a membrane. The optimum membrane is easily permeated by the contaminant of interest, but impermeable to all other contaminant species. In practice this is rarely the case, but often more complex sampling methods can be avoided by an effective permeation sample.

When selecting an instrument to measure concentration, be sure to carefully read its specifications. Instruments are valid for certain conditions and using them in situations for which they are not designed can give disastrous results. Know the accuracy of your instrument. *Accuracy* is the measure of how close the instrument reads to the actual concentration of the contaminant of concern. If you read 250 ppm with an instrument that is ±5% accurate, the real concentration may be as low as 237.5 ppm or as high as 262.5 ppm, or 12.5 ppm either way. Be careful about reading and understanding the accuracy specification. If the manufacturer says the accuracy of the instrument is ±5% FS, that means full scale and the range or error is now ±50 ppm, if the full scale is 1,000 ppm. So then the range in our previous example (250 ppm) is from a possible low of 200 ppm to a possible high of 300 ppm.

Resolution is the smallest incremental reading that can be discerned from the scale. This is generally one order of magnitude less than the tick marks on an analog scale. Resolution is another matter on a digital readout where it represents the least decimal place available. Specifications can be confusing if they report resolution in the same way as accuracy. For instance, an instrument specification may read: resolution ±0.1 ppm. On an analog scale that means the tick marks represent every 1 ppm and you can read the nearest tenth of a ppm.

Another important specification is the *minimum detection level* (MDL) or the lowest reading available on the instrument. You may not assume that a "no reading" is equivalent to zero, only that the real concentration is something less than the MDL. Now, consider the combination of a *no reading* and the accuracy of ±5% as in our example. Do you have a true zero? Or do you have 50 ppm? Shame on you if you say either zero of fifty is the right number. All you can truthfully report is that the reading was less than MDL and the actual concentration may be as high as 50 ppm.

Sampling error is created by taking a small sample from a large amount of material during a finite time from a continuous process or permanent deposit. But what is the reading in the rest of the material? What will the reading be next week? You'll never know unless you sample again. One common mistake is to conclude that all the material, and at all times, is like your sample. Sometimes that may be close to true, but not necessarily and not at all times. The best you can do is to explicitly state your assumption that the sample is a representative one. In critical situations it is a good practice to discuss the alternatives. What if your sample is not representative?

For potentially flammable or explosive atmospheres, make sure you use equipment that is intrinsically safe for the appropriate atmosphere (see Chapter 20). Instruments are either intrinsically safe by design or by certification. When the manufacturer makes the instrument according to accepted practices that are known to limit the release of heat and electrical energy, it is designed to be intrinsically safe. If the instrument was proved by being tested in a flammable atmosphere in an independent testing laboratory, it is certified to be intrinsically safe.

Confined Space Entry

The purpose of the Confined Space Entry Standard is to prevent workers from experiencing suffocation and death by toxic inhalation. A northern utility company lost two meter readers and their supervisor due to their entry into a basement where the meter was located. Unfortunately, a compost pile was also kept in this confined space. Composting is a slow combustion process that consumes oxygen. With the basement closed up, the compost pile depleted the oxygen in the room and the first meter reader suffocated before he knew what hit him. The second meter reader, who was sent out the next day to find out why the first man did not return to the office on the previous day, also suffocated. Later that afternoon, the supervisor made the rounds to find out why the second man did not return. The next morning all three were seen in the basement and the rescue squad was wisely called, preventing further deaths.

When I was in the Navy, we used to hear horror stories every few months about a sailor or yard worker being overcome in a void (an empty, unused space on a ship that enhances buoyancy). Bacteria in sludges in these voids either consume the oxygen in the space or produce the toxic hydrogen sulfide gas. Recently, two industrial workers were killed while performing maintenance in a boiler. When boilers are operating, a chemical called the oxygen scavenger is added to the water to capture and remove free oxygen molecules in order to prevent corrosion. While this boiler was down for repairs and the two men were inside, a pool of the oxygen scavenger depleted the oxygen in the boiler and the men suffocated.

A confined space, according to OSHA, is any space that was not designed for normal human occupancy and potentially contains a hazardous environment. The environment may be hazardous due to a hazardous atmosphere, such as oxygen deficiency or a toxic gas. The potential for engulfment also creates a hazardous environment. If the internal configuration of the space could cause a worker to become trapped or asphyxiated while inside, it is considered a confined space. If other recognized safety or health hazards could prevent the employee inside from escaping, a confined space is involved.

Confined space entry is not rocket science, it is just a painstakingly slow assessment of conditions and use of so-called common sense. With all the deaths that have occurred in confined spaces, however, you would think the sense required is quite uncommon. Figure 12-27 lists the steps for establishing a sound confined space entry program. The potential hazards of confined spaces are given in Figure 12-28.

A properly identified confined space has a limited means of entry and exit. It is large enough for an employee to enter and perform work, but is difficult to get out of in a hurry. Finally, a confined space is a space that is not intended for continuous occupancy. A confined space requires a permit for entry if it has the potential to contain one or more of the hazards listed in Table 12-12. Remember:

Monitor – Ventilate – Lockout

Figure 12-27
Steps Required to Safely Enter Confined Spaces

1. Identify potential confined spaces.
2. Evaluate the potential hazards in each space.
3. Label permit required confined spaces (a hazard could be present) and confined spaces that do not require a permit (no hazard present during entry). Lock the permit required spaces closed, if possible.
4. Identify employee job duties that require entry.
5. Develop and implement procedures and practices to eliminate or control hazards for necessary confined space entry.
6. Provide the necessary personal protective equipment and rescue equipment.
7. Train personnel on getting a permit, safe entry procedure, communications, and rescue activities.
8. Establish a permit system for entry.
9. Post at least one attendant outside the confined space.
10. Provide a means to communicate and summon rescue assistance to those inside the confined space.

Figure 12-28
Confined Space Hazards

Entrapment
Asphyxiation
Explosion
Engulfment

The proper testing order for confined space is oxygen content, lower explosive limit (LEL), and toxic gas concentration. Not two long ago the proper order of testing was first for explosive gases, then for oxygen content, and finally for toxic gases. Twenty and more years ago, oxygen content was measured by a device called a flame safety lantern, which had a flame that burned bright in adequate oxygen, dim in poor oxygen, and extinguished in inadequate oxygen. Not as neat as a direct reading oxygen concentration monitor available today, but just as accurate in a qualitative way. The lantern was lit, however, by a real flame—this obviously was not recommended if a combustible gas or vapor were present.

Anyway, you can use a direct reading meter now so take the oxygen reading first. This is good procedure for two reasons. First, when concentrations of gases that are immediately dangerous to life and health (IDLH) are present, oxygen deficiency is the primary criteria. Second, instruments read differently in oxygen deficient atmospheres.

Warning!

Instruct employees to evacuate a confined space immediately if they detect the presence of hazardous conditions during entry.

If this occurs, evaluate the space to determine the cause of the hazardous atmosphere. Then implement measures to protect employees from the hazard before reentry.

Oxygen Deficient Atmospheres

Oxygen deficiency is a life threatening hazard. A certain *partial pressure* of oxygen is required in order to support the metabolic processes of the human body. Normally the atmosphere has a fixed composition (See Table 12-13) and we do not think about oxygen in the air we breathe. However, when the oxygen level starts dropping, we notice the effects pretty quickly.

Table 12-13
Normal Earth Atmosphere at Sea Level

Gas	Volume (dry)
Nitrogen	78.09
Oxygen	20.95
Argon	00.93
Carbon Dioxide	00.04

Air also contains trace amounts of other gases (on the parts per million or ppm level) such as neon, helium, and krypton. Water vapor is another important constituent of air and may represent up to five percent of the total wet volume.

For physiological reasons, the partial pressure of the oxygen (percent dry volume) must be maintained at a certain level. Just where the partial pressure becomes dangerously low is a matter of interpretation and various organizations define oxygen deficient atmosphere differently (see Table 12-14).

The symptoms of oxygen deficiency depend on the exact partial pressure of oxygen in the breathing air. Table 12-15 summarizes physiological effects as the partial pressure decreases.

Under what conditions might you find deficient oxygen levels? Anywhere an oxygen consuming process is taking place, such as a decaying organic pile. Most piles of organic materials are suspect, whether composting is intended or not. Areas where internal combustion engines are operating or where they exhaust may be deficient in oxygen. The vicinity of processes

Table 12-14
Definitions of Oxygen Deficient Atmosphere

Source or Standard	Oxygen Content, %
American Conference of Governmental Industrial Hygienists (ACGIH) *(Threshold Limit Value)* (1973)	18.0
OSHA 29 CFR 1910.94 (General Industry) 29 CFR 1915.51 (Maritime)	 19.5 16.5
American National Standards Institute (ANSI) Z88.2 (Respiratory Practices) Z88.5 (Firefighting)	 16.0 19.5

Table 12-15
Physiological Effects of Oxygen Deficiency

O_2 Level, Vol. % @ sea level	Effect
16-12	Increased breathing volume. Accelerated heart rate. Impaired attention and thinking. Impaired muscle coordination.
14-10	Very faulty judgment. Very poor muscle coordination. Muscular exertion causes rapid fatigue that may lead to permanent heart damage. Intermittent respiration.
10-6	Nausea. Vomiting. Inability to perform vigorous movement or loss of all movement. Unconsciousness, followed by death.
Less than 6	Spasmodic breathing. Convulsive movements. Death in minutes.

using solvents is also suspect. Deficient levels of oxygen may also be found in areas where inert gases (such as nitrogen, methane) or the noble gases (helium, neon, argon, krypton, xenon, and radon) may displace oxygen. Welding and brazing are two more suspect processes. Anywhere that combustible gases and vapors pose a threat of fire and explosion may have deficient oxygen levels. Also, wherever toxic gases and vapors pose a real threat, the oxygen level may be deficient. Processes emitting carbon monoxide or cyanide, while the oxygen level may be normal, are dangerous because these compounds prevent the body from properly processing oxygen and suffocation ensues anyway.

Metalworking Fluids

Major changes introduced into the U.S. machine tool industry during the latter decades of the twentieth century led to a significant increase in the overall consumption of metalworking fluids (MWFs). The use of synthetic MWF increased as tool and cut speeds increased. Newer automated machines tend to be totally enclosed, which allows the use of local exhaust ventilation. Around 1975, many plants started installing recirculating air cleaners, improving existing air filtration systems, and adding comfort air conditioning. MWF exposures are declining from what they were in the mid-half of the twentieth century. NIOSH has confirmed that improvements in engineering controls and work practices have reduced concentration of MWF to less than 0.5 mg/m^3 in most plants.

Nevertheless, recently a concern among bargaining units, notably the United Auto Workers, about exposure to MWF, has led OSHA to begin the process of issuing a new standard. Many epidemiological studies have been conducted to learn whether these chemical mixtures are as dangerous as some think. A weak correlation between MWF and chronic chemical pneumonitis has been shown, however, if you factor out those predisposed (because they started to work with asthma), no one is left. This section will discuss MWFs and their generic metal removal fluids (MRFs) briefly in order to make you aware of the issue, trends, and pending action by OSHA.

Effects of MWFs

MWF (coolant) is a complex mixture of oil, detergent, surfactant, biocide, lubricant, anti-corrosive, and other additive packages. MWF is used in grinding, boring, drilling, and turning processes. The useful properties of these fluids is that they remove heat from workpieces as metal is removed, but they also serve as lubricants, and corrosion inhibitors.

Modern health studies suggest, but do not prove, exposure to MWFs increase the risk for skin diseases, respiratory effects, and cancer. The suspect respiratory effects include irritation, bronchitis, lipoid pneumonia, hypersensitivity pneumonitis, and asthma. Cancer sites under study include the esophagus, stomach, pancreas, larynx, colon, and rectum. Skin effects of concern are folliculitis, oil acne, keratosis, irritant contact dermatitis, and allergic contact dermatitis.

Effects on Skin

Perhaps as much as one-third of the workforce that uses MWFs have some dermatitis symptoms at any given time. Often these problems are not reported to the OSHA log recordkeeper. However mild, skin irritation begins at the moment and site of contact with a MWF. This irritation acts to prevent these fluids from reaching more sensitive underlying tissue. The outermost, least permeable protective layer is the epidermis. This cellular tissue is surrounded by lipids that act as mortar does to solidify a brick wall. Water-based cutting fluids start the dissolution of the essential surface lipids of the skin. *Delipidization* is an early symptom of irritant dermatitis. Emulsifiers and wetting agents destroy the fatty fraction of the cell membrane, leading to a whitening and drying of the affected area. Once the lipid barrier is dissolved, the skin becomes extremely permeable. A capillary bed lying directly beneath the dermis gives access to the bloodstream.

Oil folliculitis or oil acne is caused by the plugging of the follicular canal by cutting fluids. Hyperpigmentation, photosensitivity, and eczematous dermatitis are also caused by exposure to cutting fluids.

Industries where workers have exposure to MWFs have the highest incidence rates of skin disease. Insoluble or straight MWFs produce folliculitis, oil acne, and keratoses. Soluble, synthetic, and semisynthetic MWFs primarily cause irritant contact dermatitis and occasionally allergic contact dermatitis. Skin carcinomas have been associated with PAH content of mildly refined base oils. Severe refinery methods reduce PAH content to less than one percent.

Skin carcinomas have been associated with straight MWFs. However, changes in refinery methods have reduced the content of polycyclic aromatic hydrocarbons (PAHs), which are the suspected carcinogens. Not only that, but the MWFs are not used straight, but diluted. Therefore recent use of MWFs has not produced any skin carcinogenesis. Straight oil exposure has been found to be associated with increased risk for laryngeal and rectal cancer. Synthetic fluid exposure was associated with an increased risk for pancreatic cancer. Unfortunately, important lifestyle factors were not determined in these studies and all the suspect cancer sites are also sites of major lifestyle cancers. Studies have not shown an association between straight MWF exposure and cirrhosis or lung cancer.

Inhalations Effects

The causative agent in lipid pneumonia is respirable mineral oil. The results of many epidemiological studies of respiratory symptoms present compelling evidence that exposure to MWF aerosols causes symptoms consistent with airways irritation, chronic bronchitis, and asthma. Clinical asthma induced by MWF exposure appears to involve specific sensitizers.

Another potential exposure from MWFs is microbial contamination: bacteria and fungi. MWFs, which are water-based or water-contaminated, may be colonized by bacteria and fungi. High aerobic bacterial counts have been measured in many MWFs. Workers may be exposed to microbially contaminated MWFs by extensive skin contact and by inhaling aerosols. Some of the respiratory effects reported in exposed workers may be related to contaminating microorganisms and their products: endotoxins, exotoxins, and mycotoxins. The toxic effects of endotoxin inhalation include fever, obstructive pulmonary effects, and inflammation.

Opportunistic pathogens are more common contaminants. One MWF incident was traced to a self-limited nonpneumonic form of Legionellosis that produced influenza-like symptoms. An outbreak of hypersensitivity pneumonitis was related to Pseudomonas fluorescens contaminated MWF.

In a study reported by Woskie, *et al,* MWF exposed workers had a higher exposure to culturable bacteria in the aerosol fraction that is less than eight microns in aerodynamic diameter than in larger size fractions. The airborne bacteria associated with MWF aerosols was reported to be more likely to be *Pseudomonas,* whereas workers who were not exposed to MWF were more likely to be exposed to *Bacillus.*

Exposure Prevention and Reduction

In December 1995, OSHA established a priority planning process for rulemaking. In January 1996 EPA began screening twenty high production MWFs for health effects. NIOSH recommended a draft standard in May 1996. ANSI drafted a voluntary U.S. standard in January of 1996. Manufacturers of MWFs and the Independent Lubricant Manufacturers Association, in January 1996, disputed OSHA's claim that MWF can lead to cancer. MWF users, primarily the UAW and the American Automobile Manufacturers Association, presented symposiums on the topic in Fall 1995 and again in Fall 1997.

The old TWA exposure limit is 5.0 mg/m^3 as oil mist. A consensus is developing to introduce a new TWA exposure at 0.5 mg/m^3 as total particulate.

What can you do to prevent worker exposure? First, evaluate the MWF being used now. Can you use safer alternatives? What are the current health concerns? Next, determine the TWA in your plant. Target 0.5 mg/m^3 TWA exposure as your highest acceptable level. Mist eliminators, barriers, and electrostatic precipitators are commonly used engineering controls. Personal air-purifying respirators may be called for in extreme cases where a high residual concentration of MWF remains airborne despite engineering controls. Gloves should be worn to handle MWF soaked parts. Operators of machines using MWF should practice exemplary cleanliness. Training is need for hazard communication and PPE usage.

Splashing and mist generation can be minimized by applying the MWF at the lowest possible pressure and flow volume consistent with adequate part cooling, chip removal, and lubrication. The MWF should be applied at the tool-workpiece interface, minimizing contact with other rotating parts of the machine. Fluid delivery should cease when machining ceases. Select MWFs that have less than one percent mildly refined base-oils. That way, any PAH content will be less than one percent. Monitor the MWF and additive concentrations continually and adjust as needed.

Biocides should be used to control microbial overgrowth. During a shutdown when no circulation or aeration of the MWF occurs, the aerobic population decreases as the oxygen in the fluid is consumed. This results in the MWF overgrowth with anaerobic bacteria, which produce noxious gases and odors. Keep in mind that the use of biocides to control bacteria growth typically leads to fungi proliferation. This may lead to clogged filters and will interfere with the metalworking operation.

Another disadvantage of dosing MWFs with biocides is the production of endotoxins from the dead organisms. A study by Thorne, DeKoster, and Subramanian showed the airborne levels of endotoxin in automotive machining plants may exceed the thresholds of individuals with acute respiratory problems. The study correlated the airborne endotoxin concentration with the concentration in bulk MWF. Bulk concentration of endotoxin is particularly high when microbial growth in the MWF reservoir is high. Therefore, a program of carefully monitoring and control of *microbiota* and endotoxin in machining fluids is recommended.

Hands, *et al,* studied the mist exposure from three different kinds of machine enclosures and reported their results. They found that the state-of-the art enclosure by the original equipment manufacturer (OEM) was superior to either retrofit enclosures or little or no enclosure. OEM enclosures are capable of maintaining MWF exposures less than 0.5 mg/m^3, especially when paired with local exhaust, whereas retrofits and no protection may have higher exposures.

Compressed Gas Management

A gas or mixture of gases contained with a pressure exceeding 40 psi at 70°F is called a compressed gas. A gas or mixture of gases contained with a pressure exceeding 104 psi at 130°F, regardless of the pressure at 70°F, is also a compressed gas. A liquid having a vapor pressure exceeding 40 psi at 100°F as determined by ASTM D-323-72 (the so-called Reid vapor pressure) is technically a compressed gas.

Cylinders of compressed gases must be stored in a manner that prevents tipping, falling, or rolling. If you have ever seen a compressed gas cylinder tear through a cement block wall after someone knocked it over and the regulator valve broke off, you will understand this requirement. Chain compressed gas bottles to a wall, stanchion, or building support. Replace the protective caps on the valve when not in use. Maintain separate storage for flammable gases and oxidizers. Place guard posts around liquefied petroleum tanks to prevent vehicles from colliding with the tanks. Post an adequate number of no smoking signs. Always store acetylene cylinders with valve end up to prevent dangerous decomposition.

Have the vendor or distributor pick up bottles that have illegible labeling. The label must identify the hazardous material contained, give appropriate hazard warnings, and identify the manufacturer by name and address.

Leave compressed gas cylinders outside of confined spaces.

References

About Protecting Yourself with PPE (Personal Protective Equipment). South Deerfield, MA: Channing L. Bete Company, Inc., 1973.

Anderson, David O. "Basic IH Screening Techniques." *Occupational Health & Safety.* May 1996, pp. 116, 118.

Anderson, David O. "Dispelling the Myths of Toxicology." *Occupational Health & Safety.* May 1996, pp. 28-33.

Anderson, Rosalind C. "Indoors, the Newest Polluted Space." *Pollution Engineering.* April 1, 1992, pp. 58-60.

Baier, Edward J. "Endangered Species." *Occupational Health & Safety.* July 1989, pp. 20-23.

Brock, Melanthea. "IDLH, the Loaded Gun." *Occupational Health & Safety.* May 1996, p. 40.

Burke, Adrienne. "Chemical Exposures: Where You Least Expect Them." *Industrial Safety & Hygiene News.* September 1996, pp. 21-22.

Burke, Adrienne. "Sick Building Syndrome." *Industrial Safety & Hygiene News.* October 1997, p. 48.

Burke, Adrienne. "Under Your Skin." *Industrial Safety & Hygiene News.* July 1996, pp. 21-22.

Burke, Adrienne. "What Workers Need to Know about Contact Dermatitis." *Industrial Safety & Hygiene News.* July 1996, p. 26.

Burton, D. Jeff. "Finding Equivalent Round Duct Diameters for Square and Rectangular Ducts for Friction Estimation." *Occupational Health & Safety.* Date unknown, p. 20.

Burton, D. Jeff. "A Fix on Fans." *Occupational Health & Safety.* April 1998, p. 18.

Burton, D. Jeff. *IAQ and HVAC Workbook.* 2d Ed. Bountiful, UT: IVE, Inc., 1995.

Burton, D. Jeff. "New Equation Estimates Emission Rates from Pooled Liquids." *Occupational Health & Safety.* Date unknown, p. 32.

CBC 100: Basic Course in Fan Selection. Glendale Heights, IL: Chicago Blower Company, no date.

Cheremisinoff, Nicholas P. "Equipment Roundup: Flans, Blowers, and Compressors." *The National Environmental Journal.* January/February 1992, pp. 60-62.

Colton, Craig E. "Respirator Fit Testing." *Occupational Health & Safety.* May 1996, pp. 54-65.

Dalene, Marianne, Gunnar Skarping, and Pernilla Lind. "Workers Exposed to Thermal Degradation Products of TDI- and MDI-Based Polyurethane: Biomonitoring of 2,4-TDA, 2,6-TDA, and 4,4'-MDA in Hydrolyzed Urine and Plasma." *AIHA Journal.* August 1997, pp. 587-591.

Daugherty, Jack E. "The Safety Equation: Modeling Safely Predicts Solvent Vapor Exposure." *Parts Cleaning.* November/December 1997, pp. 21-23.

Daugherty, Jack E. "Stopping Hazardous Materials at the Door." *Occupational Hazards.* May 1996, pp. 83-85.

De Chacon, Jeffrey R. and Norman J. Van Houten. "Indoor Air Quality." *The National Environmental Journal.* November/December 1991, pp. 16-18.

Eichman, T.G. "Determining Respiratory Protection Needs." *Plant Services.* November 1993, pp. 50-53.

Eisma, Patricia Lyn. Practical Advice on Gas Detection for Confined Space Entrants' Safety." *Occupational Health & Safety.* December 1992, pp. 18-24.

Fenske, Richard A. and Shari G. Birnbaum. "Second Generation Video Imaging for Assessing Dermal Exposure (VITAE System)." *AIHA Journal.* September 1997, pp. 636-645.

Frund, Zane N., Jr. and John W. Cobes III. "Protecting against Multiple Hazards." *Occupational Health & Safety.* November 1997, pp. 32-36.

Gressel, Michael G. "An Evaluation of a Local Exhaust Ventilation Control System for a Foundry Casting-Cleaning Operation." *AIHA Journal.* May 1997, pp. 354-358.

Gupta, Ram S. *Environmental Engineering and Science: An Introduction.* Rockville, MD: Government Institutes, Inc., 1997.

Hall, Stephen K. "Health Risk Assessment of Chemical Exposure." *Pollution Engineering.* December 1988, pp. 92-97.

Hall, Stephen K. "Biological Monitoring of Metal Exposure." *Pollution Engineering.* January 1989, pp. 128-131.

Hands, David, Maura J. Sheehan, Ben Wong, and Henry B. Lick. "Comparison of Metalworking Fluid Mist Exposures from Machining with Different Levels of Machine Enclosure." *AIHA Journal.* December 1996, pp. 1173-1178.

Johnson, Linda F. "Selecting Personal Protective Apparel." *Occupational Health & Safety.* May 1996, pp. 67-76.

Joseph, Maureen. "HPLC Detector Options for the Determination of Polynuclear Aromatic Hydrocarbons." *Pollution Equipment News.* August 1992, pp. 86-87.

Keith, Michael F. "Surveying Air for Onsite Action." *The National Environmental Journal.* May/June 1992, PP. 49-52.

Keenan, Tom. "Fear of Chemicals: How to Develop Respect and Prevent Panic among Workers." *Industrial Safety & Hygiene News.* September 1989, pp. 34.

Johnson, Arthur T., Cathryn R. Doody, David M. Caretti, Michael Green, William H. Scott, Karen M. Coyne, Manjit S. Sahota, and Benhur Benjamin. "Individual Work Performance during a 10-Hour Period of Respirator Wear." *AIHA Journal.* May 1997, pp. 345-353.

Johnson, Linda F. "Choosing Where to Begin." *Occupational Health & Safety.* May 1996, p. 32.

Johnson, Linda F. "Dangerous Atmospheres: IH Concerns in Confined Spaces." *Occupational Health & Safety.* May 1996, pp. 43-47.

Korbee, Leslie, and I.L. Bernstein. "Workers Risk Occupational Asthma by Exposure to Myriad of Chemicals." *Occupational Health & Safety.* March 1988, pp. 28-31.

LaBar, Gregg. "Keeping the Skin Safe." *Occupational Hazards.* August 1997, p. 35.

LaBar, Gregg. "Substituting Safer Materials." *Occupational Hazards.* November 1997, pp. 49-51.

Lapides, Michael A. "Cutting Fluids Expose Metal Workers to the Risk of Occupational Dermatitis." *Occupational Health & Safety.* April 1994, 82-86.

Leffert, Kim A. "Employers Must Consider Legal Aspects of Reproductive Hazards." *Occupational Health & Safety.* July 1989, pp. 24-5.

Lemke, Peter T. "Compressed Gas Cylinders: Safe Transport, Storage, and Handling." *Occupational Hazards.* February 1998, pp. 43-45.

Markiewicz, Dan S. "Ranking Substitute Chemicals." *Industrial Safety & Hygiene News.* April 1993, p. 16.

McClure, Alan D. "Compressed Gas Safety and Your Facility." *Occupational Health & Safety.* May 1996, pp. 84&86.

Ness, Shirley A. *Surface and Dermal Monitoring for Toxic Exposures.* New York: Van Nostrand Reinhold, 1994.

NIOSH. *Criteria for a Recommended Standard: Occupational Exposures to Metalworking Fluid.* U.S. Department of Health and Human Services. Public Health Service. Centers for Disease Control and Prevention. National Institute for Occupational Safety and Health.

Odom, J. David, III, and Christine R. Barr. "Sick Building Litigation: The Role That Occupant Outrage Plays." *The Synergist.* October 1997, pp. 28-30.

OSHA Compliance Manual. The Merritt Company, 1988.

Plog, Barbara A., ed. *Fundamentals of Industrial Hygiene.* 3rd ed. Chicago: National Safety Council, 1988.

Polakoff, Phillip L. "Chemical Toxicity Damages Liver; Greatest Danger Is from Inhalation." *Occupational Health & Safety.* November 1987, pp. 16-17.

Rekus, John F. "Hazard Communication: OSHA's Most Frequently Cited Standard." *Occupational Hazards.* February 1998, pp. 39-42.

Rekus, John F. "Selecting the Right IH Instrument." *Occupational Hazards.* September 1997, pp. 43-46.

Sandler, Howard M. "Symptoms: The Workplace Dilemma." *Occupational Hazards.* July 1997, pp. 53-54.

Schlatter, C. Nelson. "Permeability of Chemical-Protective Gloves Plays Major Role in Safety." *Occupational Health & Safety.* February 1989, pp. 24-31.

Smandych, R. Susan, Murray Thomson, and Howard Goodfellow. "Dust Control for Material Handling Operations: A Systematic Approach." *AIHA Journal.* February 1998, pp. 139-146.

Smith, S.L. "Thin-Skinned: The Risks of Dermal Exposure." *Occupational Hazards.* September 1993, pp. 111-114.

Swanson, Sandra. "PPE and Chemical Hazards: Don't Let a Bad Marriage Burn You." *Safety + Health.* September 1996, pp. 58-60.

Talty, John T., ed. *Industrial Hygiene Engineering: Recognition, Measurement, Evaluation, and Control.* Park Ridge, NJ: Noyes Data Corporation, 1988.

Thorne, Peter S. and Jeannine DeKoster. "Pulmonary Effects of Machining Fluids in Guinea Pigs and Mice." *AIHA Journal.* December 1996, pp. 1168-1172.

Thorne, Peter S., Jeannine A. DeKoster, and Periyasamy Subramanian. "Environmental Assessment of Aerosols, Bioaerosols, and Airborne Endotoxins in a Machining Plant." *AIHA Journal.* December 1996, pp. 1163-1167.

Toth, Weldonna and Dean Liliquist. "Data Quality in Industrial Hygiene." *Occupational Health & Safety.* May 1996, pp. 39, 41.

Vandergrith, Edwin F. "Meeting OSHA Regulations on Toxic Exposure." *Chemical Engineering.* June 12, 1980, pp. 69-73.

Werely, Linda. "PPE Selection for Chemical Use: Looking Beyond OSHA's Guidelines." *Occupational Health & Safety.* November 1993, 27, 30-31.

Wilsey, P.W., J.H. Vincent, M.J. Bishop, L.M. Brosseau, and L.A. Greaves. "Exposures to Inhalable and "Total" Oil Mist by Metal Machining Shop Workers." *AIHA Journal.* December 1996, pp. 1149-1153.

Woskie, Susan R., Mohammed Abbas Virji, David Kriebel, Susan R. Sama, David Eberiel, Donald K. Milton, S. Katherine Hammond, and Rafael Moure-Eraso. "Exposure Assessment for a Field Investigation of the Acute Respiratory Effects of Metalworking Fluids. I. Summary of Findings." *AIHA Journal.* December 1996, pp. 1154-1162.

Zakrzewski, S. Ed. "How Toxic Are You?" *Professional Safety.* April 1993, pp. 26-30.

13

Human Factors and Ergonomics Management

Overview

Ergonomics is the science of measuring equipment and surroundings with respect to the people who use the equipment and inhabit the surroundings. Ergonomics is also the art of assessing work procedures as they adapt to people.

A Polish professor, Jastrzebowski, coined the word "ergonomics" around 1860. Literally, ergonomics is the measurement of work but the emphasis is on the capabilities of the worker to perform the task safely, rather than on production or quality. Ergonomics became popular during World War II when complex and confusing aircraft, radar, and other equipment created performance and maintenance problems. Such problems were addressed by engineers, psychologists, anthropologists, and physiologists, leading to a multidisciplinary approach to ergonomics. Phrases such as *human engineering and engineering psychology* were used to describe these efforts to create the most utilitarian workspace for such diverse men and women. The nascent science was called *ergonomics* in Europe, but *human engineering or human factors engineering* in America. In 1993, the Human Factors Society was renamed the Human Factors and Ergonomics Society.

To reiterate, ergonomics is the science of fitting the task to the person, not the person to the task. Ergonomics uses physiology, engineering, anthropometry, and psychology. *Anthropometry* is the science of measuring body dimensions and the capabilities of the body.

Ergonomic illnesses tend to consist of chronic trauma to the muscles and skeleton. In fact, the number one ergonomic illness is the category generically called *cumulative trauma disorder* (CTD) and specifically *carpal tunnel syndrome.* Applied ergonomics prevents illnesses such as carpal tunnel syndrome and back injury. Applied ergonomics also improves productivity and quality by improving performance and reducing errors. Employees who suffer quietly from CTD typically work slower to compensate for the pain, without even thinking about it.

Ergonomics assesses the physical and mental activities of employees and compares them to the physical and mental demands of the assigned task. The environment is also assessed, from lighting to noise level to tools being used. Ergonomics is not unlike what industrial hygienists and safety engineers have done for years, but some people specialize in ergonomics.

Musculoskeletal disorders (MSDs) are a significant health condition in many workplaces. Physical workload has been recognized as a substantial contribution in the etiology of sundry musculoskeletal complaints. Protracted physical load is implicated in distortion of biological structures, cumulative localized muscle fatigue, local inflammatory reactions, and increased levels of catecholamine releases. Table 13-1 reviews the evidence for MSD causes as determined by the National Institute of Occupational Safety and Health (NIOSH).

Human Error

Human error has been discussed before but it is critical to the art and science of ergonomics. The body tolerates a range of stresses and even heals itself without permanent effects. However, once the range of tolerance is exceeded, for whatever reason, but notably due to human error, the recovery ceases to be complete.

In design, human error contributes to building and installing machines and work stations that contribute to excessive stress on the human body. Even if the stress is within the normal range of tolerance for an individual, as that person's body ages, the range of tolerance grows tighter. Often, ergonomic design error is due to ignorance of ergonomic principles.

Often, ergonomic injuries occur due to human error in operations. For instance, an untrained employee attempts to perform a task that causes physical stress on the body when an experienced employee could have performed the task safely.

Improper maintenance is another human error that can lead to ergonomic injury or illness. Lack of maintenance on a crane hoist eventually causes a chain to slip several notches. An employee who attempts to grab and stabilize the load can suffer several strains and sprains.

Table 13-1
NIOSH Evidence for MSD Causes

Body Part	Risk Factor	Evidence
Neck/Shoulder	Repetition	Some
	Force	Some
	Posture	Strong
	Vibration	Insufficient
Shoulder	Repetition	Some
	Force	Insufficient
	Posture	Some
	Vibration	Insufficient
Elbow	Repetition	Insufficient
	Force	Some
	Posture	Insufficient
	Combination	Strong
Hand/wrist CTS	Repetition	Some
	Force	Some
	Posture	Insufficient
	Vibration	Some
	Combination	Strong
Hand/wrist tendinitis	Repetition	Some
	Force	Some
	Posture	Some
	Combination	Strong
Hand-arm vibration syndrome	Vibration	Strong
Back	Lifting/forceful movement	Strong
	Awkward posture	Some
	Heavy physical work	Some
	Whole-body vibration	Strong
	Static work posture	Insufficient

Human error in documentation, such as safety procedures, has ergonomic illness potential. One employee continued using the wrong lever to do a certain heavy lifting job and consequently was taking more force on his spine than necessary. No wonder his was a workers' compensation case for back strain.

The potential impact of human error is far reaching. Usually we think of physical accidents that lead to first aid cases or a visit to the emergency room at a nearby hospital as the only human error scenarios. More insidious is the chronic injury that in ergonomic circles is labeled an illness.

Recognition of Human Factor Situation

Ergonomic related problems in the workplace include cumulative trauma disorders (CTDs), repetitive motion disorders (RMDs), repetitive motion injuries (RMIs), repetitive strain injuries (RSIs), and musculoskeletal disorders (MSDs). These are not entirely independent of each other and the various problems do overlap. Ergonomics problems can be predicted if certain risk factors are present in the workplace. Some risk factors are job related, while others are not.

Job-related risk factors include awkward postures, repetition of strenuous tasks, force, frequent and/or heavy lifting, and vibration. Other job-related risk factors include contact stressors, low temperature, noise, and mental stress. The latter can lead to a psychological condition or as ergonomists say it, "the pain in the brain ain't necessarily due to strain." Non-job related risk factors include diabetes, arthritis, obesity, smoking, pregnancy, gender, vitamin B-6 deficiency, wrist size and shape, and the use of oral contraceptives.

The CTD issue is not new, as some would like to believe. In 1714, Ramazzini wrote:

> Various and manifold is the harvest of diseases reaped by craftsmen. All the profit they get is fatal injury to their health, mostly from two causes. The first and most potent is the harmful character of the materials they handle. The second, I ascribe to certain violent and irregular motions and unnatural postures of the body, by reason of which, the natural structure of the vital machine is so impaired that serious diseases gradually develop therefrom.

According to OSHA's Ergonomic Program Management Guidelines for Meatpacking Plants, issued in 1990, at least one physical finding such as a positive Tinel's, Phalen's, or Finkelstein's test indicates a reportable CTD illness. Swelling, redness, or deformity, or loss of motion also indicates a reportable illness. If one or more subjective symptom is presented, such as pain, numbness, tingling, aching, stiffness, or burning, and at least one of the following: medical treatment, lost work days, restricted work activity, transfer to another job, or rotation to less strenuous jobs.

Sandler cites the Sir Bradford Hill criteria for determining whether repetitive motion injuries are job-related or not. First, what is the *strength of association* between the RMI and work tasks? The second criterion is *consistency*.

In other words, is the work-related explanation consistent with the type of injury you would expect? *Specificity* is another criterion. Can the injury be tied to a specific task? *Temporality* means that the injury occurred about the time the suspect work task was performed. A *biologic gradient* means that a dose response exists. The more the suspect task is performed, the more severe the injury. What is the *plausibility* of the work-related explanation for the injury? In other words, is the explanation believable? *Coherence* is a criterion that forces us to ask two questions. First, does any *experimental evidence* exist that demonstrates that this RMI is work-related? Then, can we deduce by *analogy* that the injury is work-related? For instance, if carpal tunnel syndrome is an injury associated with keyboarding, then it follows that heavy users of keypads, such as adding machines, could also develop CTS.

Analyzing Human Factors

Certain human factors place employees at more ergonomic risk than other factors. Prolonged poor posture is a common ergonomic risk factor. Excessive force on the body and vibration are risk factors as well. Any job that uses bending, squatting, kneeling, lifting, reaching, or other body mechanics is risky. Finally, repetitive motion risks are making news in the trade and professional journals and this awareness indicates that many workers are at risk.

Many investigators use a video camera to tape employees at work. Video allows the investigator to review the task several times in order to identify risk factors and to demonstrate to others his or her findings. Table 13-2

Table 13-2
Duration of Videotaping

Job Cycle	Duration of Taping
<30 sec.	10 cycles
30 sec. - 2 min.	3-4 cycles
>2 min.	1 cycle

summarizes Stankevich's recommendations for duration of taping for this purpose. This article, by Stankevich, is highly recommended for the ergonomic videographer, but is too rich in detail to do it justice here or even summarize.

Anatomical Planes of Reference

The anatomical planes of reference, shown in Figure 13.1, are used to define movement and posture of the body. In the case of the back, flexion

Figure 13-1
Human Anatomical Planes

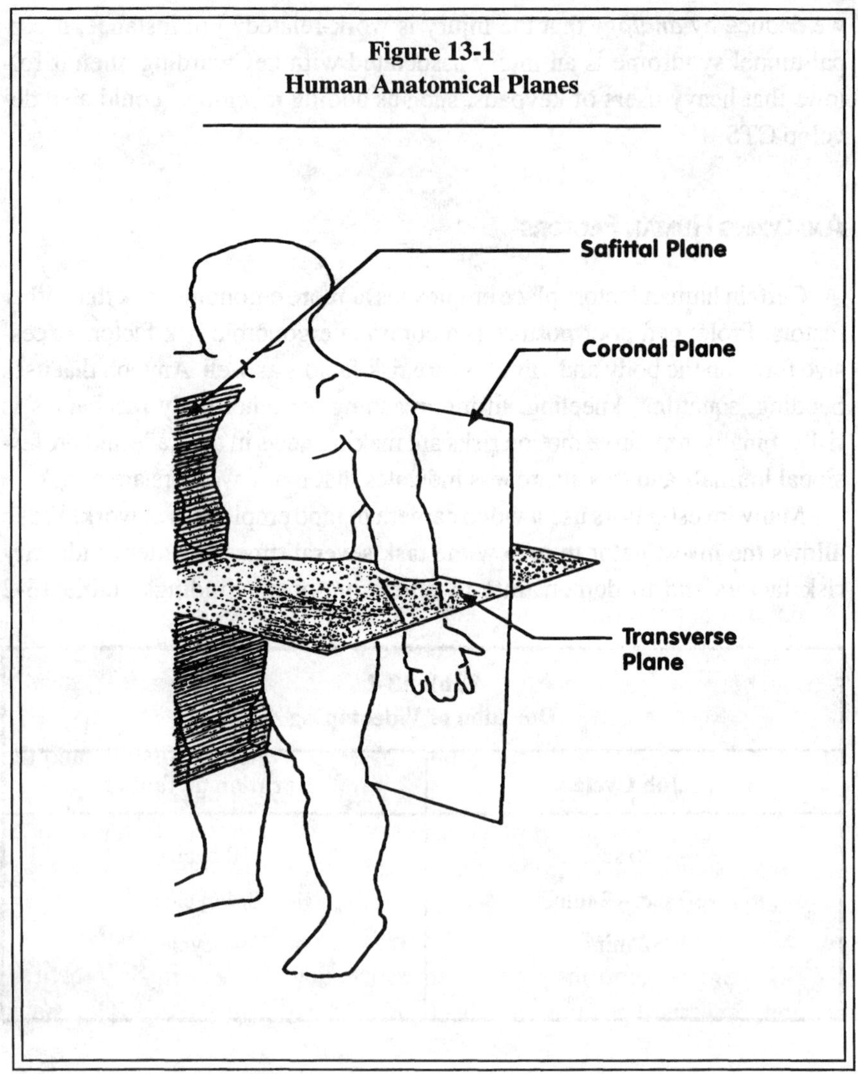

and extension is described about the sagittal plane. Lateral bending of the back is described about the coronal plane. Twisting of the back is described about the transverse plane.

Work Site Analysis

Figure 13-2
Work Site Ergonomic Analysis

1. Review injury and illness records for evidence of ergonomic problems.
2. Identify tasks that increase the risk to the worker.
3. Verify low risk tasks for light duty or restricted work assignments.
4. Target and prioritize high risk tasks for reduction or elimination of the hazard.
5. Conduct periodic follow-up surveys of the workplace.

In order to correct ergonomic problems an analysis of the work site must first be conducted. Figure 13-2 lists the steps involved. A qualified person should perform the analysis.

Integration with Manufacturing and Industrial Engineering

The science of ergonomics was established by the military, especially the Army Air Corps before it became the Air Force, during World War II. Military equipment has always been considered "one size fits all" and the soldiers and airmen being inducted were anything but one size. Somehow all these young men had to be able to expertly and deftly use the tools of war. Ergonomics as a discipline came into being. Today ergonomics is being integrated into factories and industries all over the country, tailoring workplaces to fit not only able-bodied men, but women and the disabled as well.

Ergonomics identifies physical stressors such as heavy loads, repetitive motions, awkward postures, and heat and cold in the workplace. The goal is to eliminate these stressors from the workplace or to minimize the risk of injury and illness to the exposed workers. If unable to eliminate the stressors, ergonomics fits the workplace to the worker's body dimensions. Thus, the comfort level of the worker is accommodated. Computer work stations are examples of worker accommodation. Another accommodation is worker friendly workstations in factories. Consoles, software windows, and software structures, such as pop-up forms and help boxes, are designed to match the mental model and memory capacity of the human mind. User friendly consumer products have also been designed.

Ergonomic designs are intended for the majority—fitting roughly 95 percent of all male workers in the workplace. Anthropometry is the study of people in terms of their physical dimensions. Using anthropometric data, machine and equipment designers can accommodate the 5th percentile of women to the 95th percentile of men.

Redesigning the Workplace

Alexander suggests that we first determine what we want the ergonomics program to accomplish before we decide how we will make changes. Figure 13-3 lists seven questions he considers.

Figure 13-3
Alexander's Strategic Planning Questions

1. What do you want the ergonomic program to accomplish?
2. How will you monitor results and what data will you measure to demonstrate progress?
3. What are the barriers to the program and how can they be overcome?
4. What policies are likely to be affected?
5. Who should be involved and what are their roles?
6. How important is ergonomics relative to other safety and health issues for your company?
7. What is your general plan?

Matching the worker with a job is a standard ergonomic practice. First, consider the normal worker attributes. What is his or her strength, endurance, mobility, coordination, vision, and hearing? Then consider the job requirements. What are the stressors? Reach, force, repetitions, postures, visual demands, auditory demands, motions, or speeds? Well-matching results lead to high productivity, quality workmanship, low grievance rates, and low injuries. On the other hand, poor matching results in low productivity, poor quality, and high rates of injury and grievances.

Easy Reach

Carson goes for instant results in her tips for improving work table design. For one thing, she divides the work table into three smaller areas. The chief area is Carson's *direct work area* where immediate needs are kept less than fourteen inches from the edge of the work table. Tools and parts that are needed less often are kept in the *often accessed area* about fourteen to eighteen inches from the edge of the table. Finally, the *rarely accessed area,* more than eighteen inches from the edge, contains support tools and parts that are infrequently used.

Some ergonomists go beyond this simple organizational layout. For one thing, they might reduce the size of the work table. Another alternative is to tile the surface so the farthest areas are now brought closer to the worker's reach. They might also provide a cut-out into the work surface so the worker can get closer to all the tools and parts needed.

Reaching into or working out of boxes can exacerbate the potential for muscle strain. The use of tilt-tables or inclined box stands eliminates this problem. Spring-loaded bins also eliminate the need to bend down into a box. How far or deep should an employee be required to reach? What is the reach of your smallest employee? Lazy Susan devices have been used to allow the worker to rotate far parts and tools to his or her proximity when needed. Chutes and hoppers can be used to shorten reach, also. Large containers that have removable sides help prevent back strain caused by bending down into the container to fetch parts.

Twisting and bending to make long reaches can lead to muscle strains. At the very least, twisting and bending makes the work more difficult. Therefore, this ergonomic redesign not only yields improved occupational health but also improved productivity and quality.

Work Height

Adjust chairs of sitting employees so that they are at a proper work height. A mismatch between the level of the work and the height of the employee, sitting or standing, leads to poor posture. Standing work generally ought to be at elbow height. Heavier work requiring upper body strength should be moved to a slightly lower level. Lighter precision work can be situated at a slightly higher level. Again, tilting the work surface can easily place the work at an optimum height.

One way to minimize stressful posture is to minimize the amount of bending required. All lifting should be done between the knuckle and elbow heights. (These heights are with reference to the arm/hand when they hang loose beside the body.) This ideal lifting height is typically about 30 to 42 inches from the floor when standing. If you cannot raise the work, then lower the worker. Provide a chair that is easily adjustable. Avoid horizontal reach distances exceeding fifteen inches. For standing work, the acceptable window for a horizontal reach is from 12 to 20 inches in front of the body. For seated work, frequent reaches should be limited to 8 to 15 inches in front of the body. Do not allow container dimensions to exceed 30 by 20 by 18 inches (length x width x depth).

The height of work surfaces should be as listed in Table 13-3.
Add 2 to 3 inches to the height for precision work. Subtract 2 to 3 inches from the height for heavy work. At seated workstations, the seat pan of chairs should be adjustable from 16 to 20 inches.

Table 13-3
Height of Work Surface

Work Station Type	Height, Inches
Standing	35-48
Sitting/Standing	38-47

Reduce Magnitude of Load

Minimize the weights or forces to be handled by the worker, so that he or she is handling loads consistent with his or her capabilities. Some guidelines for weight handling not mentioned elsewhere in this book are:

- Minimize forceful pinching and grasping of tools
- Suspend tools from the ceiling using a spring balancer
- Keep loads as close to the body as possible
- Reduce pushing or pulling forces
- Provide sufficient recovery time after a forceful exertion
- Provide proper handles on objects to be lifted frequently
- Educate workers about proper work and lifting techniques

An ergonomically correct load, or safe load, that suits all people and all types of work situations does not exist. The safe load for a given task depends on frequency of lifting, distance by which the load is displaced, type of grip available, and duration of lifting. NIOSH and other researchers have developed equations and tables by which safe loads can be calculated for a given lifting scenario.

Calculated safe loads are acceptable for 75 percent of the female workforce and 99 percent of the male workforce. The approved equations and tables limit energy expenditure to about 3.5 Kcal/min and do not impose a compressive load greater than 770 lb. in the lumbar/sacral region.

Mital *et al* developed Table 13-4 based on the above criteria.

The box size dimension given in the table is the width as measured out from the body. The frequency of lifting is given from one lift per shift (8 hours assumed) to 8 lifts per minutes. High frequency lifting (8/min.) should be avoided.

To reduce pushing or pulling loads, use larger wheels. Or use wheels with better bearings. The floor and the wheels should both be in good repair. Use powered assistance to tug the load. The load to be pushed or pulled should have good handles. Remember, pushing, in general, is easier than pulling.

Eliminate Awkward Postures

Awkward postures impose biomechanical loads on the spine due to the moment-arms created by the body parts when they deviate from the neutral

Table 13-4
Recommended Weight of Lift (lb.) for Two-Handed Symmetrical Lifting for 8 hours

Box Size (inches)	1/8 hr	1/30 min	1/5 min	1/min	4/min	8/min
Floor to 31.5 inches (knuckle height)						
29.5	30.8	24.2	21.5	19.8	19.8	17.6
19	35.2	26.4	22.0	22.0	19.8	17.6
13	41.8	30.8	28.6	26.4	24.2	19.8
Floor to 52 inches (shoulder height)						
29.5	26.4	19.8	17.6	16.5	16.5	14.3
19	28.6	22.0	17.6	17.6	16.5	14.3
13	35.2	25.3	24.2	22.0	19.8	17.6
Floor to 72 inches (overhead)						
29.5	24.2	17.6	15.4	15.4	15.4	13.2
19	26.4	19.8	15.4	15.4	15.4	13.2
13	30.8	22.0	22.0	19.8	17.6	15.4
31.5 inches to 52 inches (knuckle to shoulder)						
29.5	33.0	28.6	26.4	24.2	19.8	15.4
19	33.0	28.6	26.4	24.2	19.8	15.4
13	37.4	30.8	28.6	26.4	24.2	18.7
31.5 inches to 72 inches (knuckle to overhead)						
29.5	28.6	24.2	23.1	20.9	17.6	13.2
19	28.6	24.2	23.1	20.9	17.6	13.2
13	33.0	26.4	24.2	23.1	22.0	16.5
52 inches to 72 inches (shoulder to overhead)						
29.5	24.2	19.8	19.8	17.6	17.6	13.2
19	26.4	22.0	19.8	19.8	17.6	13.2
13	30.8	26.4	24.2	24.2	19.8	15.4

position. Wrists should be straight. The spine should have an S-curve. Elbows should hang naturally and not be elevated.

An awkward posture with an external load correspondingly increases the stresses on the body. Figure 13-4 shows the variation in disc pressure for various postures. Awkward postures include raised elbow, reaching behind the

Figure 13-4
Variations in Disc Pressure

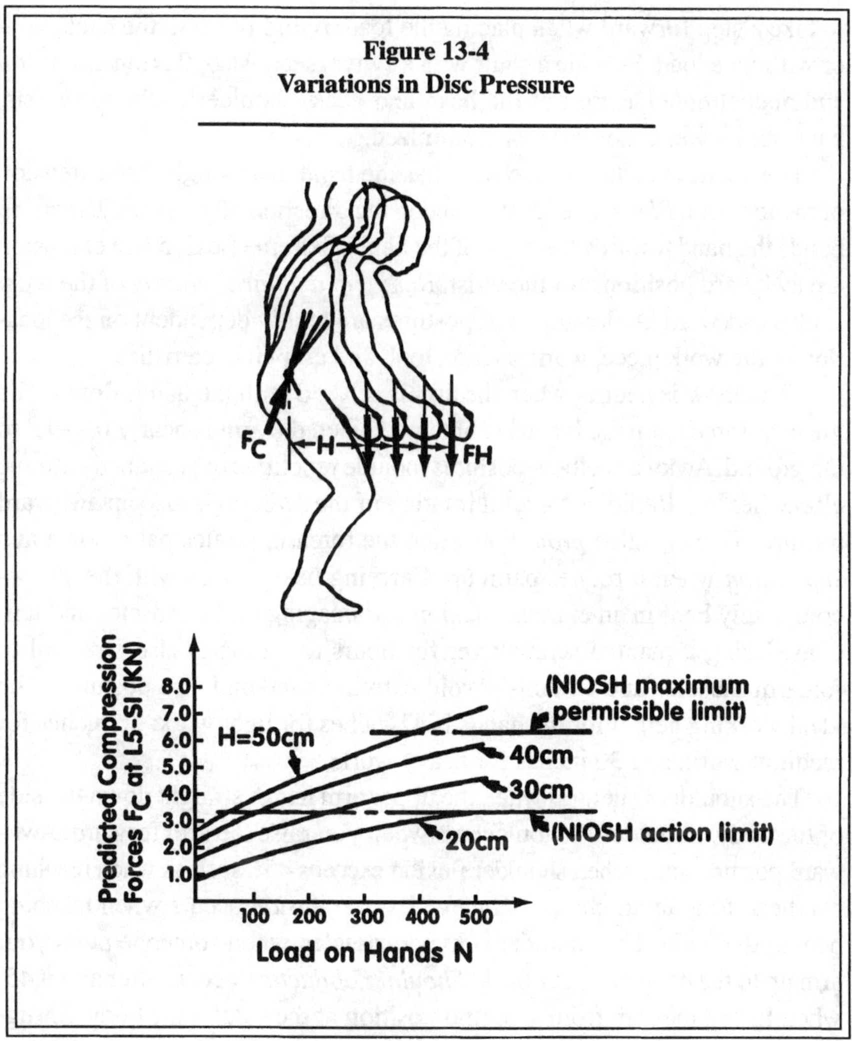

torso, extreme flexion of the arm, outward or inward rotation of the arm, flexion or extreme extension of the hand, or radial or ulnar deviation of the hand.

Modification of the workplace to accommodate an employee with functional limitations is case specific. What is the nature of the injury? Is it a fracture? Tendinitis? Carpal tunnel syndrome?

Minimizing twisting also reduces stressful posture. All objects to be lifted should be placed in front of the operator. The worker should have foot room to take a step forward when placing the load. Avoid twisting the back, with or without a load. Provide a chair with a swivel seat. Also, flexion/extension and abduction/adduction of the head and neck, shoulders, elbows, wrists, back, and lower limbs must be minimized.

The neutral position of the wrist has the hand in a straight line extension of the forearm. *Flexion* bends the hand in the direction of the palm. *Extension* bends the hand towards the back of the hand. Extreme flexion and extension are awkward positions for the wrist. *Radial and ulnar deviation* of the wrist is also awkward. Awkward wrist postures are highly dependent on the location of the work piece, work surface, tool, and user characteristics.

The elbow is neutral when the arm is relaxed with the hands down. The greatest force from the biceps is obtained when the arm is nearly parallel to the ground. Awkward elbow postures include repetitive or sustained extreme elbow flexion. Rapid or forceful rotation of the forearm is also an awkward posture. This is called *pronation* when the forearm rotates palm down and *supination* when it rotates palm up. Carrying heavy loads with the elbows completely bent in an extreme flexion is damaging to the muscles and tendons. Using a manual screwdriver for hours at a time is also stressful to forearm muscles and tendons. Avoid extreme hand and arm postures. The ideal working height for the hands is 42 inches for light work, 36 inches for medium work, and 30 inches for heavy work.

The shoulder is neutral when the upper arm hangs straight down the side of the body. Shoulder flexion occurs when you raise the arm forward. Awkward posture starts when shoulder flexion exceeds 45°, such as when reaching overhead to grab an object. *Shoulder hyperextension* occurs when reaching behind the body. The shoulder is hyperextended when someone pulls your arm up to the center of your back. *Shoulder abduction* occurs after about 45° when lifting the arm from a neutral position at the side of the body. Shrugging the shoulders increases static contraction of the trapezius muscles (the muscles in your shoulder your uncle or big brother used to pinch to bring you to your knees), causing fatigue. The shoulder is a very flexible joint, but discomfort results when flexibility and endurance are stressed. A way to avoid stressful posture is to reduce above-shoulder work.

When you bend your head down to your chest you experience *neck flexion.* Bending your head towards your back is known as *neck extension.* Lateral bending of the neck occurs from side to side, or shoulder to shoulder.

The neutral position places the head squarely upon the shoulders, straight up and down. Any movement of the head, especially lateral, greater than 20° can stress the neck and shoulder muscles. Also turning or twisting the neck left to right more than 20° rotation can lead to muscle stress.

The neutral posture of the back occurs when you are standing straight. However, the spine is not really straight in this position. Rather it forms a natural S-curve. Awkward postures include mild flexion (forward bending) for 20 to 45°. Bending more than 45° is severe flexion. Extension (backward bending) greater than 10° is stressful. Lateral bending (to either side) more than 20° and twisting the back are also very stressful.

Awkward postures of the lower limbs include repetitive or sustained knee flexion (squatting). Repetitive bending of the ankle, either by flexion or extension, is stressful.

Instruct employees to push loads if posture is the problem, but pull them if the weight is the problem. Be cautioned that pulling leads to hyperextension of the back, which results in extra loads on the spine.

Static postures, which do not change for a long duration, should be avoided. For instance, typists and keyboard entry personnel should change posture at least once per hour. Job rotation is a good way to handle industrial tasks with static postures.

Strive to keep individual body parts in neutral posture. Provide all the items necessary for the task within the worker's easy reach. Eliminate work above the shoulders and below the knuckles. Select the proper tools. Provide sufficient headroom and plenty of knee and thigh clearance under tables and desks. Insufficient clearance causes the worker to reach and apply stress to the musculoskeletal system.

The data in Table 13-5 often must be multiplied by safety factors to derive safe lifts for a given scenario. Some of the factors that must be considered are:

- Working duration (see Table 13-5)
- Headroom availability (see Table 13-6)
- Twisting (see Table 13-7)
- Couplings (see Table 13-8)
- Load placement clearance (see Table 13-9)
- Load asymmetry (see Table 13-10)
- Heat stress (Table 13-11)

Table 13-5
Working Duration Modifier

Duration (hours)			
1	**4**	**8**	**12**
1.14	1.08	1.0	0.92

Table 13-6
Limited Headroom Modifier

Stature	Fully Upright	95% Upright	90% Upright	85% Upright	80% Upright
Multiplier	1.0	0.6	0.4	0.38	0.36

Table 13-7
Twisting Modifier

Angle of turn (degrees)	Multiplier
0-30	1.0
30-60	0.924
60-90	0.848
Greater than 90	0.8

In order to determine the lifting capacity of a specific task:

$$L = A \times B \quad [13.1]$$

where

L = *safe lifting load*
A = *lifting capability*
B = *a* combined modifier equal to the products of the multipliers of each applicable modifier.

Table 13-8
Couplings Modifier

Couplings (handles)	Multiplier
Good, comfortable handles to initiate the lift	1.0
Poor quality handles/slippery	0.925
No handles	0.85

Table 13-9
Load Placement Clearance Modifier

Load Clearance (inches)	Multiplier
Unlimited to 1.2	1.0
0.6	0.91
0.1	0.87

Table 13-10
Asymmetry Modifier

Load Asymmetry	Multiplier
0	1.0
3.9	0.96
7.9	0.89
11.8	0.84

Table 13-11
Heat Stress Modifier

Heat Stress	Multiplier
Up to 90°F	1.0
90°F or hotter	0.88

The lifting capability A is determined for a given lifting height, lifting frequency, and box size, as determined from Table 13-4. B is the product of all the multipliers taken from Tables 13-5 through 13-11. In many cases, the values presented in the modifier tables do not precisely match those of a particular lifting scenario. Either the closest available value is chosen or values are interpolated.

Example. You investigate a lifting task in a machine shop. The employee must lift a casting that weighs 40 pounds from knuckle to shoulder height every 5 minutes. The width of the casting is 18 inches, measured out of the body. Lifting takes place for 4 hours of every shift. There is no place to get a grip on the casting. From Table 13-4, A is determined to be 26.4 lb. We have two adjustment factors that will modify this number: working duration and coupling. Since the lifting only takes place during half of each shift, we would expect the modifier to increase the base weight. True enough, from Table 13-5, the multiplier is 1.08. Similarly, without a good coupling, we anticipate reducing the safe load. The multiplier is found to be 0.85 in Table 13-10. Therefore,

$$L = 26.4 \times (1.08 \times 0.85) = 24.2\ lb.$$

The actual load of 40 pounds exceeds the safe load of 24.2 pounds. Therefore, the lifting task exposes the employee's musculoskeletal system to risk of injury. The back is especially at risk. If it is not possible to reduce the load to 24.2 pounds or less, provide the employee with a mechanical lifting device as shown in Figures 13-5 and 13-6.

Pulling is preferred over pushing loads since pulling has several biomechanical advantages. In pulling the weight of the body is in front of the load, whereas in pushing it is behind. Pulling therefore stresses the spine in a riskier fashion. Engineering solutions should be implemented in cases where it is not possible to reduce the load.

The physical load on a body over a work day can be estimated as follows:

$$AL_{gi} = \sum AL_{gi} * \left({}^{t_{ij}}\!/_{T_i} \right) \qquad [13.2]$$

Figure 13-5
Mechanical Lifting Devices
(Part A)

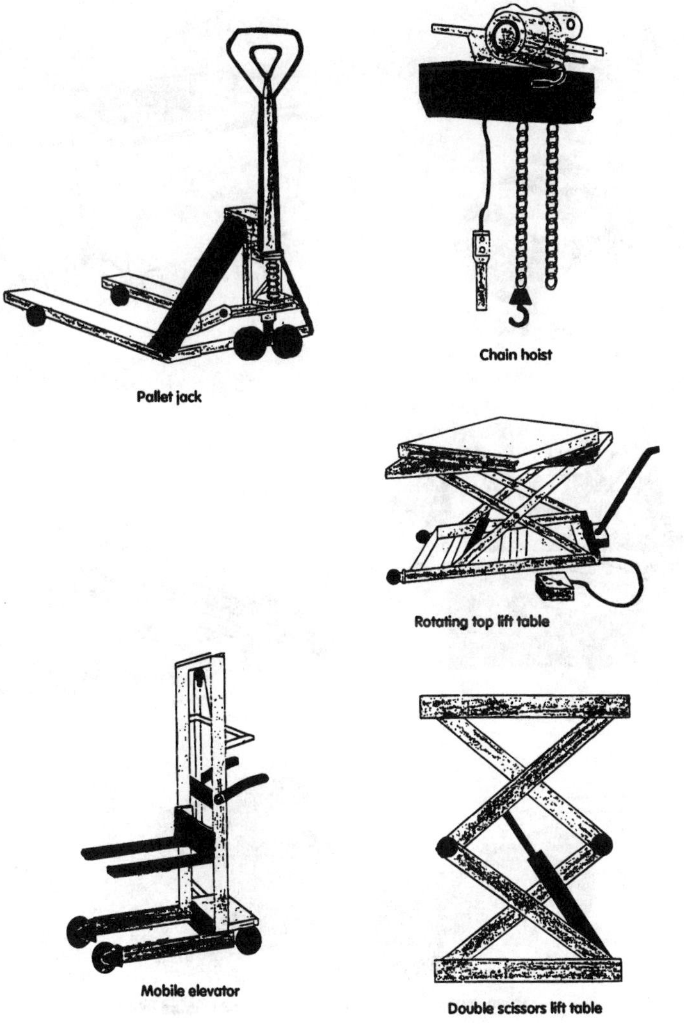

Chain hoist

Pallet jack

Rotating top lift table

Mobile elevator

Double scissors lift table

Figure 13-6
Mechanical Lifting Devices
(Part B)

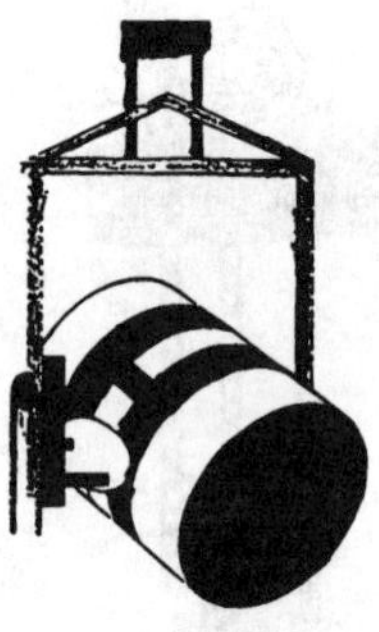
Drum lifter

Utility truck

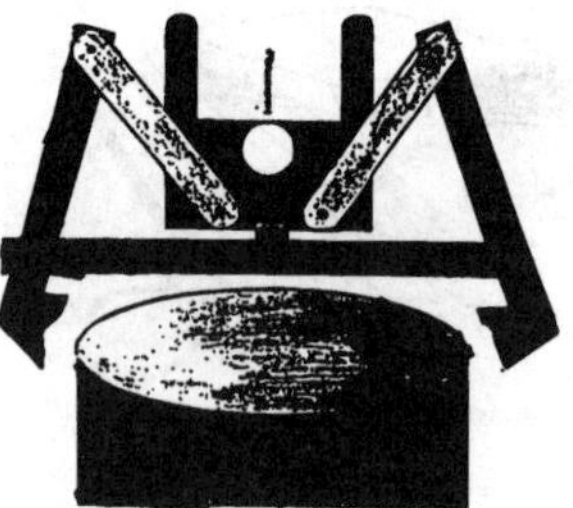
Vertical drum lifter

Bar cradle truck

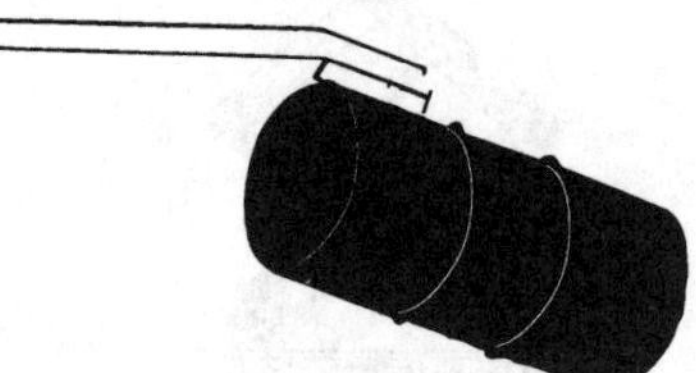
Drum upender

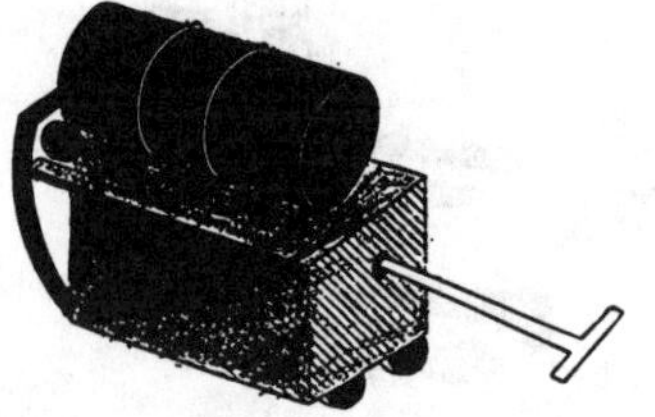
Portable drum rotator

Table 13-12
Complaints by Type Work

Complaint	Office	Operators	Miscellaneous
Back Pain	30%	37%	50%
Neck Pain	10%	29%	19%
Shoulder Pain	5%	20%	13%
Elbow Pain	2%	19%	6%
Wrist Pain	3%	14%	6%
Knee Pain	5%	19%	6%

where

AL_{gi} = average load over a shift for subject i in exposure group g
Al_{gj} = average load during task j in exposure group g
t_{ij} = duration of task j for subject i
T_i = total duration of work day for subject I

Table 13-12 lists the prevalence of complaints by type by work in a large tank farm operation. Table 13-13 gives maximum acceptable forces on the wrist.

Minimize Repetition

Minimize the number of motions in a given task. Let your tools do the work. Design the task and the work station for motion efficiency. Use the most efficient technique to do the task.

Table 13-13
Maximum Acceptable Forces on the Wrist

Force Factor	Repetition 15/min	Repetition 20/min	Daily Range hr^{-1}	Weekly Range day^{-1}
Torque, Nm	0.92-2.70	0.86-2.76	0.77-2.85	0.67-2.94
Isometric strength, Nm	2.94-9.02	3.19-9.37	3.07-9.19	2.64-9.55
Duration, sec	0.51-1.13	0.38-0.56	0.36-1.16	0.36-1.22

Minimize Fatigue

Designing the job to prevent fatigue also prevents overloading the worker's physical and mental capacities, the debilitation of which can lead to accidents, unacceptable work quality, and lost production.

One factor that contributes to fatigue is holding the same position for a period of time. This is called *static load.* Combined with strong force and awkward posture, static force can be particularly stressful. Here are some tips to combat static load.

- Provide jigs and fixtures to support the work piece.
- Use straps or handholds to grip.
- Change the handle design.
- Design the task with natural postures.
- Change postures frequently during the task.
- Minimize the force necessary to complete the task.

General weariness due to overload of physical capacity can also be addressed.

- Limit the intensity and duration of effort.
- Distribute peak work loads.
- Rotate to less demanding jobs.
- Allow frequent short breaks.
- Minimize environmental extremes.
- Minimize psychological stress.

Minimize Direct Pressure

Contact stress is direct pressure exerted on a body part during a task. Contact stress is uncomfortable and inhibits nerve function and blood flow. The most common body parts affected are the palms, forearms, and thighs:

Reduce palm pressure by changing the size, shape, contour, and covering of tool handles. These changes can distribute more evenly the pressure required to hold the tool.

Pressure is put on the forearms by leaning them against sharp edges or thin hard edges for support. Minimize forearm pressure as follows:

- Pad the edges.
- Round the edges.
- Provide arm rests.

- Redesign the task to eliminate the need to lean on a support or distribute contact over more surface area.

Standing for long periods or leaning over work surfaces for long periods decreases circulation of blood to the legs and feet. Reduce pressure on the legs and feet by:

- Providing floor mats to stand on.
- Encouraging cushioned insoles in shoes.
- Reducing the size of the work table.
- Changing the layout to eliminate the need to lean across obstacles.

Provide Adjustability

Adjustability helps the worker maintain good posture, avoid reaching, and minimize contact pressure. Some good ways to provider adjustability are:

- Use a scissors lift as a work surface.
- Provide hydraulic or pneumatic legs for tables and machines.
- Provide a good chair.
- Provide adjustable platforms.

Though correct posture is important, the body needs to change posture and move around from time to time. The optimum work station has the following features:

- It provides opportunities to change positions, move around, and alternate between sitting and standing.
- It has footrests.
- It has standing back rests.
- It provides sit-lean stands.

Clearance and Access

Make the workplace easy to work in by having plenty of room to move freely and providing easy access to everything the worker needs to complete his or her task. Too often layouts are considered only in two dimensions. Adequate clearance is required for the head, arm reach and movement, feet, torso, and knees. Determine clearance based on the largest person who will need to be accommodated, or who is expected to be accommodated.

Obstructions that prevent work from being accomplished and present potential injury to the body are not good for production or quality, much less safety. Determine accessibility based on both extremes—large and small persons. Reorganize equipment, filing cabinets, shelves, and storage lockers. Increase the size of openings. Eliminate physical barriers. Optimize the size and shape of maintenance ports. Eliminate line-of-sight barriers that interfere with visual access to important tasks.

Make the Environment Comfortable

Provide adequate lighting for the task (see Chapter 14), avoid temperature extremes (see Chapter 15), and isolate vibration (see Chapter 14).

Displays

Displays and control panels can assist or hinder the safe, productive, and quality operation of a work station. Therefore, enhance the clarity and understanding of displays and control panels, starting with design.

When precise information is needed, digital displays are better than analog. However, analog displays (moving pointers) better communicate relative information. Spatial displays are effective for organizing complex information, such as process controls. Signs and labels should contrast with their background. The size of signs and labels should be large enough to read easily and consistent with the importance of their message. Emergency information, for instance, should be very large and visible. Icons convey information more quickly than wordy signs, particularly for warning information.

Use stereotypes to reduce errors when operators interact with control panels and displays. Stereotypes represent the expectation of what the worker thinks is supposed to happen when he or she activates a control. For instance, workers expect a device to come on when the ON-OFF switch is flipped up. Also, people expect a device to speed up when a control lever is pushed forward. If you make the ON position of switches down, or to the left, or if a device slows when a velocity lever is pushed forward, the worker involved will get confused because he or she expects something different.

Standardizations have become so common in industry that it is confusing to deviate. For instance, stop buttons are RED. Using another color confuses people. The touch telephone buttons are arranged in a particular order.

If you sold a telephone that had the buttons arranged in any other order, people would be confused and it would not sell very well. An example of this is the new keyboard arrangement that was offered for sale a few years ago. Although it probably has a better arrangement of keys than the standard QWERTY arrangement, it has not sold well to date—not even to novice key punchers. What if you were to put the numbers on the analog clock in reverse order and run the hands counterclockwise? Wouldn't that work just as well? Sure, it would, but who are you going to sell it to? Not even the military-type 24 hour digital clock, which is easy to read, has caught on. People are confused and dismayed if standard arrangements are changed.

Organization of Work

Improve work organization. Organization and disorganization are two faces of the same coin that contribute to mood and morale, one way or the other. Organization, more to the point, also contributes in various and subtle ways to the physiological health of employees—in the form of little aches and pains that may not even be attributable to work. Many aches and pains may go unnoticed and unreported. So, why be concerned about these, then, if they are not going to make you accountable to OSHA? Because they will cost you. The employee can't tell you why, perhaps, they feel these aches and pains, but they may take a sick day off every now and then because they don't feel well. The less motivated the employee is, the more likely he or she will be tempted to stay home and nurse these nondescript aches and pains. Often even the employee does not recognize the true reasons for playing hooky. He or she, being in an agitated mindset, may stay home to get even with the boss, or because fishing is more appealing than making gadgets, or for some other reason. In the long run, then, disorganization will cost you. The savings of organization may never be measurable, but it is money in the bank, nevertheless.

Organization evolves from meticulous planning. Brainstorming, collective anticipation, thinking ahead as a team effort—all these are part of the preparation that goes into bringing about effective organization. Get more people involved in the planning phase of any project. Consider all ideas and points of view.

Communication also is important. Information must be shared or it is of no use. Issues need to be discussed.

Today, with the corporate downsizing trend, stretch and enlarge jobs. Make the job more challenging but not so overloaded that the worker cannot safely perform all of its required tasks. Combine tasks and responsibilities. Cross train as many workers as possible so that any one of them can perform the tasks of another. A broad division of labor that includes thinking as well as doing makes a content, motivated workforce which is also inclined to work safer. Such a workforce is also more aware and considerate to the effect their work has on those following them in the work sequence. Therefore, they can help make the next person's job easier and safer, as well as their own.

Condition the Worker

Employees who are involved in strenuous material handling should be fit for the job. Even though pre-employment screening must stand before the scrutiny of federal equal employment opportunity laws and the Americans with Disabilities Act (another federal law), make physical fitness part of the job description of those jobs. First of all, such workers should not have any serious back problems, including a history of herniated discs or torn ligaments. Preferably, the employee will have no known personal risk factors, such as smoking. The employee whose job includes strenuous material handling should be lean and muscular and proportionally built. Men are generally stronger and able to perform strenuous tasks with lower risk of injury. This does not preclude females or even certain disabled individuals from demonstrating that they can perform the required tasks. Physical fitness training should be encouraged for all such employees.

The safety department, in cooperation with the human resource or training department, has the responsibility for instructing employees in proper work methods and procedures. Worker training and education is another method of preventing back injuries. Training and education alone cannot guarantee a reduction in back injuries. Training is no substitute for poor ergonomics. Workers can be trained through live demonstrations, video tapes, and posters.

Back belts are popular in many material handling workplaces, but they do not reduce the load on the spine or offer any kind of protection to the back. They should not be considered as PPE. Employees should be made aware of their ineffectiveness and be discouraged from using them as the only solution to lifting strain.

Figure 13-7
Body Part Discomfort Survey

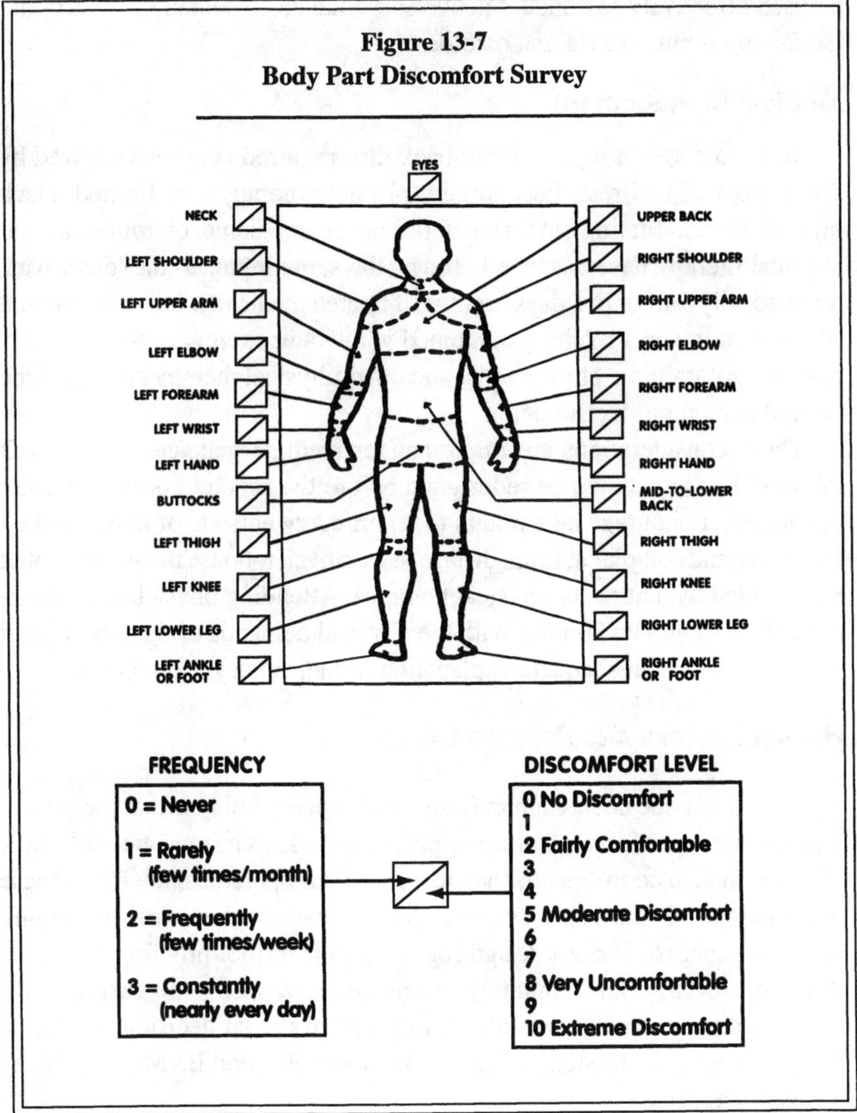

Active Surveillance

The use of interviews and questionnaires to identify back problems in the pre-clinical stage is good ergonomic practice. Passive surveillance through the OSHA 200 log and injury/illness record does little to correct the problem. A Body Part Discomfort Survey as shown in Figure 13-7 can be used effectively to conduct active surveillance, especially for early detection of cumulative trauma disorders.

Medical Management

Ten years ago, a ruptured low-back disc required surgery, followed by six months of bed rest. Post-surgery physical therapy was limited. Back injuries are debilitating but fortunately science, medicine, chiropractic, and physical therapy have advanced. Today, the same injury to the lower back requires only four or five days bed rest, if treated medically. Intensive muscle strengthening and stretching combined with water exercises, aerobic exercise, weight training, chiropractic, and other physical therapy can make the injured person productive again.

These considerations are important for medical management of back injuries. The job should be redesigned before the problem occurs in other employees. Encourage all workers to report every episode of musculoskeletal injury and complaint. Immediate, conservative, in-house treatment should be provided by a nurse or physical therapist. Attending physicians and chiropractors should be familiar with the physical demands of the job. Injured workers deserve an adequate explanation of their injury and recovery.

Human Factors and Process Safety

The chief issue about human factors and process safety is that the human factor is the central issue. True, equipment can fail on its own, but did a lack of maintenance contribute to that accident? Did operator inattention cause the accident? In the Navy, we tried to eliminate human error to varying degrees of success. Process design engineers call this fool-proofing, we called it sailor-proofing. Unfortunately, no matter how well a process is fool-proofed, someone will come along, make an irrational decision, do something inexplicable, though perhaps quite innocently, and BAM!—a process accident happens.

A high state of training and positive motivation are the two best strategies to combat the effect of human factors on process safety. Fool-proofing is necessary but not adequate by itself. Train, train, train. Drill, drill, drill. You must make proper procedures become second nature. The employee must be able to perform the required tasks of each job without thinking, but you must also impel the workers to think about safety issues before executing each task. Workers who have to concentrate on the mechanics of their tasks do not have time to think about safety as well.

Select process equipment that meets the needs of the application. When the worker has to assist the equipment in its function, he or she has no time for thinking about safety. Safety must also be kept in mind when designing and selecting the process equipment.

The process equipment must be rugged. No employee can be expected to behave safety when the equipment he or she uses is inherently dangerous and could fail at any moment. Each component of the equipment must have structural integrity. The method of assembly must also have structural integrity. Bearings at pivot points and rollers should be self-lubricating. Legs and roller pins should be chromed. Controls, as discussed above, should be designed to acceptable standards.

Cumulative Trauma

Repetitive motions cause trauma to muscles and ligaments of the upper and lower arms, elbows, wrists, hands, fingers, shoulders, neck, chest, and abdomen. They may even affect the legs and the back. Often the damage is accumulated over a period of time before it manifests itself. Harmful motions include lifting, twisting, squeezing, hammering, pressing with the fingers, handling small objects in the hands, and pushing or pulling. Repetitive motions are classified as:

- Low repetition: once every 30 minutes or more;
- Moderate repetition: once every 2 minutes or more; and
- High repetition: once every 15 seconds or more.

Typical high repetitions of concern are job cycles that last less than 30 seconds, movements that are repeated 7,600 to 12,000 times per shift, or continuous keyboard entry. Repetitive motions should particularly be matched to the functionally impaired employee. Reduce the cycle time and/or the duration of exposure. You can also have the employee alternate between repetitive and non-repetitive tasks.

Cumulative trauma disorder is any disease caused by a force that acts repeatedly on some part of the body until the body compensates for it by manifesting a recognizable set of signs and symptoms. Figure 13-8 lists some CTDs.

CTDs may take weeks, months, or even years to develop into recogniz-

Figure 13-8
Some Cumulative Trauma Disorders

Carpal Tunnel Syndrome (CTS)
Cubital Tunnel Syndrome
DeQuervain's Disease
Epicondylitis (Tennis Elbow)
Gamekeeper's Thumb
Guyon's Canal Syndrome
Hypothenar Hammer Syndrome
Myalgia
Myofacial Pain Syndrome
Peritendinitis
Raynaud's Phenomenon
Tendinitis
Tenosynovitis
Thoracic Outlet Syndrome
Trigger Finger

able diseases. Once diagnosed and interdicted they may take weeks, months, or even years for recovery. Before the full-blown disease is clearly recognizable, symptoms are often poorly localized, nonspecific, and episodic. Another complication for diagnosis is that CTDs may have more than one causal factor.

Employees with carpal tunnel syndrome and other nerve disorders should not be exposed to vibrating tools and cold environments. Sharp edges that the wrist, forearm, or leg contact should be rounded off or padded with foam. Forearm rests should be provided for continuous precision work. Workers should avoid sudden acceleration and deceleration (jerky movements) of the arms. Arm movements should be smooth. Illumination and noise levels can also affect repetitive motion. Address them if they are a concern.

The risk of CTD increases with constant repetition, exertion of force, awkward posture, and lack of rest. Sharp or hard objects increase the risk. Cold temperature increases the risk. Improperly designed tools, work areas, or controls increase the risk.

In the workplace, the human back is the most abused and injured part of the body. Eighty percent of all people experience some back pain in their lives, although only a fraction are reported as occupational injuries. After the back, the arms, legs, multiple body parts, eyes, neck, and muscle systems are most injured. On any given day, 6.5 million Americans stay in bed because of back pain. Annual estimates of new cases of low back pain range from 10 to 15 percent of existing cases. Low back pain accounts for 40 percent of all lost work days. The annual cost to U.S. industry for back injuries is $20 billion. One company found that back injuries accounted for 30 percent of all cumulative trauma disorders (CTDs) reported in 1994. Another company found that the back accounted for 45 percent of its diagnosed CTD cases. The average cost of a back injury was about $6 thousand.

Spine

The back or spine is the framework on which the whole body is built and consists of 33 rigid blocks, or vertebra, and flexible soft tissue, or discs, between the upper 24 vertebra. The discs give our backs the flexibility to bend and twist on several planes at once. They are the shock absorbers of the spine. This flexible structure allows you to walk, sit, stand, lift, work, and play. The spine is also a conduit for 31 nerve branches, which pass from the brain to the various parts of the body. Nerves are a network of electrochemical connections that pass orders from the brain to activate specific regions of about 400 muscles and the limbs.

Back problems are common in the workforce, causing pain, inconvenience, and disability. The injury-prone back creates problems that are expensive. The individual can best serve his or her spine by having regular checkups and maintaining an exercise program to avoid overweight and to strengthen the back muscles. The employee should consciously practice good posture.

Vertebra

The vertebra (singular vertebrae) are short, cylindrical bones. The cervical region of the spine, see Figure 13-9, contains 7 vertebra, the thoracic region 12, and the lumbar region normally contains 5 vertebra. The exact count of vertebra in the lumbar region varies some from person to person. A few persons even have an extra half vertebra, which typically leads to lower back complaints.

Figure 13-9
Spine and Vertebra

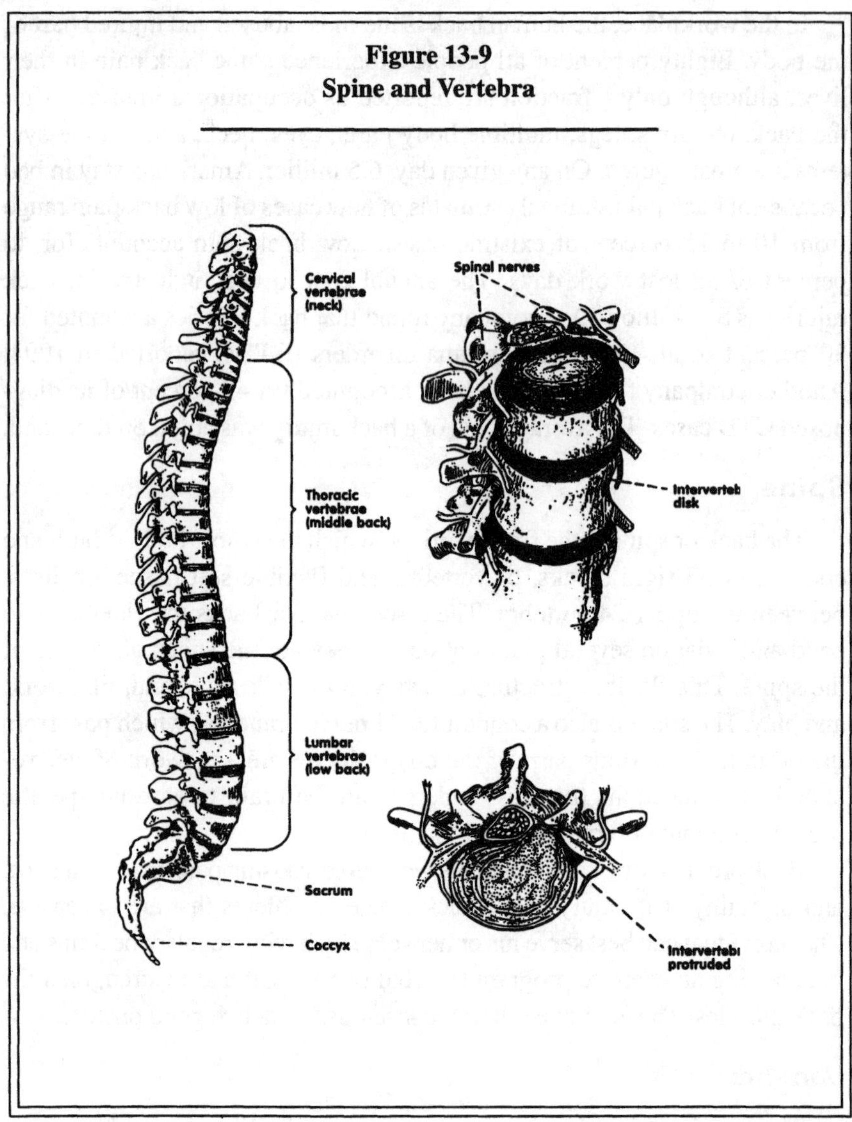

Intervertebral Discs

The disc between each set of vertebra is composed of a fibrous material called *collagen.* Discs provide the primary articulation for the vertebra and the spine as a whole. The major role of the discs is to bear the weight of the body and its external loads.

Ligaments

Spinal ligaments support and link the vertebra. These ligaments are passive, elastic stops that prevent excessive motion.

Muscles

Muscles that are under voluntary control are the motors that power the spine. Spinal musculature consists of flexors for forward bending and extensors for backward bending. Tension neck syndrome is a common work-related muscle disease.

Back Injuries

To understand back injuries, we must first talk of pain. Pain is a physiological mechanism for ensuring system safety. When an abnormality of some kind exists, such as a tissue injury, nerves transmit distress signals by electrical impulse to the brain. This signals the injured organism by opposite path to take precautionary and preventive measures. The origins of back pain are linked to the various structural elements that make up the back. Examples of back pain are listed in Figure 13-10. Origins of back pain with a strong association with work-related factors are shown in **bold** letters.

Loading

Overexertion loading occurs when in a single event, the structures of the back are overloaded beyond their strengths. This leads to direct trauma, which may result in the fracture of the vertebral bodies or to the rupturing of the muscle or ligaments at a single site. Possible causes of overexertion loading are slips and falls, underestimation of a load to be moved or lifted, or to improper coordination in team lifting. In the latter case, team members may release the load but for the person who was injured.

Static loading occurs when a person holds a load without any movement. When this occurs due to postural fixity, the worker has to maintain a fixed posture for long periods of time, leading to muscle fatigue. A static posture interferes with the circulation of blood in a muscle, and thereby with oxygen supply and removal of breakdown products.

Dynamic loading occurs due to repetitive lifting. The fatigue life of the tissues in the spine assume an important role in dynamic loading since tissue lose strength due to intermittent loading. Involuntary dynamic loading

Figure 13-10
Potential Origins of Back Pain

Muscular/Ligamentous

- **Trauma:** acute or cumulative injury.
- **Sprain:** injury to ligaments holding bone to bone.
- **Strain:** small tears in muscles.
  Acute strain: caused by sudden stress (single overexertion).
  Chronic strain: caused by repeated stress.
- **Postural imbalance:** uneven stresses on musculoskeletal system.
- **Spasm:** muscle contraction due to sudden movement.
- **Myofascitis:** inflammation and tenderness of muscle.

Structural

- Spondylolysis: congenital defect of the bony segment.
- **Spondylosis:** disc degeneration due to mechanical stress.
- Spondylolisthesis: displacement of one vertebrae over another.
- Osteoporosis: loss of bone content.
- Scoliosis: abnormal curvature of the spine.
- **Compression fractures**
- **Dislocation**
- Osteoarthritis: degenerative disease.
- Spinal Stenosis: degenerative narrowing of spinal canal.
- **Trauma:** load force exceeds the strength of tissue.

Discogenic/Neurological

- Disc herniation: slipping, bulging, or rupturing of disc. Occurs mostly in youth in the lumbar or sacral regions.
- Nerve Irritation
- Tumors

Nonstructural Disorders

- Infection: bacterial infection through blood supply.
- Metabolism: nutrient deficiencies such as osteoporosis.
- Congenital Disorders
- Inflammatory Disorders: mostly rheumatoid arthritis.
- **Psychoneurotic Disorders:** stress, hysteria, hypochondriasis, malingering.
- **Toxicity:** poisoning due to pollutants, industrial waste, or radiation, leading to cancer, and/or nerve irritation.

due to occupational vibrations may also cause back pain. A back injury that occurs under involuntary dynamic conditions is a Cumulative Trauma Disorder (CTD). It occurs due to prolonged exposure of the workplace factors, not a single overexertion. The acceptable compression force on the L5/S1 disc is 770 lb. Any more force than that requires some action to reduce the force. The maximum permission load on the L5/S1 disc, in terms of pounds of force, is 1430. Only 25 percent of men and less than one percent of women have the muscle strength to lift more than MPL anyway, but exceeding MPL puts unacceptable forces on the spine. Below 770 pounds, the action level, the load is a minimal risk to most industrial workers.

Personal Risk Factors

Several factors directly relate to the person, or the person-work system, which may increase or decrease the probability for back injury. Percentage body fat is one such factor. Fat lacks the ability to contract and develop force like a muscle. Therefore, an obese person has less load lifting capability compared to a lean person.

Another factor, metabolism, may affect availability of oxygen. If the oxygen supply is deficient due to smoking, lack of exercise, or other personal risk factors, the worker will fatigue rapidly. Muscles can be made stronger by training. Therefore, a worker who trains with weights or performs regular strenuous exercise is physically better suited for manual material handling.

Finally, muscle strength decreases with age. By age 40, muscle strength is typically 90 percent of what is was in the individual's late 20s. By age 50, it is 85 percent, and only 60 percent by age 60. From age 40 on, the spine experiences a gradual decrease in compressive strength—1474 lb. at 40 and 748 lb. at 60.

Back injuries may occur due to a single overexertion leading to a mechanical failure of one or more of the structures supporting the back. The mechanical failure may be a compression fracture, a torn ligament, a torn muscle (sprain), or micro tears to muscle (strain). A mechanical failure may occur due to prolonged exposure of the work-related risk factors, which may lead to trauma, or myofascitis (muscle inflammation). The risk for injury can be increased due to effect-modifying factors such as aging, onset of disease, and personal lifestyle.

Prevention

Prevention of back injuries is divided into three categories: primary, secondary, and tertiary. *Primary prevention* aims towards the initiation of a disease. In the case of back injury, ergonomics, teaching of work techniques, and pre-placement screening are all recommended. *Secondary prevention* involves efforts to modify the progression of a disease. This implies disease detection by a plant nurse, or a physician, or the worker himself or herself—followed by treatment or exposure modification or both. *Tertiary prevention* minimizes the consequences of a disease or injury once it is manifest. Optimal medical care and workplace modifications reduce disability from disease. The major focus of the safety manager is primary prevention—the things done by the facility to eliminate work-related risk factors through ergonomics.

A method used by athletic weight lifters is an excellent practice for occupational material handling as well. Go into the squat position by bending the knees. Keep the load as close to the body as possible. Also, if possible, keep the entire load between the thighs. Imagine an athlete about to pick up a barbell. Size up the load—push on one of its edges. If it feels very heavy, get help. Do not stoop—that is, bend your back—when lifting. Do not twist your back while placing the load—rather, take a step in the desired direction. Keep your back straight. Place heavy loads on shelves that are located between 30 and 42 inches from the floor. Relative to the body, this is the level between the knuckles and elbows of most people when their hands are hanging by their sides in the standing position.

Carpal Tunnel Syndrome

Carpal tunnel syndrome (CTS) is a nerve disease caused by the repeated flexing of the fingers until the tendons in the wrist swell and put pressure on nearby nerves. Predisposing factors are aggravation of compression effects within the carpal tunnel. For instance, the pressure you feel after wringing a towel dry is such an aggravation of compression effects. Some drivers who firmly grip the steering wheel also have this aggravating predisposition for CTS. Also, nocturnal accentuation of CTS may occur from unconscious flexion at the wrist during sleep.

The wrist is particularly susceptible to cumulative trauma because it contains muscles, nerves, blood vessels, tendons, ligaments, and bones. The wrist section shown in Figure 13-11 shows these body parts labeled while

Figure 13-11
Cross Section of Wrist

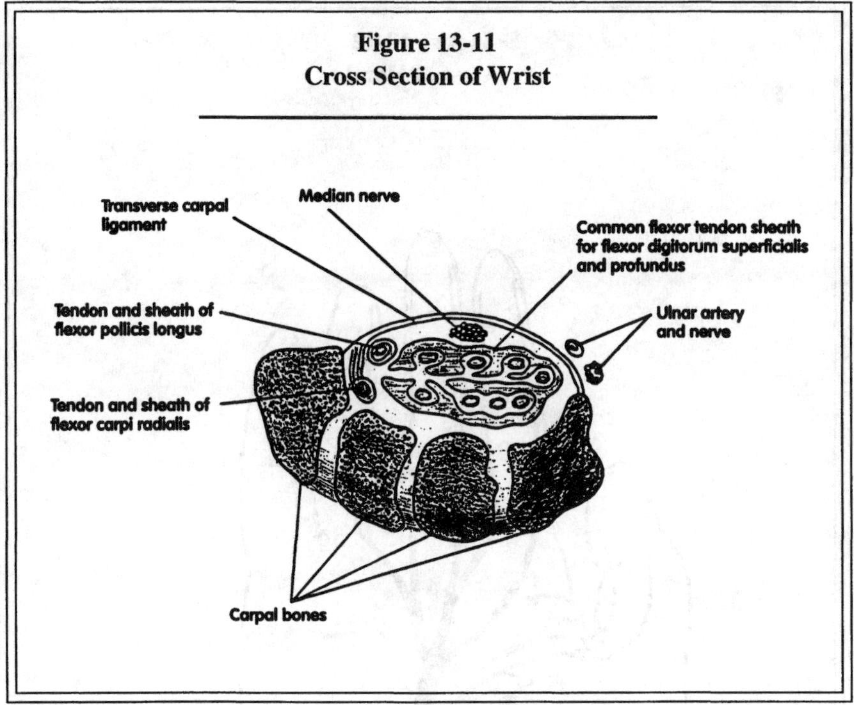

Figure 13-12 shows a clearer picture how the median nerve is pinched in the carpal tunnel.

Early warning of CTS comes from sensory complaints. Pain or tingling may be noted in the fingers supplied by the median nerve. The complaints may predominate in a single finger. The chief characteristics of the pain is burning or tingling that sometimes becomes intense enough to be distracting, if not disabling.

The next warning sign is general weakness. Loss of thumb strength may be noted when attempts are made to raise it perpendicularly against resistance. Forced flexion of the wrist for even a short period will result in reproduction of pain or exaggeration of the symptoms of CTS.

Soon afterwards, the proximal thenar pad begins to atrophy. The distribution of the median nerve includes a branch to the thenar muscles, particularly to the abductor pollicis brevis. Tinel's sign occurs when parethesias radiate to middle or index fingers, or both, when the midvolar wrist is lightly tapped at the level of the flexor crease. The CTS develops when any of a

Figure 13-12
View of Hand

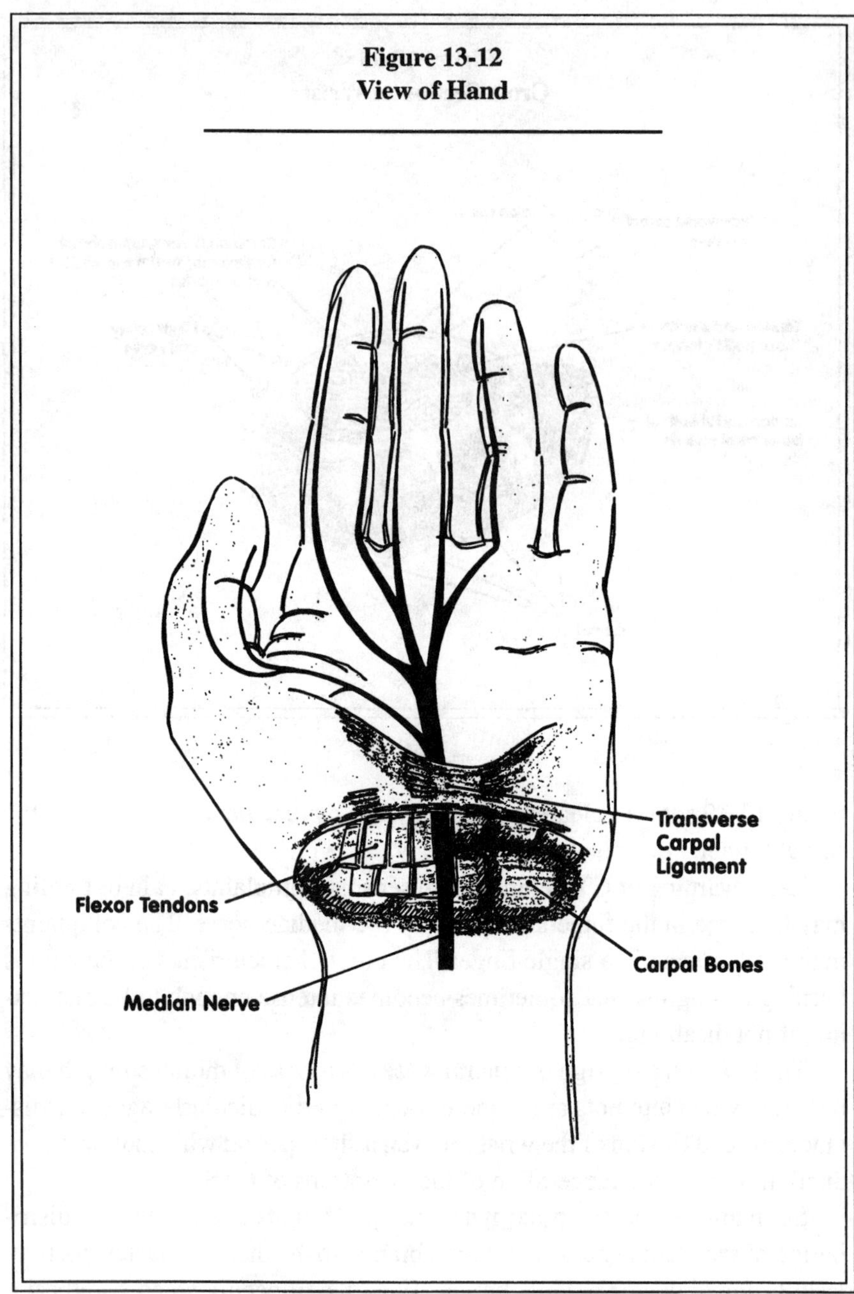

number of widely varying conditions produces compression of the median nerve at the wrist. Pseudo-CTS refers to compression of the median nerve proximal to the carpal tunnel by the flexor digitorum superficialis muscle.

Tendinitis

Tendons are inflamed by repeated tensing of muscle and tendon. Tennis elbow (lateral epicondylitis) and pitcher's elbow (medial epicondylitis) are examples of elbow tendinitis. Shoulder tendinitis is manifested in rotator cuff tendinitis, not to be confused with adhesive capsulitis, and bicipital tendinitis. Hand-wrist tendinitis includes flexor tendinitis, extensor tendinitis, and De Quervain's syndrome. Dupuytren's contracture is a tendinitis of the hand. Achilles tendinitis occurs in the heel.

Tenosynovitis

When repeated movement inflames the covering of a tendon, the disease is known as tenosynovitis. Trigger finger (stenosing tenosynovitis) is one example of this disease. Flexor tenosynovitis and extensor tenosynovitis in the hand-wrist are other examples.

Trigger finger (sometimes trigger thumb) is a disorder of the flexor tendons of the hand in which a nodule in the tendon catches on the edge of the tendon sheath through which the tendon is supposed to glide. Usually the ring finger (4th) or thumb (1st) is involved, but any of the fingers could be. Anyway, the finger locks in a flexed position until the nodule is forced past the obstruction. Straightening of the finger is difficult. A distinct click can be heard when straightening the finger. The palm at the base of the finger will be tender and a nodule is generally felt.

Typical treatment of mild to moderate occupational trigger finger includes rest, splinting, and nonsteroidal anti-inflammatory drugs. The doctor may inject cortisone if no improvement is seen over time. For severe, recurring, or persistent symptoms, the doctor may rely on surgical release of the proximal part of the tendon sheath. After surgery, a hand and finger exercise program is needed.

The employee disabled with trigger finger will miss work or be restricted, depending on the severity of symptoms and which digits are affected. If the dominant hand is affected, the disability will be more restrictive. The employee's lifestyle and underlying systemic disease will also impact the

Table 13-14
Length of Trigger Finger Disability with Medical Treatment and Injection

Type Work	Expected Disability
Sedentary	0-3 weeks
Light	0-3 weeks
Medium	0-3 weeks
Heavy	0-4 weeks
Very Heavy	0-4 weeks

Table 13-15
Length of Trigger Finger Disability with Surgery

Type Work	Expected Disability
Sedentary	1-8 weeks
Light	1-8 weeks
Medium	1-8 weeks
Heavy	2-8 weeks
Very Heavy	2-8 weeks

length and restrictiveness of the disability. The type of treatment required also affects lost or restricted workdays.

Complications of surgery may lengthen lost or restricted days. Infection is another risk after surgery. Adhesions of the tendon sheath may occur. The surgery may also damage digital nerves. Tables 13-14 and 13-15 summarize the expected lengths of disability based on treatment strategy and type of work.

Circulatory Problems

Repeated trauma may also cause circulatory problems due to pressure on blood vessels. Vibration may also interfere with blood flow. White finger is an example of this type of illness. Another is hypothenar hammer syndrome.

Eyestrain

Eyestrain due to inadequate lighting conditions and vibration is discussed in the next chapter. Poorly devised control labeling and annunciator panels may also lead to eyestrain. Pay attention to these factors when designing control panels and instrumentation.

Headache

Two other nerve diseases are worthy of mention. One is radiculopathy (cervical syndrome) and the other is thoracic outlet syndrome. Both affect muscles and nerves in the neck and shoulders and lead to headaches, emotional stress, and fatigue.

Fatigue

Joints osteoarthroses can also be a workplace musculoskeletal disease.

Bursae

Bursitis is a common disorder that may or may not be work-related. Knee bursitis, for example, may be due to heavy lifting at work, but may be complicated by off-work activities such as contact sports.

Hand Tools

To prevent gradual exhaustion of muscles in the arm and hands, static load, such as holding a hand tool, should not exceed 10 percent of the worker's maximum muscle strength capability. Dynamic loads, such as carrying a load while walking, use large groups of muscles so up to 40 percent of the worker's maximum muscle strength capability may be safely used. Robinson and Lyon cite research that supports these limits.

Avoid wrist deviations. Grip strength decreases when the wrist is deviated from neutral. Wrist deviation greater than 30° modifies the quantity of force conveyed from hand to tool. Use better tools, such as those which

Table 13-16
Maximum Power Grip (N) for Women

% Population	15/min	20/min	25/min
90	4.5	4.5	4.3
75	9.0	8.9	8.7
50	14.0	13.9	13.5
25	19.0	18.9	18.3
10	23.5	23.3	22.6

Table 13-17
Maximum Acceptable Forces on Wrist at 15 Motions/Minute

	Torque Nm	Handle Lever m	Force N
Power flexion	2.11	0.081	26.0
Pinch flexion	1.97	0.123	16.0
Power extension	1.22	0.081	15.1
Ulnar deviation	1.96	0.140	14.0

have a Bennett (bent to fit the grip) handle. Bend the tool, not the wrist. Reorient the workpiece or the work surface. Use jigs or fixtures to take some stress off the grip. Make sure tools are balanced. Avoid pinch grips whenever possible, except for during precision work. Grip strength is maximized for a power grip and exertion is minimized. Table 13-16 gives the maximum acceptable power grip for women. Finally, Table 13-17 gives forces on wrist at 15 motions per minute.

For power grips, use a cylindrical cross-section inline handle or a pistol grip that is offset 70 to 80° from the work axis. The diameter of the power grip handle should be from 30 to 45 mm, or 1.25 to 1.75 inch. The diameter for pinch grips for precision work should be limited to 5 to 15 mm, or 0.2 to 0.6 inch. The surface of the grip should be slip-resistant and nonconductive.

The length of a handle should range between 115 and 125 mm, or 4.5 to 5 inches. As far as practical the grip should be interchangeable for use in the dominant hand of the worker.

Power tools heavier than 5 pounds should be counterbalanced. In precision work, tools heavier than one pound should be counterbalanced.

For hand tools such as pliers, scissors, or screwdrivers, the handle length should be held to a minimum length of 4 inches. The optimal tool handle length is 5 inches. Add a half-inch when tools are used with gloves. If a spring assisted opening action is provided, the tool will be easier to use.

The design of the work station should allow the worker to lay down the hand tool, or hang it on a holder, between work cycles.

References

The ABCs of Moving and Lifting Things Safely. South Deerfield, MA: Channing L. Bete Co., Inc., 1986.

About Ergonomics. South Deerfield, MA: Channing L. Bete Co., Inc., 1993.

About Repetitive Motion Injuries. South Deerfield, MA: Channing L. Bete Co., Inc., 1991.

Alexander, David C. "Planning a Successful Ergonomics Program." *Workplace Ergonomics.* May/June 1995, p. 12.

Burdorf, Alex, Marcel van Riel, and Teus Brand. "Physical Load as Risk Factor for Musculoskeletal Complaints among Tank Terminal Workers." *AIHA Journal.* July 1997, pp. 489-497.

Carson, Roberta. "Key Ergonomic Tips for Improving Your Work Area Design." *Occupational Hazards.* August 1994, pp. 43-46.

"Health Profiles: Trigger Finger or Thumb." *Ergonomics News.* May/June 1997, p. 12.

Mital, A., A.S. Nicholson, and M.M. Ayoub. *A Guide to Manual Materials Handling.* Washington, DC: Taylor & Francis, 1993.

Musculoskeletal Disorders and Workplace Factors: A Critical Review of Epidemiologic Evidence for Work-Related Musculoskeletal Disorders of the Neck, Upper Extremity, and Low Back. National Institute of Occupational Safety and Health. July 1997.

Pederson, Paul. "Work Station Positioning Equipment Keeps Employee Safety within Reach." *Occupational Health & Safety.* July 1994, pp. 36-39.

Robinson, Fred, Jr., and Bruce K. Lyon. "Ergonomic Guidelines for Hand-Held Tools." *Professional Safety.* August 1994, pp. 16-21.

Stankevich, Bernard A. "Guidelines for Videotaping and Evaluating Cumulative Trauma Disorders." *Professional Safety.* May 1994, pp. 37-40.

"Twelve Principles of Ergonomics. Principle 1: Keep Everything in Easy Reach." *Safety Update.* Comprehensive Loss Management, Inc. January 1993.

"Twelve Principles of Ergonomics. Principle 2: Work at Proper Heights." *Safety Update.* Comprehensive Loss Management, Inc. April 1993.

"Twelve Principles of Ergonomics. Principle 3: Reduce Excessive Forces." *Safety Update.* Comprehensive Loss Management, Inc. June 1993.

"Twelve Principles of Ergonomics. Principle 4: Work in Good Postures." *Safety Update.* Comprehensive Loss Management, Inc. August, September 1993.

"Twelve Principles of Ergonomics. Principle 5: Reduce Excessive Repetition." *Safety Update.* Comprehensive Loss Management, Inc. October, November 1993.

"Twelve Principles of Ergonomics. Principle 6: Minimize Fatigue." *Safety Update.* Comprehensive Loss Management, Inc. December 1993, January 1994.

"Twelve Principles of Ergonomics. Principle 7: Minimize Direct Pressure." *Safety Update.* Comprehensive Loss Management, Inc. February, March 1994.

"Twelve Principles of Ergonomics. Principle 8: Provide Adjustability and Change of Posture." *Safety Update.* Comprehensive Loss Management, Inc. April, May 1994.

"Twelve Principles of Ergonomics. Principle 9: Provide Clearance and Access." *Safety Update.* Comprehensive Loss Management, Inc. June, July 1994.

"Twelve Principles of Ergonomics. Principle 10: Maintain a Comfortable Environment." *Safety Update.* Comprehensive Loss Management, Inc. August, September 1994.

"Twelve Principles of Ergonomics. Principle 11: Enhance Clarity and Understanding." *Safety Update.* Comprehensive Loss Management, Inc. October, November 1994.

"Twelve Principles of Ergonomics. Principle 12: Improve Work Organization." *Safety Update.* Comprehensive Loss Management, Inc. December 1994, January 1995.

Sandler, Howard M. "The Cal/OSHA 'Ergo' Standard: Let the Games Begin." *Occupational Hazards.* September 1997, pp. 49-52.

Snook, Stover H., Donald R. Vaillancourt, Vincent M. Ciriello, and Barbara S. Webster. "Maximum Acceptable Forces for Repetitive Ulnar Deviation of the Wrist." *AIHA Journal.* July 1997, pp. 509-517.

Take Care of Your Back. South Deerfield, MA: Channing L. Bete Co., Inc., 1993.

Neelam, Sridhar R. "Using Torque Arms to Reduce CTDs." *Ergonomics in Design.* October 1994, pp. 25-8.

Werrell, Marjorie and Zachary J. Koutsandreas. "Ergonomics: A Good Place to Start." *Occupational Hazards.* September 1997, pp. 37-40.

Work Practices Guide for Manual Lifting. Revised. Cincinnati, OH: NIOSH, 1991.

You and Your Back: How to Prevent Injury and Maintain Health. South Deerfield, MA: Channing L. Bete Company, Inc., 1977.

14

Sight and Hearing Conservation Management

Light

Light is nonionizing radiant energy with the capacity to produce visual sensation in humans. A few quanta, or energy packets, can stimulate the human retina and thus be seen as light. To see an object, light of suitable quality and intensity must form an image on the retina of adequate size, contrast, and duration for the retina to transform the energy into nerve energy. The nerve impulses are conducted to the brain and integrated into consciousness.

Physics of Light

Light is a flux of energy particles called photons traveling through space. Photons travel at 3×10^8 m/s, the speed of light. Light speed is related to frequency, f, and wavelength, l, as follows:

$$c = \lambda f \qquad [14.1]$$

The wavelength and frequency of the different parts of the electromagnetic spectrum are given in Table 14-1.

The photon energy of nonionizing radiation is insufficient to produce ionization of the atoms it strikes. The energy required to ionize simple atoms, such as hydrogen, is as much as 12 eV, which is the approximate dividing line between the two types of radiation—ionizing and nonionizing. The minimum amount of ionizationenergy (12 eV) roughly corresponds to 100-nm wavelength in the UV range.

Biological molecules absorb nonionizing radiation and changes are induced in the vibrational and rotational energy of ions and polar molecules. Generally at longer wavelengths in the microwave range, this energy is dissipated as heat, affecting tissue. At shorter wavelengths in the UV range, fluorescence is produced.

Table 14-1
Electromagnetic Spectrum

Spectrum Portion	Wavelength, μm	Frequency, Hz
Radio Frequency	$>10^4$	$0\text{-}3\times10^{12}$
Infrared	$10^3\text{-}0.7$	$3\times10^{11}\text{-}4.3\times10^{14}$
Visible	0.7-0.4	$4.3\times10^{14}\text{-}7.5\times10^{14}$
Ultraviolet	0.4-0.01	$7.5\times10^{14}\text{-}3\times10^{16}$
Gamma and X-rays	<0.01	$>3\times10^{15}$

Visible Light

The human eye is sensitive to radiant energy from 380 to 740 nanometers (nm) in wavelength. We call this range of electromagnetic energy *visible light.* For young eyes, the limits of visibility are from about 313 to 900 nm. For all practical purposes, the average range is used in the workplace. Age, glare, state of adaptation, and visual acuity modify vision, or the ability to translate radiant energy packets into consciousness. In the retina, rods respond with peak efficiency to 0.51 μm. The cones tend to respond better at 0.555 μm.

Color that is seen by a human is a somewhat subjective perception, yet not entirely. Color as seen depends on the color of the light source, the color of the object seen, and the combination of the eye and the brain of the viewer. Shade or tint depends on such things as brightness of background light, intensity of the source, and absorption or reflectance of the object seen. The eye responds to a very small portion of the electromagnetic spectrum—the visible spectrum, made up of red, orange, yellow, green, blue, indigo, and violet. However, color response is not uniform. The best response is around 550 nm, green to yellow, while the response at the low end, approaching ultraviolet, and at the high end, the near infrared, is poor.

Light in the green-yellow wavelengths is the brightest, providing the most lumens per watt of power. However, the red and blue wavelengths in a light source are required to give the best color, though the brightness may suffer.

Infrared

The infrared (IR) spectrum begins where the visible light spectrums leaves off. At wavelengths around 740 nm the infrared spectrum begins and ends around 1,500 nm. IR is emitted from the sun, hot metals, electrical appliances, incandescent bulbs, furnaces, welding arcs, lasers, and plasma torches. Energy and wavelength are highly dependent on the temperature of the source. IR exchange between two hot bodies acts in accordance with the Stefan-Boltzmann Law, which states that energy is exchanged with the fourth power of the absolute temperature, as seen in Equation 14.2.

$$Q_R = E_1T_1^4 - E_2T_2^4 \tag{14.2}$$

where

E = emissivity of the hot body.

Table 14-2
Ultraviolet Ranges

Specific Region	Wavelength, µm	Photon Energy, eV
Vacuum UV	<0.16	7.7
Far UV	0.16-0.28	7.7-4.4
Middle UV	0.28-0.32	4.4-3.9
Near UV	0.32-0.4	3.9-3.1

Table 14-3
Biological Effects of UV Radiation

UV Region	Wavelength, µm	Biological Effect
UV-A	0.4-0.315	Black light; induced fluorescence
UV-B	0.315-0.28	Sunburn
UV-C	<0.1-0.28	Germicidal

Infrared causes eye damage by energy absorption in the epithelium. Iris tissue absorbs infrared at wavelengths shorter than 1.3 µm. Prolonged or intense infrared exposure will create lesions of the iris from tissue heating.

Ultraviolet

The ultraviolet band of the electromagnetic energy spectrum ends around 380 nm where visible light begins. It starts around 10 nm. Table 14-2 lists the UV ranges.

Its biological effects are summarized in Table 14-3.

Sources of UV radiation include the sun (which is the major source of all wavelengths) high and low pressure mercury corona discharge lamps, plasma torches, welding arcs, quarts lamps, xenon discharge lamps, lasers, and fluorescent lamps. The penetration depth of UV is shallow compared with the thickness of the skin. The chief biological effect, then, is sunburn, an erythermal response. Shorter UV wavelengths are absorbed by the cornea of the eye and this has the potential to produce cataracts. UV wavelengths between 0.295 and 0.32 µm especially are thought to be cataractogenic. The erythermal threshold is about 0.5×10^{-6} W/cm^2 for an exposure lasting seven hours. The ACGIH established a TLV for UV in the 0.315 to 0.4 µm range as 1 mW/cm^2 for exposures lasting longer than 1,000 seconds and 1 J/cm^2 for briefer exposures.

When employees work outdoors, a serious health hazard they are exposed to is ultraviolet radiation. The skin has a limited ability to protect itself from the sun's rays and, after that, problems arise. First, a slight redness develops on the skin. Anywhere from two to six hours later, a full-blown sunburn becomes evident. The pain and acute damage peak in 12 to 16 hours. If no further exposure occurs, the sunburn fades in a few days. Unfortunately, outside workers are repeatedly exposed to harmful ultraviolet rays, increasing their risk of skin cancer.

Ultraviolet protection for outdoor workers includes:

- Wear long pants, long sleeve shirts, and broad brimmed hats.
- Clothing material should be a tightly woven fabric.
- Generously apply to exposed skin a sun block having a sun protection factor (SPF) of at least 15.
- Apply the sun block thirty minutes before going outside.

Table 14-4
SPF Protection Levels

Skin Type	1 hr	2 hr	3 hr	5+ hr
Very fair/extremely sensitive (never tans, always burns)	SPF 15	SPF 30	SPF 30	SPF 45
Fair/sensitive (tans slowly, burns easily)	SPF 15	SPF 15	SPF 30	SPF 45
Fair (tans gradually, usually burns first)	SPF 15	SPF 15	SPF 15	SPF 30
Medium (tans well, burns minimally)	SPF 8	SPF 8	SPF 15	SPF 30
Dark (tans easily, rarely burns)	SPF 4	SPF 8	SPF 8	SPF 15

Sun block is required, even on cloudy days because ultraviolet rays penetrate cloud cover.

SPF represents the incremental time period an individual can stay in the sun without getting burned. The time to burning, t_{burn}, determines the time protected, t_{prot}:

$$t_{prot} = t_{burn} \times SPF \qquad [14.3]$$

For example, if an individual normally burns with 10 minutes of exposure, then a sun block with an SPF of 45 will give him or her 450 minutes of protection. If he or she eats lunch in the shade, the sun block ought to last throughout the shift. Table 14-4 summarizes the benefit of sun block protection.

Outdoor workers should perform self-examinations for skin cancer frequently. The warning signs are:

- Asymmetry—half of a skin mole is not the mate to the other half.
- Irregular border—the edges of the mole are ragged, notched, or blurred.
- Color—the mole has a molted look due to shades of tan, brown, black, red, or blue.
- Size—any mole or skin growth larger in diameter than 6 mm (a diameter of a pencil eraser) or growing rapidly is suspect.

The employee should contact his or her physician if any of these warning signs are noted.

Laser

The term LASER is an acronym standing for Light Amplification by Stimulated Emission of Radiation. This describes a physical process that produces the phenomena we have come to know as laser light. Ruby lasers produce light emissions in a three stage pumping system: ground state, broad energy absorption band, and upper laser level. Gas lasers and rare earth lasers operate in a four level system. Carbon dioxide lasers are commonly used in industry for cutting, drilling, and welding. These lasers operate on a wavelength of 10.6 mm in one of three modes: continuous, pulsed, or Q switched. Switched is a method that enhances storage and immediate discharge of energy to produce extremely high peak power pulses.

Table 14-5 lists common laser systems.

Table 14-6 lists the lasing-action of other ions.

Table 14-5
Common LASERS

LASER	Wavelength, μm	Photon Energy, eV
CO_2	4.8-8	0.258-0.155
Dye	0.36-0.65	3.5-1.9
GaAs	0.85-0.9	1.47-1.38
H_2O	27.8-118.6	0.044-0.01
HCN	33.7	0.037
HeCd	0.325-0.4416	3.82-2.808
He-Ne	0.6328-3.39	1.96-0.366
N	0.337	3.68
Nd-Glass	1.06	1.17
Nd-YAG	1.06	1.17
Ruby (Cr^{3+})	0.6934	1.79

Table 14-6
Ions Capable of Producing LASER

Active Ion	Wavelength, μm
Nd^{3+}	0.9-1.4
Ho^{3+}	2.05
Er^{3+}	1.61
Cr^{3+}	0.69
Tm^{3+}	1.92
U^{3+}	2.5
Pr^{3+}	1.05
Dy^{2+}	2.36
Sm^{2+}	0.70
Tm^{2+}	1.12

The chief biological effect of lasers is eye damage. After an acute exposure, the eye is sensitive to glare. Massive damage to the foveal region of the retina is the worst damage that lasers can cause. Most damage occurs in the pigment epithelium of the retina due to absorption of energy in the wavelength range from 0.7 to 1.4 μm. The visible and IR wavelengths cause thermal damage. UV and blue light cause photochemical damage. Laser also causes

Figure 14-1
Summary of LASER Damage to Tissue

Wavelength Range	Tissue Damaged
R-Far	Cornea
IR-Near	Retina ($\lambda < 1.4$ μm)
UV	Cornea
Visible	Pigment of retina

Figure 14-2
LASER Classification

Class 1	Incapable of producing radiation levels; exempt from control measures.
Class 2	Visible wavelength lasers or lasers with low potential for injury.
Class 3	Intrabeam viewing of laser or specular reflections of the beam may be hazardous.
Class 4	Direct or diffusely reflected radiation may be hazardous to eye; direct beam is a skin and fire hazard.

some mechanical damage of the eye. Figure 14-1 summarizes tissue damage caused by lasers. Figure 14-2 summarizes the classes of lasers.

Light Calculations

The primary design considerations for lighting are:

- Achieve a pleasant visual environment.
- Provide sufficient light to allow the efficient performance of tasks requiring visual acuity.
- Control nonionizing radiation to prevent damage to the body and skin.
- Ordinary light is not a problem even at intense levels.
- Avoid short pulsed intense light that causes flash blindness.
- Shield high intensity light that causes retinal burns: lasers, welding arcs, and spotlights.

Luminous intensity is the indication of how much light is given off by a source in a given direction. The *candela* is the unit of measurement. The *lumen* is the light output from a source of light. The *illumination level* is the amount of light falling on a surface, measured in *footcandles.* One hundred lumens on one square foot surface equals one hundred footcandles. A surface one foot from a source with an intensity of one hundred candelas has an illumination level of one hundred footcandles. The term *lux* is the metric equivalent of the lumen where the unit surface is one square meter.

Photometric brightness is called *luminance,* which is the amount of light emitted or reflected from a certain surface area. Luminance is measured in *footlamberts.* A surface that emits one lumen per square foot has a luminance of one footlambert. A bare 40-watt fluorescent lamp, for instance, has a luminance of 2,400 footlamberts. If the surface area is measured in meters, the luminance is measured in candelas per square meter.

Reflectance measures how much light is reflected from a surface and is the ratio of luminance of the surface to illumination on the surface:

reflectance = luminance/illumination

The reflectance of a black surface is zero. A perfectly white surface has a reflectance of one hundred percent, or 1.0. The reflectance of most surface finishes ranges between 5 and 95 percent. Use 0.5 if the reflectance is unknown and the surface is neither black nor perfectly white.

Effects of Light on Human Sight

Seeing is a learned ability. Training, in our formative years, improves our seeing and expands the limits of our eyes and nervous systems. Seeing is a perceptual process that is affected by our emotions, other sensations, and learned associations. Our ability to see varies greatly over our lifetimes.

Many conditions and diseases disturb the eye and vision. Emmetropia refers to an average, normal eye. A defective eye is labeled *ametropia.* An eye with little or no vision that appears normal is called *amblyopia.*

Far-sightedness, or *hyperopia,* is due to the axial length of the eye being too short, which places the image behind the retina, or the focusing mechanism being too weak to place the image on the retina. Far-sighted eyes are corrected by placing a plus lens in front of them to replace the image on the retina. In *near-sightedness,* or *myopia,* the image is formed in the vitreous because the eye is too long or the focusing mechanism is too strong. Near-sightedness is corrected by using a minus lens. *Astigmatism* results from the deformation of the cornea and is corrected by a cylindrical lens. *Presbyopia* is the decline of focusing ability with age.

Aniseikonia is a spatial distortion in vision caused by size differences in images due to unequal magnification by the two eyes. Some aniseikonia is present in most eyes, but if the size difference exceeds one or two percent then one eye needs a lens to boost magnification. *Aphakia* is the lack of a lens. A spectacle lens must substitute for the eye lens. *Glaucoma* is a dis-

Table 14-7
Snellen Eye Chart Organization

Line Number	Acuity Score
1	20/200
2	20/100
3	20/70
4	20/50
5	20/40
6	20/30
7	20/25
8	20/20
9	0/15
10	20/13
11	20/10

ease that increases pressure in the eyeball, leading to mechanical damage and loss of sight, unless stopped.

Visual Acuity

Visual acuity is measured on a Snellen eye chart with naked eyes or as corrected with either spectacles or contact lenses. The standard distance from the chart is 20 feet (6.1 m). Different charts are available with different letter order to guard against memorization. As the test subject reads the chart, he or she is allowed to proceed to the next line if no more than one letter is incorrectly identified per line. Table 14-7 lists the Snellen Eye Chart Organization. Subjects continue reading the eye chart until they incorrectly read two or more letters on a line.

When brief flashes of light alternating with darkness illuminate the eye, the eye sees a flickering until the rate of alternation reaches from ten to thirty cycles per second. At this point, the images fuse and appear continuously as one image. The rate of fusion of images is the *critical flicker frequency* (CFF). CFF increases with increased illuminance.

Talbot's Law

Fluctuating and steady lights of the same energy content appear equally bright.

For brief exposures, intermittent light is less efficient. For long exposures, fluctuations help.

Glare is unwanted light in the field of view that causes annoyance, discomfort, loss in visual performance, and loss of visibility. The glare of the morning sun may hinder your ability to drive safely to work on an east bound street. An overhead lamp behind your computer may be a source of glare as you attempt to read your morning electronic mail. Both of these are examples of *direct glare. Reflected glare* is not so straightforward to detect, as it consists of images of lamps or luminaires reflected from the task. If an annoying lamp is behind you as you look at the computer screen to read your mail, the phenomena of reflected glare is involved.

Volatile Organic Chemicals (VOCs)

A common work-related complaint is eye irritation. The office eye syndrome manifests itself in itchy, watering eyes, redness, throbbing under the eyes, excessive dryness, and with various forms of headache, from a dull ache to sharp, intense pain. Low humidity, excessive dust, elevated carbon monoxide concentration, elevated carbon dioxide concentration, environmental tobacco smoke, mold spores, pollen grains, dust mites, animal dander, poor lighting, noise, and psychological stress have all been thought to contribute to office eye syndrome. The primary cause, recently identified by Scandinavian researchers, is a very low, but measurable, concentration of volatile organic chemicals (VOCs) in the air.

Low levels of VOCs act as a surfactant, degrading the tear film that hydrates the eye. The tear film not only waters the eye, but also protects the eye from dust impact, prevents biological infection, and lubricates the eye as the eyelid passes over it. That which breaks down the tear film, then, leads to eye irritation in any of several forms. The eye irritation is the source of the headaches due to the strain of reading or focusing on a work task.

The moisture barrier created by spreading tears through blinking normally lasts 45 to 55 seconds. Then the eye blinks again, forming a new moisture barrier. In the presence of low-level VOCs, the barrier only lasts ten to fifteen seconds, requiring blinking anywhere from three to six times as often to maintain the moisture barrier. This extra blinking and the failure to moisten the upper part of the eye anyway leads to irritation.

VOC concentrations on the order of 1,000 to 1,500 $\mu g/m^3$ or approximately 0.3 to 0.4 ppm, depending on the molecular weight of the particular VOC species, breaks down the tear film fast enough to cause problems. Office activities intensify this process of irritating the eye. Hence, the office eye syndrome is named. Contact lenses also worsen the problem. The syndrome can also occur in other occupational settings where eye strain and VOC concentration are both feasible.

Where do the VOCs come from in the office? Particle board, carpet, wall coverings, floor tile mastics, paints and varnishes, cleaning compounds, and improper chemical storage are the most obvious possibilities. VOCs are also part of the metabolic process of fungi. Certain species of mold, with the right nutrients and ambient temperature, produce VOCs.

To reiterate, common lighting problems encountered in the workplace are:

- Glare-shines in the worker's eyes
- Shadows-obscures details
- Contrast-blends the work into the background

Lighting Strategies

Several options are available for improving lighting.

- Use diffusers or shields to minimize glare.
- Place lights differently to avoid glare.
- Use task lighting or indirect lighting to soften shadows.
- Use back lighting to enhance contrast.

Foreign Bodies in the Eye

Foreign body damage to the eye is a more common eye injury than injury due to nonionizing radiation. In fact, it is one of the more common accidents encountered in the industrial setting.

Types of Foreign Bodies

Foreign bodies come in all shapes, sizes, and kinds. Anything that does not belong in the eye is a foreign body. Technically, even chemical splashes are included, but since we covered that in chapter 12, we will limit our discussion here to solid objects as foreign bodies.

Sources of Foreign Bodies

A chip from a machining process is probably the most common foreign body removed from the eye in the occupational setting. A sliver of metal, metallic kerf, dust, some other particulate—anything that is propelled through the air by a mechanical device is a potential foreign body. Blowing dust is another source of foreign bodies in the eye.

Protective Measures and Equipment

Eye protection is necessary for dusts, grit, metal chips, wood chips, stone chips, or other flying or airborne articles.

Industrial safety glasses provide impact resistance for particles flying through the air. Prescription spectacles that meet ANSI specification Z87 may be worn as safety glasses, but other glasses, even if termed "impact resistant," may not be. Most sunglasses, by the way, do not meet the ANSI specifications. The appropriate glasses have a specification number stamped or printed on them. Side shields and/or brow shields are required where flying objects could enter the eyes from the side or over the glasses.

For liquids and vapors, industrial safety glasses are inadequate. Flexible- or cushion-fitting goggles, called chemical splash goggles or goggles for short, are required. Goggles may also be used in lieu of impact glasses with side and brow shields. Prescription glasses can be worn under the goggles. Goggles can be uncomfortable, resulting in some common abuses. Watch out for employees who remove the ventilator screens from goggles to make them more comfortable. Such goggles no longer offer adequate protection.

For heaving chipping jobs, such as using a chipping hammer to remove paint, chipping goggles are highly recommended. These have impact-resistant eye pieces set into goggle-like frames. Ordinary glasses will not fit under these goggles, but they can be purchased with corrective lenses on an individual basis.

Special goggles are required for oxy-acetylene torch cutting and use of laser beams. Welding requires a special helmet with an eye protection shield inset.

Immediate flushing with potable water is the most effective means of treating chemical contamination of the eyes. Immediate access to an emergency eye wash within ten to fifteen seconds is critical to saving the contaminated eye. Therefore, the distance to the nearest eyewash or deluge shower must be less than one hundred feet from the work station. Placement within ten to fifteen feet of highly acidic or caustic chemicals is preferred. The duration of flushing is also critical and should be a minimum of fifteen minutes. Twenty to thirty minutes of flushing with water is recommended by medical experts for optimum effectiveness. OSHA requires emergency eyewashes at workplaces where there is a reasonable risk of exposure to corrosive chemicals, such as acid or alkaline materials, or other liquids hazardous to the eyes. Emergency eye wash fixtures, though required, may not substitute for protective equipment or engineered protective systems.

What eye wash devices are available? Plumbed eyewash stations, portable eyewash stations, showers, manual drench hoses, and combination units are all available.

A plumbed eyewash station can be free standing, mounted on a wall, or mounted on a counter top or work table. Caps or covers should cover the eye nozzles when not in use and should be loose enough that water pressure alone will remove them. The station should be easy to operate for a person with a chemical in his or her eyes. Foot treadle operation allows the victim to use his or her hands to hold open eyelids for flushing.

Portable eyewash stations are used for those hard to get to areas where permanent plumbing is not feasible. Be sure to use eyewash additives that kill bacteria yet do not cause eye irritation or are not toxic themselves.

Deluge showers used for chemical safety can also be used to flush the eyes. Pull the chain or cord and stand under the shower with face tilted up. The shower should have a minimum thirty gallon per minute flow so it needs a suitable drain for the nine hundred gallons or more that will be used in a thirty minute period.

Hand-held or manual drench hoses are used to provide a gentle stream to augment an eyewash or shower. The hoses are not allowed to replace an eyewash or shower.

Noise

Noise is unwanted sound. For the most part, my music may be your noise, and vice versa. But, when sound begins to interfere with voice communication and cause hearing impairment, we can both generally agree that it is noise.

A qualitative indication that a noise level is too high is when you have to raise you voice from normal speaking volume to be heard. If you cannot hear someone less than two feet away, it is probably too noisy. If, after you leave an area of high sound level, sounds are muffled or dull in a quiet area, the noise is probably too high. If you experience ringing in your ears after exposure to a loud sound, that is a good indication that it was high level noise you were exposed to.

Physics of Sound

When air (or water or some other medium) is mechanically disturbed, sound waves are produced. Sound, then, is nothing more than rapid pressure variations in, this case, air. Anything that can induce pressure variations in another medium is a *sound source. Frequency* is the high or low pitch of sound proportional to the number of complete wave cycles that pass a given point each second. One cycle per second (cps) is called a *Hertz,* abbreviated as Hz. High-frequency noises are more damaging to the ear than low-frequency noises. Noise in an industrial facility, as any sound, is a combination of frequency and energy level, or loudness, or intensity. Noise is measured in decibels, dBA.

The transmission velocity of sound is dependent on the mass and elastic properties of the material through which it travels. In a homogeneous medium sound travels at the same speed in all frequencies. At normal temperature and pressures, the speed of sound through air is about 1,130 feet per second, or 770 miles per hour. The speed of sound in water is 4,700 feet per second, or 3,200 mph! The speed of sound in steel is 16,500 feet per second, or 11,250 mph!

Sound with frequencies, f, less than 20 hertz is *infrasound. Ultrasound* has a frequency greater than 20,000 hertz. As safety managers we are concerned with frequencies in the heard range which lies between these two extremes.

Sound frequencies that we can hear are commonly divided into eight octave bands. The top frequency of an octave band is two times the lowest frequency in the band. The octave band is generally referred to by its center frequency.

An object or volume of air *resonates* or strengthens a sound at one or more particular frequencies depending on the size and shape of the object or air volume.

The average amplitude, A_0, of air pressure changes are directly related to the sound energy or power. Humans can detect sound pressure changes, between the lowest perceptible sound and sound that is so loud that it causes pain), over a range of ten million decibels. Since this absolute number is too huge to manipulate, a logarithmic scale for sound pressure is used and measured in decibels (dB).

The weakest sound pressure that the human ear can hear under ideal hearing conditions is the *threshold of hearing,* P_0. P_0 is 0.0002 dynes/cm^2 or 0.00002 Pascal. Any measured sound pressure, L, is a quantity greater than P_0.

$$L = 20\log\left(\frac{P}{P_0}\right) = 10\log\left(\frac{P^2}{P_0^2}\right) \qquad [14.4]$$

If the reference pressure P_0 is as stated above, L = 20 log P + 94. A measured pressure of 1 N/m^2 means that L = 94 dB. A sound reduction of 50 percent is represented by a six dB reduction in pressure level. Conversely, a six dB increase in sound pressure level is a doubling of the sound pressure.

The average sound level, L_{avg}, is based on the average taken over the time of measurement. A time-weighted average (TWA) is based on eight hours, regardless of the time of measurement. If the sampling time is less than eight hours, L_{avg} is always less than TWA. L_{avg} is useful for making projections from short-term samples, assuming the environmental sound level persists without radical variation.

The sound intensity per unit area on a surface is called the sound power level, L_W, and is measured in watts. The formula is

$$L_W = 10\log\frac{w}{w_0} \qquad [14.5]$$

The reference power, w_0, is 10^{-12} watts. Two sound sources having the same power level add three decibels to the sound level. For instance, if a machine has a sound power level of 84 dB the operator does not need to be in the Hearing Conservation Program, as defined by OSHA. But suppose a second machine, which is an identical sound source, is moved into the work area. The new sound level is 87 dB. Now both operators must be in the Hearing Conservation Program. As the difference between sound levels of two sources increases, the contribution to the overall sound level decreases. Table 14.10 gives the correction factor to be added to the higher of two sources contributing to sound level. To continue our example, suppose a machine that contributes 72 dB is moved near the other two. What is the overall sound level expected to be now? The contribution of the new machine will be insignificant, so the overall sound level will still be 87 dB. But if an 82 dB source were moved in, the new sound level would be 89 dB.

A *pure tone* is a sound having a single frequency. This does not occur often, but examples include a tuning fork or pitch pipe. *Broad-band* noise is spread over several frequencies. When spread over a small range of frequencies, the noise is *narrow-band.* Many industrial sources are broad-band, but specific equipment such as a high-speed drill or activities such as aiming an air hose at a narrow opening produce narrow-band noise. Devices that create an impact generate repetitive impulse noise. Impulses greater than 200 per minute are considered continuous noise.

Noise that is spread across a broad-band with random or uneven sound pressure levels is called *white noise. Pink noise* is also spread across a broad-band, but with fairly even pressure levels.

The area over which sound waves travel is called a field. Field characteristics influence the amplitude and nature of sound at various points within the field.

A *free field* is a space without walls or barriers to sound pressure waves. A very large room, such as an auditorium, can approximate the outdoors as a free field. The sound pressure level, L, from a single noise source varies according to the distance from the source.

$$L_2 = L_1 + 20\log\left(\frac{d_1}{d_2}\right) \qquad [14.6]$$

where

d_2 = new distance from source
d_1 = distance from which L_1 was measured

> Sound pressure level is reduced by 6 dB for every doubling of distance and increased by 6 dB for every halving of distance.

Table 14-8
Adding Sound Pressure Levels

Difference in sound pressure, dB	Add to the larger
0	3.0
1	2.5
2 or 3	2.0
4 or 5	1.5
6 to 8	1.0
9 to 12	0.5
over 12	0.0

When a wall is struck by sound, only a small portion of the sound is transmitted through the wall. Most of the sound is reflected. Some of the sound is absorbed, especially by porous materials. The ability of the wall, or any barrier, to block the transmission of sound is called its *transmission loss rating* (TL), measured in decibels. The TL of a barrier does not vary regardless of how it is used. *Noise reduction* (NR) represents the number of decibels of sound reduction actually achieved by a particular enclosure or barrier. NR is determined by comparing the noise level before and after installing an enclosure over a noise source. NR and TL are not necessarily equal.

Noise Calculations

When two or more noise sources are located in the same field, the effect is additive, but not algebraically, due to the logarithmic scale for sound pressure. The logarithmic addition is shown in Table 14-8.

For example, say that you have a machine generating 88 dB sound level in a room by itself and Manufacturing tells you they want to locate a second, identical machine in the room. What noise level will both machines make? The difference is sound levels will be zero (88 minus 88 dB), so 3.0 decibels are added to the greater (88). The combined noise level will be 91 dB. In this scenario, abatement is required per OSHA.

Suppose you have five machines generating 81 dB, 67 dB, 83 dB, 89 dB, and 83 dB, respectively. What is the combined noise level? 1) Arrange the measurements from low to high: 67, 81, 83, 83, 87 dB. 2) Start adding each two in order. 81-67=14; add 0 to 81 = 81. 83-81=2; add 2 to 83 = 85. Careful! 85, not 83, -83=2; add 2 to 85 = 87. 87-87=0; add 3 to 87 = 90 dB. So your total combined noise level will be 90 dB.

Effects of Noise on Human Hearing

The ear has three parts: outer ear, middle ear, and inner ear. The only function of the outer ear is to channel sound pressure waves into the covering of the middle ear, the ear drum. The three smallest bones in the human body—the hammer, anvil, and stirrup—are located in the middle ear. These bones pass vibrations of the ear drum, caused by sound pressure, into the inner ear. The inner ear consists of a fluid-filled mechanism called the *cochlea.* Thousands of tiny hair cells inside the cochlea respond to the vibrations and transmit the sensation, caused by the vibrations, on to the fibers of the auditory nerve. A sound is thus heard.

Noise, which is undesirable sound, is by definition a nuisance. But noise may also damage hearing, temporarily or permanently. In addition, noise may create stress that affects both physical and mental wellbeing. Noise can also be the cause of accidents when an employee fails to hear proper instructions or warning signals. Noise that exceeds 85 dBA over an 8-hour day may cause hearing loss.

Typically, brief exposure to loud noise produces a temporary hearing loss or auditory fatigue. After a few hours of relative quiet, hearing is restored to normal. Each individual has a limit on the number of recoveries he or she can make from auditory fatigue. Continuous or frequent exposure to loud noise will permanently damage the hearing.

When permanent hearing loss occurs, the injured person is unable to hear high-pitched or soft sounds. With this type hearing loss, one may have trouble understanding conversation, especially in the midst of background noises or other conversations, and have difficulty with telephone conversations. He or she may also experience *tinnitus*—ringing or roaring in the ears. Unfortunately, no cure exists for loss of hearing caused by noise. Contrary to popular belief, hearing aids do not restore noise-damaged hearing.

Healthy, young ears can hear a frequency range of 20 to 20,000 Hz. *Presbycusis* is hearing loss induced by aging and the use of drugs. *Nosacusis* is hearing loss caused by injury or illness. Impairment of sound transmission due to infection or ear wax accumulation is called *conductive hearing loss*. *Sensorineural hearing loss* is the poor transmission of sounds by nerve pathways due to noise damage to either the hair cells in the cochlea or the auditory nerve. If exposure to excessive noise is stopped, and after sufficient rest, the hair cells recover. Continual excessive noise, however, causes the hair cells to lose their resiliency and permanent hearing loss ensues.

Noise-induced losses and presbycusis are differentiated by the *4,000 Hertz shift* or the *notch*. Hearing losses in and around the notch are associated with exposure to loud noise. If hearing loss is in the 4,000 Hz range and the workplace exposure is greater than 90 dB, the presumption is that the loss is industrial related.

Overall health may also be damaged. Mental and physical stress are common in individuals subjected to excessive noise exposures. Stress, in general, is created when muscle reflex activities increase due to noise, causing a continuous tightening of muscles, which leads to even greater tension levels. Certain illnesses are possible from the exposure. Small blood vessels in the extremities are susceptible to excessive noise. Damage to these vessels can increase arterial blood pressure as the heart pumps harder to overcome the increased resistance to blood flow. Changes in vision can occur, such as the narrowing of the field of vision or modification of color perception, due to excessive noise. The typical vision damage is a partial deficiency in the perception of the color red. Excessive noise can also lead to the loss of touch sensitivity or slowed response to pain, an important defense mechanism. Finally, as mentioned above, excessive noise can lead to accidents. One way this may happen is when an exposure to sudden unexpected sound triggers a startle-response, resulting in secondary injuries such as bumps, scrapes, or contusions. Figure 14-3 lists some effects of noise on humans.

How much noise can damage hearing and cause other effects? The louder the noise, the shorter the time taken to cause damage for any given frequency. For many years, the standard has been set by OSHA, as in Table 14-9.

Noise Measurement

OSHA requires that measurements of the workplace be made using the A-weighted circuit of the sound level meter (SLM) or an A-weighted dosimeter. The A-weighted device has a filter that excludes most of low-frequency sound

Figure 14-3
Effects of Noise on Humans

Destruction of inner ear
Dilation of pupil
Secretion of thyroid hormone
Heart palpitations
Secretion of adrenaline
Secretion of adrenaline cortex hormone
Movements of the stomach and intestines
Muscle reactions
Constriction of blood vessels

Table 14-9
Duration of Exposure for which Hearing Loss Is Feasible

Sound Level (dBA)	Hours of Exposure
90	8
92	6
95	4
97	3
100	2
102	1.5
105	1
110	0.5
115	0.25

from the measurement. The A-weighted measurements correlate with hearing loss in many industrial circumstances.

What noise levels are dangerous for hearing? Any level over 80 dB is potentially hazardous. OSHA requires hearing protection if the level is greater than 85 dB and, if the noise level is greater than 90 dB, engineering controls are required to abate the sound level to 90 dB or less. The American Speech-Language-Hearing Association reports the sound levels in Table 14-10 as average noise levels for everyday sounds, from Dessoff as modified by data from Myers.

Modern dosimeters provide data logging, are programmable, allow several thresholds to be selected, and download to computers and printers. The OSHA threshold is set at 90 dB for measurement of most workplaces. However, if the dosimeter has multi-threshold capability, then resetting to a threshold of 80 dB allows you to evaluate noise exposures in those workplaces where sound level is in the high eighties. You cannot do this if the threshold is set at 90 dB.

When OSHA measures noise exposure in the workplace, the compliance health officer will place a dosimeter on each employee to be moni-

Table 14-10
Average Noise Levels of Everyday Sounds

Sound	Noise Level, dB
Whisper	30
Quiet room	40
Moderate rainfall	50
Conversation, dishwasher	60
Busy traffic, vacuum cleaner	70
Alarm clock	80
Computer room	80
Lawnmower	90
Light tractors	90
Sheet metal shop	95
General workshop noise	95
Power equipment	95
Snowmobile, chain saw	100
Printing press	100
Heavy tractors	100
Outboard motors	100
Diesel locomotives	105
Highways	105
Riveting	105
Engine rooms, on ships	105
Forestry machines	110
Rock music	110
Payloaders	110
Concrete industry	110
Forges	110
Textile industry	110
Pneumatic drills	115
Jet-plane taking off	120
Mining	125
Jackhammer	130
Airfields, jets running	135
Firearms, air-raid siren	140

tored and also use a sound level meter. Before a citation can be issued for results, the compliance health officer must calibrate the instrument before and after taking the readings.

Effective Hearing Conservation Programs

Hearing conservation is about protecting something that is priceless: the ability to hear. An OSHA Hearing Conservation Program has five parts or steps.

The first step in establishing a good Hearing Conservation Program is to measure the noise level in the workplace. Not only is this necessary to determine which employees must be included in the program, but the sound level is needed in order to engineer noise abatement and select appropriate hearing protection.

Next, hearing tests must be performed on all affected employees. Any employee exposed to 85 dBA or greater for a shift (100 percent dose or more) must have a baseline and annual hearing test. The baseline test indicates whether a preexisting hearing loss is present. The test must be conducted by a licensed or certified audiologist, otolaryngologist, another physician, or hearing conservationist who is certified by the Council on Accreditation in Occupational Hearing Conservation or has otherwise demonstrated competence in providing audiometric examinations.

Third, the employer must provide initial and annual training. The effects of noise on hearing must be explained to employees in the program. Hearing-protection devices that are made available to them must be introduced and their advantages and disadvantages must be explained. Rationale for proper selection of these devices and techniques for fit, use, and care must be demonstrated. Finally, training must explain the purpose and procedures of audiometric testing.

Noise Control Techniques

In hearing protection, Personal protective equipment should only be relied upon: 1) as temporary protection until engineering or administrative controls can be implemented and as a last resort in two instances: first, when the time-weighted average exposure is 85 dBA or more and, secondly, when the employee has a standard threshold shift or greater hearing loss. A *standard threshold shift* is a 10 dB average loss from the baseline test at frequencies of 2,000, 3,000, or 4,000 hertz (the speech range) in either ear. Three types of personal protective gear are available for hearing protection: earplugs, canal caps, and ear muffs.

When properly fitted into the outer part of the ear, canal plugs can be effective hearing protection. Earplugs come in three types: formable, premolded, and custom-molded. Formable earplugs are the once-size-fits-all variety made of soft, pliable material that is twisted or wadded into shape before being inserted into the outer ear canal. This category may be further divided into disposable and semi-disposable brands. Waxed cotton or acoustical fiber plugs are used once and thrown away. The semi-disposable plugs are made of molded sponge or foam and are rolled tightly before insertion. The semi-disposable plug then expands in the outer ear canal to give a good seal against noise energy. These types can be worn more than once before being thrown away—hence, the name semi-disposable. Pre-molded earplugs are typically made of a soft, silicone rubber or plastic. These plugs are reusable and come in two types: universal and multi-size. The universal type fits most ear canals. The multi-size type comes in several sizes for better fit. Finally, custom-molded earplugs are molded in the exact shape of the user's ear. A silicone rubber or plastic molding compound is poured in each ear and allowed to set. These molds are either used to make the final plugs or are sometimes used as the earplugs themselves.

Canal caps close off the ear canals at the opening. Made of a soft, rubber-like substance, these caps are kept in place over the ear canal opening by a light band under tension.

Ear muffs consisting of cushioned cups and a headband fit over the entire ear to keep out noise. Some ear muffs are fitted to safety hats. Electronic ear muffs keep out noise but amplify desirable sounds. Dielectric ear muffs have no metal parts to protect workers exposed to electrical shock. Finally, folding ear muffs allow easy storage where full-time protection is not required.

Table 14-11 compares the three general types of ear protection.

Noise Reduction Rates

Noise reduction rate (NRR) is a term used by ear plug and muff manufacturers to communicate the level of reduction their devices offer. Typical NRRs range from 5 to 30 dB. NRRs are meant to adjust the C-weighted sound level, not the A-weighted sound level, required by OSHA. Therefore

Table 14-11
Comparison of Ear Protectors

Type	Effectiveness	Fit	Maintenance
Earplugs	up to 30 dB reduction	must be right size; properly inserted	wash reusable plugs; store in bag or case
Canal Caps	up to 25 dB reduction	fits all	wash; store in bag or case
Ear Muffs	15-25 dB reduction	fits all	wash; store in bag or case; repair or replace cracks; replace worn cushions

you cannot simply take the sound level of an area or personal exposure and subtract the NRR to find the new exposure level. You can convert A-scale to C-scale, make the adjustment, then convert back to A-scale. While this can be done, it is a tedious process, so OSHA has approved an approximation:

$$NRR_{adj} = {(NRR_{mfr} - 7)}/{2} \qquad [14.7]$$

where

NRR_{adj}= the new NRR as adjusted, dB
NRR_{mfr}= the manufacturer's NRR, dB

Sound is produced by changes in force, pressure, or speed. The greater the change the louder the noise.

More noise is produced if a task is carried out with great force for a short time than with less force for a longer time.

Airborne sound is usually caused by vibration in solids or turbulence in fluids.

Reduction of vibration in solids reduces the noise generated.
Reduction of turbulence in fluids reduces the noise generated.

Vibration can produce sound after traveling long distances. Such vibrations cause distant structures to resonate.

Stop vibration as close to the source as possible.

Frequency Reduction

The slower the repetition, the lower the frequency of noise. Frequency is determined primarily by the rate at which changes in force, pressure, and speed are repeated. The longer the time interval between changes, the lower the frequency of noise generated. The level of noise is dependent on the amount of change.

High frequency sound is strongly directional and more easily reflected. When it strikes a hard surface, high frequency noise is reflected. Sound absorbing material is also a good control measure for high frequency noise.

A sound insulating enclosure can be constructed over the source of high frequency noise.

High frequency sound is greatly reduced by passing through air.

If the low frequency noise source does not cause problems in its immediate vicinity, try shifting the sound toward higher frequencies.

Low frequency noise travels around objects and through openings. It radiates in all directions at about the same power level, much as the ripples on a pond surface produced by a pebble. Low frequency sound travels around barriers unless they are very large.

> A complete enclosure of damped material lined with sound absorbent is needed for low frequency noise abatement.

Low frequency noise is less disturbing than high frequency noise. The human ear is less sensitive to low frequencies.

Vibration Reduction

Small vibrating surfaces generate less noise than large surfaces.

> Machines always vibrate to some extent.
> Keep machines as small as possible.

Densely perforated plates produce less noise than solid plates. As a surface vibrates, it pumps air back and forth like a piston. This causes the sound to radiate from the plate. When the surface is perforated, the pump functions poorly. Other alternatives perforations in a plate are mesh, gratings, and expanded metal.

A long, narrow plate produces less sound when vibrating than a square plate will. Vibrating plates form excess air first on one side then the other. Sound is generated from both sides. Close to the edges of the plate the pressure differences are balanced out.

Plates with free edges produce less low frequency noise. If a plate with free edges vibrates, the pressure equals out between the sides of the plate. Clamping the edges of a plate prevents this pressure equalization, resulting in the generation of a greater level of low frequency sound.

> To the extent feasible, leave the corners and edges of plates free to vibrate.

When a plate is struck by an object, the plate vibrates and makes noise. Light objects and low speed produce the least impact noise. The sound level produced by the vibrating plate depends on the weight of the impacting object and the speed with which it strikes the plate.

> Reducing the initial height of a dropping object from five meters to five centimeters reduces the impact sound level by 20 dB.
>
> ---
>
> Reducing the weight of a falling object from one kilogram to one hundred grams reduces the impact sound level by 20 dB.

As vibration moves throughout a plate, it gradually decreases as it travels. This *attenuation* is rather small in most plates, which have low *internal damping*. The internal damping of steel is notably poor. A damped surface generates less noise.

> Good damping can be achieved by surface coating or adding intermediate layers of material that have better internal damping.

Resonance greatly increases noise from a vibrating plate, but it can suppressed by damping the plate.

> Sometimes effective noise reduction is achieved by damping only part of a surface or even a single point.

Resonance can be shifted to a higher frequency to more easily *dampen* noise. Large vibrating plates especially have low frequency resonance that is economically infeasible to dampen.

> A reinforcing grid added to a plate will increase the resonance frequency.

Reduction of Tones

When air or gas passes by an object, a strong pure tone, the *Karman tone,* is produced. Prevent Karman tone by lengthening the object in the direction of flow, referred to as adding a tail, or by making the object irregular in shape. This is the purpose of the spirals on large smoke stacks. The pitch of the spiral, if you look closely, is not constant. Thus the wind encounters an irregular object, no matter from which direction it blows, and a Karman tone is avoided.

Avoid air or gas flow past hollow openings. Air or gas blowing across the edge of an opening produces a loud, pure tone. Wind instruments use this principle to make beautiful music. Air hoses in machine shops, on the other hand, make a loud screech. The volume of the hole is inversely proportional to frequency. The number of openings is proportional to frequency.

Reduction of Jet Stream Noise

Jet noise can be reduced by adding air streams. A noisy jet stream is created when a gas at standard temperature and pressure flows through a narrow opening at 325 feet per second or more (the *Mach number*).

By using an air stream with a lower speed outside the jet stream, the noise is greatly reduced. This is the principle used in the tubular mouthpieces with dual air flow on compressed air nozzles. Part of the compressed air

> Reducing gas velocity by half, when the original velocity is greater than the Mach number, will reduce the jet stream noise by as much as 20 dB.

moves at a slower speed outside the central jet stream, reducing the noise from the nozzle.

> Low frequency jet noise is easier to abate if converted to high frequency jet noise.

Increasing jet stream frequency can be accomplished by reducing the diameter of the outlet. A common practice is to replace one opening with several smaller ones. This is the principle used in noise diffusers.

Reducing Pressure Drop Noise

Rapid pressure changes in liquids increase the noise of a piping system. Turbulence, the sound producer, is increased when the pressure in a liquid system is dropped quickly. This pressure drop produces gas bubbles from boiling liquid, resulting in a roaring noise. This is called *cavitation.* One way cavitation is created is when the volume of the system suddenly increases by a large amount.

> Cavitation noise is avoided by slow changes in volume.

Control valves, for instance, frequently have small valve seats, resulting in large velocities with large pressure drops. Twisted flow paths and sharp edges in the valve body can produce intense cavitation. Sound radiates directly from valves and pipes and is conducted to walls. To remedy this situation, use valves with larger cone diameters, straighter flow pathways, and rounded edges in those pathways. Cavitation can also be controlled by reducing pressure in several small increments instead of the large pressure drop experienced in a control valve. Insert pressure reducing orifices upstream to the place where a certain pressure is required. Choose each orifice plate so that the pressure does not drop sufficiently to cause cavitation, but so that the total pressure drop across all the orifices equals the desired amount.

Reducing Noise by Design

Do not locate sound sources near the corners of rooms. The closer the source is to a reflecting surface, the more noise will be measured from it. The worst case location of a sound source is in a corner with three reflecting surfaces—that is, next to two walls and the floor, or two walls and the ceiling. Move noise generators away from walls and corners.

Thick, porous layers of sound insulating material absorb both low and high frequency sounds, but not equally as well. Thicker insulation is required to control low frequency. Any porous material through which air can pass makes a good sound absorbent: felt; foam rubber; foamed plastic; textile fibers; sintered metals; and ceramics. Some companies specialize in manufacturing sound absorbing insulation so you are safest using material recommended by them. However, in a pinch, improvise. Thinner absorbents handle higher-pitched tones. Below 100 Hz, the required thickness may not be economical or even physically feasible, but some thickness will abate the noise. One way to enhance the absorption of low frequency noise and cut down on the required thickness is to place air gaps between layers of insulating material.

In machine shops with intense low frequency noise, two practices are common. Absorption baffles are often hung from the ceiling to provide low frequency sound absorption. Another practice that may be more common is to install horizontal absorbing panels about eight inches from the ceiling—much like ceiling panels in an office, but higher.

Another way to handle low frequencies is to mount absorbent panels on studs. The absorption achieved is only effective over a narrow range of frequencies, though. The range is dependent on the stiffness of the panels, the distance between fastenings, the distance from the panel to the wall, the damping ability of the material, and the thickness of the material. Fasteners must be spaced further apart for low tones. Panel thickness must be thinner for low tones. The absorption panel must be placed further from the wall for low frequency sound abatement. Since all these factors are critical for success, it is recommended that you hire the services of a good sound engineer to design your abatement project. The absorbent material company may provide that service for you.

High frequency noise is typically abated by using sound shields. The taller the shield, with respect to the noise source, and the closer it is to the source, the more effective it will be. Sound shields are more effective when used in conjunction with ceiling panels. This is because uninsulated ceilings can reflect the noise around the barrier.

If the noise of an exhaust pipe or duct has a narrow frequency range, a reaction muffler may take up less space than an expansion chamber (see below for controlling duct noise). Several reaction mufflers in a series are often used to abate noise that covers a large frequency range. The length of the reaction chamber determines the frequencies that are controlled, while the diameter of the muffler determines the extent of control.

Any unutilized area of a building can be turned into an absorption chamber. Quite simply, an absorption chamber is a huge muffler. For instance, a loud air conditioning duct can be run to an unused room with the duct continuing to its destination from the room's return grille. The change in velocity, due to the room's volume, will quiet the duct. Adding sound-absorbing material to the walls of the room will reduce noise even further. Locate the inlet and outlet to the room in offset locations to prevent high frequency sound from being directed downstream. The larger the room and the thicker the absorbing material, the lower the frequency at which the absorption chamber is effective. The shape of the chamber is of little significance in abating sound.

Manufactured absorbent panels even come decorated. Perforations, ribs, and other ingenious schemes to make panels attractive, generally do not hinder the absorption of sound.

HVAC Noise Control

Heating, ventilation, and air-conditioning (HVAC) systems contribute to the noise in spaces they heat or cool. HVAC noise consists of low frequency fan noise, mid-frequency duct system component air flow noise, and high frequency damper and diffuser noise. Fan noises decrease about 40 decibels from a high level at 63 Hz to a low level at around 250 Hz. Airflow noises range from about 250 Hz to 2,000 Hz with the highest level at the low end of the band and dropping 10 decibels over the band. The highest level diffuser and damper noises start around 750 Hz and drop about 10 decibels to about 8,000 Hz.

Fan Noise. Manufacturers commonly use the terms total fan sound power level and ducted fan sound power level. *Total fan sound power level* is the noise generated by the inlet and discharge of the fan, motor, drive train, and casing. *Ducted fan sound power level* is noise that is transmitted into the duct system from the fan. The ducted fan sound power level is required to analyze the acoustics of a duct system.

Figure 14-4 summarizes the procedures to use when evaluating the manufacturer's information.

> The ducted fan sound power level may be estimated by subtracting three decibels from each octave band of the total fan L_W levels.

Fan noise, when the fan is attached to a duct, is increased by the total pressure of the system, a phenomena called *static regain,* so design ducting to achieve the lowest possible operating pressure. The static regain, L_S, is

$$L_S = 20\log_{10}(P) \qquad [14.8]$$

where

P = system operating pressure

Fans that are oversized and operated at less than design speed or undersized fans that are operated above design speed generate more noise than necessary.

Figure 14-4
How to Use Manufacturer's Information

1. Is the manufacturer using total or ducted fan L_W levels?
2. Evaluate manufacturer's installation recommendations.
3. Review the fan manufacturer's potential fan choices.
4. Obtain fan L_W from the manufacturer whenever possible.

Select the fan to operate near its maximum efficiency.

Fans also make less noise when placed in smooth, undisturbed flow streams. If turbulence is already present in the air stream, the fan will create a more intense sound. To reduce this effect, move the fan farther downstream from the disturbance, such as a barrier, or elbow, or vanes.

Silencer Noise. A few common types of duct silencers are available. One type is the rectangular silencer with internal sound-absorbing baffles. Another is round silencer, which comes with or without internal sound-absorbing baffles or bullets. The flow of air through the silencer is obstructed by the reduction of cross-sectional air flow area created by the baffles and bullets. Downstream, as air expands, turbulence generates noise. If this noise level is within 10 dB of the attenuated sound power level, the two noise levels are combined to yield a greater noise level. See Section 14.3.2 for adding two sounds together. The effective use of silencers can abate noise by 10 to 15 dB.

Damper Noise. Noise from dampers is a function of the air velocity flowing past the damper. Poor balancing creates high velocities, which create excessive noise.

Diffuser and Grille Noise. Noise from diffusers and grilles is generated by turbulence created from air flow through these devices. A lower exit velocity produces a lower sound level.

Locate dampers sufficiently upstream from diffusers and grilles to eliminate noise.

A well-designed system does not require balancing dampers.

> For an air velocity less than 325 feet per second, reducing the speed by half makes the noise produced 15 dB weaker.

Design the system with minimum air flow and select diffusers and grilles from manufacturer catalogs based on the published sound power level data with reference to noise criteria (NC) levels.

Duct Component Noise. Elbows, tees, crosses, and other components in the duct system generate noise as air flows through them. Turning vanes help in elbows, but do not eliminate noise. Noise level in these components is a function of air velocity, duct size, number of vanes, radius of bends, location of dampers, location of elbows, and location of branch takeoffs. Ducts without impediments to airflow produce the least amount of noise. Undisturbed flow also produces the least amount of exit noise. The design engineer needs to follow the American Society of Heating, Refrigeration, and Air-conditioning Engineers (ASHRAE) recommendations on placement of these components.

As sound waves travel through a duct, some energy is transmitted to surrounding surfaces reducing the sound level. This is called *natural attenuation.* Significant attenuation is found in long single-walled duct runs, expressed in decibels per foot, dB/ft. Attenuation is a function of duct shape and size. Round duct attenuation is related to diameter. Flat oval duct attenuation is related to the width of the minor axis.

Noise from turbulence is increased proportional to the elbow angle. However, when sound enters a sheet metal elbow, some of the energy is reflected. This reduces the amount of sound energy scattered downstream. In fact, all changes in a duct reduce sound transmission more than turbulence adds noise because of the noise that is reflected back down the original duct upstream of the change. This attenuation applies to elbows, branches, and changes in volume, shape, or wall materials. The attenuation is rated as dB/elbow or other fitting or change. When the angle of the elbow is less than 90°, the attenuation is proportional to the actual angle divided by 90. Elbow attenuation is also related to the diameter of round elbows and the minor axis of flat oval elbows.

The most significant attenuation in ducting comes from power splits. Sound energy splits with the air at divided-flow fittings and is proportional to the cross-sectional areas of the downstream paths. The attenuation from splits applies at all frequencies. Expansion chambers are another method that is useful for reducing low frequency noise in ducts.

> The lower the frequency that must be abated, the more volume that is required for the expansion chamber.

Absorption mufflers are effective over a broad band. The simplest form of absorption muffler is absorbing material placed on the wall of the duct. Lower frequencies are reduced by thickening the material. The space between absorbing walls must be close together to abate high frequency noise. Therefore, large ducts must be divided into several small ducts by placing sheets of absorbent materials lengthwise in the duct with small spaces for air flow between them. This is called a packed muffler.

A significant change of area at the end of a duct run reflects low frequency sound back into the duct.

> End reflection is practically eliminated when a variable air box, diffuser, or register is placed at the duct opening.

Augmented Attenuation. Lining duct is an effective means of reducing duct noise. Rectangular duct lining is made with a solid outer shell with a layer of insulation adjacent to the interior wall of the duct. A coating on the insulation prevents erosion of the insulation by airflow. Performance data on duct lining is available from ASHRAE.

Another noise abatement strategy with ducts is to use a double-wall duct. These ducts and fittings have a solid outer shell and a perforated inner shell with as much as three inches of insulation between them. Acoustical data in dB/ft is available from the manufacturer. Performance of double-wall ducts is a function of diameter or minor axis, insulation thickness, and air velocity.

Baffle and bullet silencers placed downstream of fans also enhance attenuation. Performance is measured in terms of *insertion loss*, the change in sound pressure level at a specified location when a sound attenuating device is placed between the source and the measuring device. Silencers are selected according to percent open area to produce the desired noise control while maintaining an acceptable pressure drop. *No-loss* silencers have no baffles and bullets and introduce no pressure drop. The latter are effective devices to abate low frequency noise.

Reverberation

A problem in rooms with hard surfaces such as concrete floors and walls is *reverberation.* The term refers to the lifetime of reflected acoustical energy as measured according to the time it takes to reduce the sound energy by a factor of 60 dB. Excessive reverberation time causes the sound energy to build until it impedes communication. Reverberation is controlled by adding materials to the surfaces that dissipate rather than reflect sound energy. The Sabine equation, which relates the reverberation time in a room to its volume, surface areas, and materials of construction.

Enclosure

One way to deal with noise is to isolate the source by enclosing it. The enclosure, regardless of material of construction, is insulated with sound-absorbing material. Ventilation may have to be provided if the source also generates heat, particulate matter, aerosols, gases, or vapors. The effective use of enclosures can reduce noise level in the 3 to 20 dB range.

The TL of a homogeneous single layer wall is estimated from its surface weight, expressed in kg/m^2 or lb/in^2.

$$TL \propto W_S \times f \qquad [14.9]$$

where

W_S = surface weight, kg/m^2
f = frequency, Hz

Table 14-12
Transmission Loss for Common Building Materials

Material	TL @ 1000 kg/m² Hz	TL @ 5000 kg/m² Hz	Limiting TL
Chipboard	12	26	30
Plaster	12	26	35
Glass	12	26	35
Aluminum	12	26	38
Steel	12	26	46

Based on data supplied by OSHA, the TL is calculated as:

$$TL = 8.5 + \left(0.0035W_S \times f\right) \quad [14.10]$$

TL is independent of material until a limiting value is reached. Table 14-12 gives the TL range for common building materials with limiting values.

A single wall provides poor sound insulation at its resonance frequency. The TL at this frequency is less than estimated. This *coincidence valley* disappears only if the wall has good internal damping. Table 14-13 shows coincidence valleys for steel plate, according to OSHA data.

The thickness of a material determines the frequency at which the coincidence valley occurs.

> Cavitation noise is avoided by slow changes in volume.

In most 20 cm thick single layer walls, the coincidence valley occurs around 100 Hz. Rigidity and weight are both important in thick walls. At higher frequencies, TL increases proportionally to both rigidity and weight. Figure 14-5 provides a comparison of some walls.

Two light walls separated by an air gap furnish good TL. TL increases proportionately with the gap, up to about 15 cm. Beyond that distance, sound-absorbing material must be added for further increases in TL. Double walls

Table 14-13
Coincidence Valley for Steel Plate

Plate Thickness, mm	Resonance Frequency, Hz	TL, dB
1.5	6700	28
10.0	1000	28
80.0	125	28

can provide the same TL as single walls that are five to ten times heavier. However, the double walls should have few connections. The closer the studs are together, the less TL is available. The thicker the layers, the farther apart the studs must be.

Substitution and Repair

Another strategy for noise control is to substitute the noise source. For instance, a loose bolt may be causing a vibration that generates excessive noise. Replacement of the bolt and/or tightening may be all that is needed. Noise abatement structures are no substitute for good maintenance practices.

Figure 14-5
Comparison of Wall TL

The following walls have a 30 dB TL at low frequencies, a 60 dB TL at high frequencies, and a mean TL of 55 dB.

15 cm cast concrete

23 cm cement block

28 cm cinder block

Vibration Isolation

Noise is often generated by vibration of equipment, which is transmitted through floors and walls to other spaces or the far field. Placement of the equipment on resilient pads prevents such transmissions. A good way to isolate an extremely heavy piece of equipment with low frequency vibration that produces noise is to place it on a concrete base plate that rests directly on the ground. Additional vibration/noise protection is achieved by separating the base plate from the rest of the building by means of a joint. A clay layer in the ground can retransmit the vibration back into the building. In the latter case, place pilings beneath the base plate. The pilings should penetrate the clay layer into any amorphous material below. Vibration isolation of a particular machine may be an ineffective noise control measure if the sound is transferred through a utility connection, such as a connection for oil, water, electrical power, or compressed air. Even process inlets and outlets can transmit sound/vibration. Make utility connections and process inlets and outlets flexible.

Noise Cancellation

An interesting strategy is the cancellation of the noise by picking the noise up by a microphone and feeding it to a computerized signal processor. A wave form identical to the original but 180° out-of-phase is generated, amplified, and fed back to the source via a speaker system. The two sounds combine and cancel each other out.

Mufflers

Equipment with noisy exhausts can be controlled by adding a muffler. An effective muffler can reduce noise by around 10 dB.

Recordkeeping

Noise measurements taken by a sound level meter or by a dosimeter measuring area noise in the workplace must be retained for two years. However, if a dosimeter is placed on an employee to measure personal exposure to noise, the results must be retained for as long as the worker is employed, then thirty years beyond that time.

Vibration

Although vibration is an integral factor of noise—which is the vibration of some medium transferred to our bodies' hearing sensors—let's turn our attention now to vibration as a direct hazard. Vibration acting directly on the body can cause considerable discomfort and injury. Vibration also has negative psychological consequences: aggravation, irritability, short temper, malcontent, and various stages of depression.

Physics of Vibration

Resonance determines how much kinetic energy the body absorbs from vibration. Each part of the body has a distinct resonance level. Resonance for the entire body is induced by the 5 to 11 Hertz (Hz) range. The head is affected from 20 to 30 Hz. Thirty to 40 Hz affects the hands. The eyes and visual acuity are affected from 60 to 90 Hz. Over 100 up to 200 Hz affects the lower jaw.

Displacement is the magnitude of motion of a body. This term is more significant in rigid bodies and is important in the study of deformation and bending of structures. *Velocity* is the time rate of change of displacement. For sinusoidal vibration, velocity is proportional to displacement and frequency. *Acceleration* is the time rate of change of velocity. For a certain mass, its acceleration is proportional to the applied force.

Effects of Vibration on the Human Body

Two types of vibration phenomena are observed with respect to the human body: segmental and whole body vibration.

Segmental Vibration

Segmental vibration is most often encountered in the use of hand tools. The most common form of segmental vibration is hand-arm vibration (HAV). The effect of HAV on the body depends on frequency, acceleration, direction, and duration.

What is the frequency of vibration? A frequency range of 25 to 150 Hertz is associated with Raynaud's phenomenon or vibration white finger (VWF). What is the acceleration? An acceleration of 1.5 to 80 g is associated with Raynaud's phenomenon or VWF. One g of acceleration is approxi-

Table 14-14
Effects of Vibration on the Hand-Arm

Frequency Range, Hz	Effect
30-480	Vascular tissue damage
250-350	Neurological disruption
60-480	Decreased blood flow to fingers
280-600	Raynaud's syndrome
10,000	Vibration white finger

mately 10 m/sec^2. Direction is measured in Cartesian coordinates (x,y,z). Duration is the segment of time over which the vibration lasts.

Disorders associated with HAV effect are vascular, bone and joint, and neurological and muscular systems. Vascular disorders are *Raynaud's phenomenon* or VWF. Bone and joint disorders include *decalcification, degeneration,* or *deformity* of metacarpal bones, and *osteoarthritis* from repetitive shocks. Neurological disorders include *sensory loss, paresthesia,* and *carpal tunnel syndrome.* The peripheral nervous system is particularly affected. *Muscle atrophy* occurs due to a combination of nervous and muscular disorders. *Dupuytren's contracture* involves thickening and shrinking of fibrous tissue under palmar skin. *Tendinitis* and *tenosynovitis* are also muscle related. Table 14-14 lists the effects of various frequency ranges on the hand-arm.

Whole Body Vibration

Whole body vibration affects the whole body as opposed to the arm and hand, or leg and foot. Martens cites research that has linked whole body vibration to low-back pain. An example given by Martens is that of the occupational truck driver. The very least effect whole body vibration causes is mild discomfort. The worst case may be a herniated disc that leads to *sciatica,* spinal degeneration.

Low frequency exposure up to about 10 Hertz causes stomach and chest discomfort and causes some muscles to contract. Breathing is affected as low as 2 Hertz. Vibrations over 10 Hertz, due to resonance, tend to affect the head and limbs more than the whole body.

Table 14-15
Safe Vibration to the Hands

Duration	Frequency Range, Hz	Upper Limit of Acceleration, G
30 minutes	8-15	250
	80	
	250	
4 to 8 hours	8-15	
	80	

The upper limit is the level at which performance decrements can be seen.

Vibration Measurement

Safe vibration to the hands is described in Table 14-15.

The acceleration of a chipping hammer is about 30 g with high frequency. A portable grinder has an acceleration of less than one g with high frequency. The weight, size, and design of the tool determine its vibration

Table 14-16
Whole Body Vibration

Vertical Vibration, Hz	Affected Body Part
2	Legs with Knees Flexing
4-5	Shoulder Girdle
4-8	Abdominal Mass
10-12	Spinal Column (Axial Mode)
18-30	Lower Arm
20+	Legs with Rigid Posture
25	Head (Axial Mode)
30-80	Eyeball, Intraocular Structures
50-200	Hand Grip
60	Chest Wall, Lung Volume

Table 14-17
Threshold Limits for Vibration Force on the Hand

<1	1.22
1-2	0.82
2-4	0.61
4-8	

characteristics. Cold temperatures exacerbate the problem. Anything that forces the employee to make a tighter grip on the tool also exacerbates the problem. Table 14-16 lists affected body parts to certain frequencies of whole body vibration.

For many tasks that apply forces to the arm, such as forces that cause repetitive ulnar deviation of the wrist, a higher tactile sensitivity is noted at the end of the work day. Repetitive motions that harm the wrist do not necessarily affect tactile sensitivity in the finger tips.

The procedure for measuring vibration is established by ISO 5349 (1986), *Guide for the Measurement and the Assessment of Human Exposure to Hand-Transmitted Vibration,* for international compliance and ANSI S3.34-1986, *Guide for the Measurement and Evaluation of Human Exposure to Vibration,* for national compliance. First, determine the acceleration of a vibrating handle or work piece in three mutually perpendicular directions (x,y,z-coordinates) close to where the vibration enters the hand. Mount a lightweight accelerometer and record several vector components of the vibration force. An accelerometer that measures vibration in the frequency range of 5 to 1,500 Hz should be used. Assess the vibration force in each direction. Compare the largest to Table 14-17, which gives the threshold limits for hand vibration in terms of acceleration.

Figure 14-6 lists the strength of evidence found in a NIOSH study linking vibration with various musculoskeletal disorders.

Figure 14-6
Evidence for MSD Causation by Vibration per NIOSH

Neck/shoulder disorders
Insufficient evidence Shoulder disorders
Insufficient evidence Carpal Tunnel Syndrome
Some evidence Hand-arm Vibration Syndrome
Strong evidence Back problems (whole body vibration)
Strong evidence

Vibration Control Strategies

Ideally, eliminate all unwanted vibration at the source. Control methods include isolation of the source by shock and vibration insulators, reduction of shock and vibration by absorbers, or vibration damping. In general, vibration effects are reduced by three methods. Isolating the source from the radiating surface is one method. A second method is to reduce the response of the radiating surface. Third, you can reduce the mechanical disturbance that causes the vibration. Specific measures are discussed below.

The most effective vibration safeguard is to remove the worker from the vibration. Can the operator perform the task remotely? If the operator's job is merely observation, install closed-circuit television and let him or her observe a video monitor in a remote location, free of vibration. If the task involves observation and control but no direct involvement with the process, can remote controls be used?

The second most effective way to deal with vibration is to isolate the vibration. Machines that vibrate should be mounted on heavy, rigid bases. Sometimes it is possible to move the part of the machine that vibrates to a location near the machine. Vibration is an inherent factor in other machines, and this cannot be done. Accumulators installed in hydraulic or pneumatic lines to store pressure will reduce vibration. Vibration isolators are made of various materials and come in various shapes. Rubber or plastic foam material may be used to isolate the machine from a floor. Dense rubber or plastic gasket material may also be used. Mineral wool, cellular material, and cork may also be used. Sometimes machines may be mounted on horizontal wire coils, spiral springs, leaf springs, or plate springs. Passenger cars are partially vibration isolated by a combination of spiral springs, leaf springs, and shock absorbers.

Spiral springs are made of either long, thin wire or short, thick wire. Improperly selected springs can increase vibrations. An object placed on springs has a *fundamental frequency.*

> Vibrations at or close to the fundamental frequency are greatly intensified in the object spring system.

The object may even break away from its moorings.

> Vibrations that have a lower frequency than the system's fundamental frequency are not blocked by the use of that particular spring.

If the base is very heavy, or very rigid, the fundamental frequency is determined entirely by the combined weight of the isolated object and the base together with the rigidity of the spring. The lighter the weight of the object and the more rigid the spring, the higher the fundamental frequency is. Avoid this situation by using springs with good internal damping.

A rigid floor may be required in order to isolate objects with low natural frequency. A stiff (extra heavy) or pile reinforced floor may be required. Place support pilings directly beneath the isolation springs.

Another way is to use materials that generate less vibration. Mounting tools to a fixture is a good strategy. An example would be a torque action arm. Make sure that tools are balanced. Select tools that have less vibration or a less harmful frequency.

Bureau, *et al,* studied the effects of wrapping a resilient material around various tools by measuring the hand-arm vibration of workers in a shipyard. Some earlier studies had shown that such tool wrappings could increase vibration most likely due to resonance-induced vibration in the 5-200 Hz range. The shipyard study confirmed this phenomena, finding that at least one wrapping material did, in fact, increase vibration in this frequency range. However, the cumulative effect of the wrappings in the 5 to 1,400 Hz range was 35-45 vibration attenuation. The vibration levels of the unwrapped tools ranged from 3.44 to 5.62 m/sec^2 in the 5 to 1,400 Hz range, using the ANSI weighted amplitudes. Wrapped tools ranged from 3.46 to 5.20 m/sec^2 in the same frequency range and weighting. Using the unweighted NIOSH method over the 5 to 5,000 Hz frequency range, the unwrapped tools measured 441 to 629 m/sec^2. The same method yielded 227 to 395 m/sec^2 for wrapped tools. This seems to indicate to me that the wrappings do decrease overall vibration, but so much not in the lower frequency range where the hand and lower arm are affected.

Reducing Exposure

Limit exposure time by rotating employees through the task or by more frequent rest breaks. Maintain equipment properly. For instance, replace worn parts that may cause excessive vibration. Implement any measure that allows the employee to relax his or her grip on the tool. Also, any measure that reduces forces that push against the body will pay dividends. Reduce the contact area on the body. Finally, control temperature. Remember, these problems are exacerbated by cold temperature.

Anti-vibration gloves provide hand and wrist support under conditions of hazardous vibration. These gloves absorb the shock that could lead to injuries.

For whole body vibration, allow affected employees to stop and stretch once per hour. Provide lumbar support where possible, such as for truck drivers. Sometimes shock absorbing material can be used to reduce the vibration.

Summarizing vibration control:

- Change equipment speeds.
- Change feed rate.
- Perform routine maintenance.
- Mount equipment on vibration dampening pads.
- Provide cushioned floor mats for standing operators.
- Provide tools that are vibration dampened.

References

About on-the-Job Hearing Conservation. South Deerfield, MA: Channing L. Bete Company, Inc., 1992.

About Protecting Yourself with PPE (personal protective equipment). South Deerfield, MA: Channing L. Bete Company, Inc., 1973.

Banach, Jim. "The 'ABCs' of Noise Measurements Set the Stage for Responsive Controls." *Occupational Health & Safety.* October 1994, pp. 75-78.

Bureau, Paul J., Ronald P. Guarneri, and Myron A. Robinson. "A Method for Measuring the Vibration Characteristics of Resilient Tool Wrap Materials in a Shipyard Production Environment." *AIHA Journal.* July 1997, pp. 518-520.

Burton, D. Jeff. "Noise Control Begins with Calculations Simplified to Add Sound Pressure Levels." *Occupational Health & Safety.* November 1993, pp. 44-45.

Burton, D. Jeff. "Noise's Temporary, Long-Term Effects on Hearing Loss Must Be Quantified." *Occupational Health & Safety.* February 1994, pp. 36 & 43.

Burton, D. Jeff. "Simple Rule Helps Estimate Weighting and Sound Levels at Various Distances." *Occupational Health & Safety.* January 1994, p. 50.

Champagne, Roger. "A Deluge Can Save the Eyes." *Occupational Health & Safety.* October 1994, pp. 69-74.

Dessoff, Alan L. "Hearing Protection: No Noise Is Good News." *Safety + Health.* February 1995, pp. 54-57.

Hawkins, Harold D. "Solving Specific Acoustical Problems in Industrial Settings: Reverberation and Attenuation." *Industrial Hygiene News.* July 1993, p.4.

Hearing Protection: A Sound Practice. Virginia Beach, VA: Coastal Video Communications Corporation, 1992.

The Industrial Environment—Its Evaluation & Control. U.S. Department of Health, Education, and Welfare. Public Health Service. Center for Disease Control. National Institute for Occupational Safety and Health, 1973.

Industrial Hygiene study Guide. 4th ed. New Jersey Section American Industrial Hygiene Association, 1989.

Johnson, Arthur T., Cathryn R. Dooly, Manjit S. Sahota, Karen M. Coyne, and M. Benhur Benjamin. "Effect of Altered Vision on Constant Load Exercise Performance while Wearing a Respirator." *AIHA Journal.* August 1997, pp. 578-582.

Martens, Mary. "How to Reduce Vibration." *Safety + Health.* May 1996, pp. 54-56.

Myers, Brian C. "Sound Strategies: How to Establish an Effective Hearing Conservation Program." *Professional Safety.* October 1994, pp. 20-23.

"NIOSH Report Links MSDs and Workplace Risk Factors." *Occupational Hazards.* August 1997, p. 16.

OSHA 3048. *Noise Control: A Guide for Workers and Employers.* U.S. Department of Labor, Occupational Safety & Health Administration, Office of Information, 1980.

Pinto, Michael A. "Scientists Call It Office Eye Syndrome." *Industrial Safety & Hygiene News.* February 1998, p. 33.

Rekus, John F. "Engineering Measures Can Muffle Hazards Heard in Noisy Work Areas." *Occupational Health & Safety.* September 1993, pp. 82-89.

Robinson, Fred, Jr., and Bruce K. Lyon. "Ergonomic Guidelines for Hand-Held Tools." *Professional Safety.* August 1994, pp. 16-21.

Sanders, Roy E. "Bad Vibrations." *Workplace Ergonomics.* May/June 1995, pp. 22-29.

Schaefer, Paul D. "Successful Noise Abatement Program Relies on Engineering Controls." *Occupational Health & Safety.* October 1992, pp. 82-86.

Snook, Stover H., Donald R. Vaillancourt, Vincent M. Ciriello, and Barbara S. Webster. "Maximum Acceptable Forces for Repetitive Ulnar Deviation of the Wrist." *AIHA Journal.* July 1997, pp. 509-517.

Sofra, John. "Computer-aided HVAC Acoustical Analysis." *Plant Engineering.* January 1995, pp. 22-24, 37.

Tsimberov, Dmitry. "Guidelines Warn When the Rattling Can Cause Harm." *Workplace Ergonomics.* October 1994, pp. 27-32.

"Twelve Principles of Ergonomics. Principle 10: Maintain a Comfortable Environment." *Safety Update.* Comprehensive Loss Management, Inc. August, September 1994.

15

HEAT AND COLD EXPOSURE MANAGEMENT

HEAT EXPOSURE

Physical labor in warm atmospheres increases body temperature and sweat rate. The loss of body fluids leads to dehydration and dangerously high core body temperatures. Dehydration has a key role in heat-related illnesses.

Usually, the body maintains a core temperature between 97° and 99°F regardless of external temperature. Typically, humans wear more clothes in cold climates and wear less clothes in hot climates to help maintain this core temperature. The body's metabolism also helps maintain this core temperature by balancing heat loss with heat gain.

Working in a hot temperature defeats the body's ability to balance heat loss and gain. This situation is caused by the fact that heat is being produced in muscles doing work and also being taken in from the environment. Without a way to lose most of the heat generated by or conducted into the body, the core temperature rises, sometimes rapidly. The risk of heat injury is thus greatly increased. The heat balance for the human body can be expressed as

$$q_s = q_m \pm q_r \pm q_c - q_e \quad [15.1]$$

where

q_s = rate of heat storage, BTU/hr
q_m = metabolic heat gain, BTU/hr
q_r = heat gain (or loss) by radiation, BTU/hr
q_c = heat gain (or loss) by convection, BTU/hr
q_e = heat loss by evaporation of sweat, BTU/hr

The storage of heat may be positive or negative. If positive, the body temperature increases and, as discussed above, the blood flow increases to periphery tissue and sweating begins. The metabolic rate is always positive and ranges from about 400 BTU/hr when the body is asleep to about 1,200 BTU/hr for sustained work. Heat released by metabolism during sleep is sufficient to boil a pint of water, by the way. Radiation is often gained, rather than lost, in industrial settings. Anytime the temperature of a surrounding surface exceeds 105°F, thermal radiation has a significant effect on the heat balance. Skin temperature is typically near 95°F. When air temperature exceeds skin temperature, convective heat gain becomes significant. On the other hand, the body loses heat when air temperature is less than skin temperature.

The partial pressure of water vapor on the skin at 95° is saturated at 42 mmHg. The greater the difference between the moisture content of air and this skin moisture the faster evaporation occurs. Thus the skin is cooled, allowing heat to be lost to the environment.

Sweating is therefore the chief way the body loses heat. Evaporating sweat carries away heat from the surface of the skin. The lowering of the skin surface temperature cools the underlying blood. Heavy clothing or high humidity interferes with sweat evaporation and reduces the potential for cooling and thus contributes to heat disorder.

The capacity for sweating is impressive, as the body is 90 percent fluid. Working for an hour in the heat can evaporate up to two quarts of fluid, weighing about four pounds. That is why you can lose weight quickly in a sauna. The dry heat helps to evaporate a lot of sweat. When sweat loss becomes great, however, the body shuts down the sweating mechanism to conserve blood and cellular fluid. When this happens, body temperature rises dramatically. Fluid losses as small as one or two percent of body weight can result in impaired work performance and increased body temperature. For a two hundred pound worker, this amounts to two quarts of sweat. For a one hundred pound worker, the loss of one quart of sweat is dangerous.

Heat Illnesses and Injuries

Transient *heat fatigue* is a temporary sense of psychological and physical discomfort that is relieved when the victim is removed to a cooler environment. The usual symptoms are irritability, depression, and impaired alertness.

Heat cramps are characterized by painful muscle cramps, particularly in the abdomen or in fatigued muscles. These symptoms are caused by an electrolyte imbalance due to prolonged sweating without adequate fluid and salt intake.

With the loss of more fluids as sweat, *dehydration* has no early symptoms; however, fatigue, weakness, and dry mouth eventually show up. At this point, the fluid loss by sweating has become excessive.

Fatigue, weakness, rapid pulse, profuse sweating, blurred vision, dizziness, fainting, and severe headache indicate *heat exhaustion.* Dehydration is continuing and blood is distributed to the periphery in an attempt to cool the body. Heat exhaustion is exacerbated by poor acclimatization and a low level of fitness.

The symptoms of *heat stroke* are chills, dizziness, and irritability. This potentially fatal illness is caused by excessive exposure to heat and subnormal tolerance to heat, which may either be genetic or acquired. Sweating typically ceases, but not always. The victim will probably be mentally confused, and may become unconscious. The skin will be dry and hot. Fifty percent of heat strokes end in fatality, so be prepared. The elderly, very young, obese, dehydrated persons, persons unaccustomed to heat, and persons with a history of heat stroke are particularly susceptible.

Some people are more susceptible to heat than others. A healthy male worker who is less than 40 years of age and acclimatized is the least likely to have a heat illness. Female workers are more at risk. It is especially risky to have a pregnant woman exposed to heat. Other persons at risk include workers over forty, overweight workers, out-of-shape workers, workers on medications, alcoholics, and chemical substance abusers. Incompatible medications include such common items as nonprescription cold and flu medicine, over-the-counter headache relievers, tranquilizers, pain medications, antihistamines, and decongestants.

Thermal burns arise from contact of human flesh with hot surfaces. Surface temperatures or radiant energy 140°F or greater can cause first degree burns (reddening of the skin with pain). Above 180°F the burn will progress to second degree (blistering) if certain body parts are exposed long enough.

How hazardous is temperature? Table 15-1 rates thermal burns according to temperature and Table 15-2 rates the hazard according to accessibility.

Measuring Heat Exposure

Thermal load in the body is the sum of the ambient temperature, T_E, which represents the external heat load, and the metabolic temperature, T_M, which represents the internal heat load. The core temperature, T_C, represents the thermal load on the body.

Table 15-1
Heat Rating

Temperature Range, °F	Degree of Hazard
00-140	Low
141-180	Medium
181+	High

Table 15-2
Accessibility Rating

Accessibility	Degree of Hazard
Remote, no normal access	Low
Reach required for contact	Medium
Contact too easy	High

$$T_C = T_E + T_M - Q_C \quad [15.2]$$

where

Q_C = heat removed by cooling mechanisms

This equation is related to 15.1, above.

Heat Loss

Normally, T_C is 98.6°F for humans. The core temperature varies only slightly, ±0.9° normally and ±5.4° maximum. The body works hard to maintain a heat balance to make heat production equal heat losses.

Conduction is the direct transfer of heat to or from the body from or to an object. For instance, when we sit, we warm the seat. Conduction is not a very efficient means of heat transfer relative to the body and represents less than three percent of heat loss.

Convection transfers heat through gas or liquid in the body to air or water. Urination removes heat from the body. So does exhalation. Convection accounts for 12 percent of heat loss.

Radiation is the most effective means of heat loss from the body, accounting for about 60 percent. When the ambient temperature is 80°F or more, radiation, as well as conduction and convection, play only a minor role in heat balances.

Evaporation of sweat dissipates more heat than any other means, especially if the ambient temperature exceeds about 80°F. As much as 95 percent of the heat load is lost through evaporation during heavy work. About 25 percent of heat loss from a resting body is attributable to evaporation. Evaporation of sweat depends on the work load and also the relative humidity. Evaporation rate is indirectly proportional to relative humidity and directly proportional to work load.

Another important factor in heat loss from the body is the state of hydration. A fully hydrated body losses sweat much more readily than a dehydrated body. In fact, dehydration compromises the sweating mechanism and less cooling takes place, raising the core temperature. The risk of heat illness is increased in a dehydrated state. Loss of more than one percent of the body's water is considered dehydration. Loss of two percent or more is considered dangerous. Unfortunately, the thirst mechanism does not activate until one percent of the water is lost. So when you feel thirsty, you are already dehydrated. This complicates heat illness prevention when heavy labor may cause a worker to lose four to five liters per hour of moisture. You just cannot drink enough water to keep pace with the losses.

Working muscles have high metabolic demands for energy. This requires high blood flow to the muscles and increased heart rate to supply blood and energy fast enough to keep pace. When cooling is required, a large amount of blood must go to the skin for cooling. Both demands cannot be met. The capacity to do work decreases and core temperature increases.

Thermal imbalance occurs when the cooling mechanism fails to keep up with the heat load. Core temperature climbs rapidly. If the core temperature exceeds 111.2°F, cell membranes deteriorate. Cell function ceases, causing cellular damage. Brain and nerve cells are particularly damaged.

Measuring Heat Load

The temperature of a hot surface is easily measured. Several kinds of thermometers are available. Where approach to the surface is a problem, infrared sensors may be employed. A simple temperature stick will give a qualitative indication of temperature excess. For instance, a stick that melts at 180°F gives a quick indication whether a second degree burn may be a hazard.

Important environmental measurements in determining a more rigorous heat load include dry bulb temperature, psychrometric wet bulb temperature, natural wet bulb temperature, globe temperature, and air velocity. Each of these factors are discussed below.

The *dry bulb temperature* (t_a) is the normal ambient temperature we are used to in our everyday lives. Simply look at a thermometer and you are reading the dry bulb temperature. Dry bulb temperature is a simple indication of the warmth of the environment. Since other factors besides air temperature influence heat stress, the dry bulb temperature is not adequate for predicting heat stress.

The *psychrometric wet bulb temperature* (t_{wb}) is determined by either a sling or battery operated psychrometer. The bulb of the thermometer has a wick wetted with distilled water—hence, the term "wet bulb." The wet bulb, which is shielded from direction radiation from the sun, gives a reading that is tied to the evaporation of the water in the wick, which is, in turn, related to the moisture content of the air. Therefore, this temperature is useful for situations where the metabolic rate is high and the evaporation of sweat maintains homeostasis. When t_{wb} is 80°F or greater, sweat is not completely evaporated. As a result evaporative cooling does not effectively reduce heat stress. A wet bulb temperature of 88°F or greater is stressful despite other factors. The *natural wet bulb temperature* (t_{nwb}) is merely a regular thermometer with a cotton wick dipped in a reservoir of distilled water.

The *globe temperature* (t_g) is measured by a thermometer that is inserted inside a six-inch, thin-walled, copper sphere painted flat black. A period of time, generally twenty minutes, is required to reach thermal equilibrium.

Air velocity (V) is measured in the work station in feet per minute (fpm). Thermoanemometers are used to measure V in locations where no flammable or combustible vapors are present. A combustible gas indicator can be used to verify that it is safe to use the thermoanemometer. A vane anemometer can be used where fans or blowers are used to direct air flow at workers.

The *effective temperature index* (ET) indicates the relative comfort of individuals exposed to varying combinations of dry bulb temperature, humidity, and air movement. This subjective index was devised from data collected by the American Society of Heating, Refrigeration and Air Conditioning Engineers (ASHRAE) for fully clothed sedentary workers. The ET may predict the acceptability of environmental conditions for office workers or clerks in factories.

When the effective temperature is corrected for radiation it is labeled the CET. Radiant heat makes an otherwise acceptable ET unacceptable. By using a globe thermometer instead of dry bulb thermometer and the dew point temperature instead of the wet bulb temperature, the pseudo wet bulb temperature is determined. This indicates the acceptability of the work environment for individuals engaged in light physical activity.

The *predicted four-hour sweat rate* (P4SR) predicts the amount of sweat that a healthy acclimatized individual doing a particular task would lose over a four-hour period. P4SR, which considers the metabolic rate, clothing, and environmental factors, is compared to established limits of tolerance.

The *wet bulb globe temperature index* (WBGT) was developed as a simple substitute for the CET. WBGT is determined from temperature measurements alone and does not require special skills or equipment. This index is useful indoors or outdoors by applying separate predictive formula. When indoors or when sunlight is not present outdoors,

$$WBGT = 0.7t_{nwb} + 0.3t_g \qquad [15.3]$$

In the sun,

$$WBGT = 0.7t_{nwb} + 0.2t_g + 0.1t_a \qquad [15.4]$$

As you see, you will not be far off, if you use equation 15.3 to estimate the outdoor WBGT all the time, though t_a is the least difficult measurement to obtain.

If several sets of measurements are taken during the work day, a time weighted WBGT can be determined as follows:

$$WBGT_{TWA} = \frac{(WBGT_1)(t_1) + (WBGT_2)(t_2) + \mathrm{L} + (WBGT_n)(t_n)}{(t_1) + (t_2) + \mathrm{L} + (t_n)} \qquad [15.5]$$

Include the various work and rest periods during the shift. The sum of the time periods should equal the duration of the shift. The TLVs suggested by the ACGIH for various metabolic rates, M, in BTU/r, are given in Table 15-3. The New Jersey section of AIHA has provided the information in Table 15-4 on metabolic rates, compiled from various literature dealing with heat stress.

Heat Stress Index

A practical engineering method used to express the factors that determine the heat balance is the heat stress index (HSI). The HSI is particularly useful for identifying the factors that need to be controlled in order to reduce heat stress potential. The index estimates work rate (the metabolic rate M) and uses simple environmental measurements: t_a, t_{wb}, t_g, and V. Heat transfer is determined by radiation, convection, and conduction and storage is calculated. The index is the ratio of the evaporation of sweat required to offset the accumulation of heat (q_e) to the maximum evaporative capacity (E_{max}) of the environment.

$$HSI = \frac{q_e}{E_{max}} \times 100 \qquad [15.6]$$

Table 15-3
Heat Stress TLV (per ACGIH)

M, BTU/hr	Intermittent Work-Rest			
	Continuous Work	75% Work 25% Rest	50% Work 50% Rest	25% Work 75% Rest
800	86	87	89	90
1400	80	82	85	88
2000	77	79	82	86

Table 15-4
Metabolic Energy (M) Required for Various Tasks

Activity	M, BTU/hr
Assembly, light	430
Bricklaying	950
Concrete mixing	1100
Forge work	1500
Machine fitting	1000
Machining, light	575-800
Printing	525
Seated work, unspecified	360
Sheet metal work	725
Shoveling	1300-2500
Typing	375-425
Walking 3.5 mph	1150-1600
Wheelbarrow pushing	1100-1600

The maximum evaporative capacity, E_{max}, is derived from the difference in partial pressure of atmospheric moisture (p_a) and the partial pressure of saturated skin at 95°F (42 mmHg). E_{max} is calculated as

$$E_{max} = kV^{0.6}(42 - p_a) \qquad [15.7]$$

where V is wind speed. The upper limit for evaporative loss from human skin is 2,400 BTU/hr. How often this limit is reached is unknown. A practical upper limit for sustained work over an eight hour period is 1,200 BTU/hr and this can be used as a default assumption if you have no other data. For a clothed worker—that is, one who is wearing long pants and a shirt—the following calculations apply:

$$q_r = 15(t_w - t_s) \qquad [15.8]$$

$$q_c = 0.65V^{0.6}(t_a - 95) \qquad [15.9]$$

and k = 2.4 in the determination of E_{max}, see Equation 15.7. The term t_s is the skin temperature, 95°F, and the wall temperature is

$$t_w = t_g + 0.13V^{0.5}\left(t_g - t_a\right) \qquad [15.10]$$

For semi-clothed workers—that is, where males are wearing a pair of short pants and no shirt, or females are wearing short pants and a halter-top—the equations used are:

$$q_r = 25(t_w - t_s) \qquad [15.11]$$

$$q_c = 1.09V^{0.6}(42 - p_a) \qquad [15.12]$$

and k = 4.0 for the E_{max} equation.

Table 15-5 provides an evaluation of the HSI.

The *wet globe temperature index* (WGT), based on the measurement of the wet globe temperature using a wet globe thermometer, integrates the thermal effects of air temperature, air movement, humidity, and thermal radiation into a single reading. The values obtained generally correlate well with the WBGT for the same conditions. In fact, adding 3°F to the WGT typically yields the WBGT.

Effects of Workloads

Several early researchers classified workloads. In Table 15-6, the system of Christensen as reviewed by Bedny and Seglin is summarized. BPM refers to heart rate, or pulse, in beats per minute. Table 15-7 summarizes a system by Wells and coworkers. Table 15-8 summarizes a system developed in Russia by Rosenblat. The Bedny and Seglin article discusses the Lehmann equation for calculating additional minutes break times (L_p) needed to reduce the strenuousness of work.

$$L_p = \left(actual\,expenditure/4.17\right) - 1 \qquad [15.13]$$

Table 15-5
Heat Stress Index

HSI	Physiological Implication
-20 to -10	Mild cold strain frequently exists in areas where workers recover from heat exposure.
0	No thermal strain.
+10 to +39	Mild to moderate heat strain. Subtle to substantial reduction of mental function.
+40 to +69	Severe heat strain threatens health unless physically fit. Some decline in physical ability. Unsuitable conditions for cardiovascular patients or those with chronic dermatitis. Unsuitable conditions for sustained mental efforts.
+70 to +99	Very severe heat strain that only a small percentage of population can tolerate.
100	The maximum heat strain that can be tolerated by fit, acclimatized young men on a daily basis.

where the actual energy expenditure is in kcal/min. Lehmann uses the German standard of 4.17 kcal/min (2,000 kcal/shift). Equation 15.14 gives the Murrell formula for calculating additional break time.

$$R = T(K - S)/(K - 1.5) \qquad [15.14]$$

Table 15-6
Christensen's Workload Classification System

Very Easy	Easy	Average	Heavy	Very Heavy	Unusually Heavy
<75 bpm <2.5 kcal/min	75-100 bpm 2.5-5 kcal/min	100-125 bpm 5-7.5 kcal/min	125-150 bpm 7.5-10 kcal/min	150-175 bpm 10-12.5 kcal/min	>175 bpm >12.5 kcal/min

Table 15-7
Wells' Workload Classification System

Easy	Moderate	Optimal	Taxing
<100 bpm	100-120 bpm	120-140 bpm	>140 bpm
<4 kcal/min	4-7 kcal/min	7.5-10 kcal/min	>10 kcal/min

Table 15-8
Rosenblat's Workload Classification System

Easier Work	Average Physical Intensity	Heavy Physical Intensity	Very Heavy Physical Intensity
< 90 bpm	90-99 bpm	100-119 bpm	120+ bpm

where

R = rest requirement, min
T = total work time, min
K = average work, kcal/min
S = standard work, 4.5 kcal/min (2,160 kcal/shift)

The constant 1.5 represents the resting level in kcal/min. The Russians calculate break time (Bt_{cal}) as a percentage of shift time based on pulse rate (PR) against a standard pulse of 100 bpm.

$$BT_{cal} = 100(PR_W - 100)/(PR_W - PA_{br}) \qquad [15.15]$$

where

PA_{br} = average pulse rate during break

Resting

When the body is resting the energy expenditure ranges one hundred Kcal/hr or less.

Light Workload

Light work may be performed sitting or standing. Sitting may involve light hand work, hand and arm work, or arm and leg work. Light hand work includes writing, typing, drafting, drawing, sewing, or bookkeeping. Hand and arm work includes working with small bench tools, inspecting, assembly, or sorting of light objects. Arm and legwork include driving a car under average conditions, or operating a foot switch or pedal. Standing involves drill press operations on small parts, milling of small parts, coil taping, small armature winding, machining with light power tools, and casual walking up to two mph. Light workloads such as these examples expend from 101 to 200 Kcal/hr.

Moderate Workload

Hand and arm work such as nailing and filing is considered moderate work. So is arm and leg work such as off road operation of trucks, tractors, or construction equipment. Moderate arm and back work includes air hammer operation, tractor assembly, plastering, intermittent handling of moderately heavy objects, weeding, hoeing, and picking fruits or vegetables. Finally, pushing or pulling light-weight carts or wheelbarrows, or walking at speeds from 2-3 mph are moderate tasks. This workload level burns from 201 to 300 Kcal/hr.

Heavy Workload

Heavy workloads burn require the body to burn more than 301 Kcal/hr. Tasks considered heavy are arm and back work with heavy objects, transferring heavy materials, shoveling, sledge hammer work, sawing, planing, chiseling hardwood or stone, hand mowing, digging, ax work, climbing stairs or ramps, jogging, running, walking more than four mph, pushing or pulling heavily loaded carts or wheelbarrows, chipping castings, or laying concrete block.

Protective Clothing

Protective clothing prevents the body from utilizing its natural cooling mechanism. Therefore workers who are required to wear protective clothing, because of the hazardous nature of their work, are exposed to another hazard: extreme body heat. Table 15-9 summarizes bodily response to vapor barrier clothing by healthy men, based on the work of Reneau, Bishop, and Ashley. Table 15-10 shows typical responses at two air temperatures.

Managing Heat Stress

As obvious from Tables 15-9 and 15-10, strong, healthy workers are required when protective clothing must be worn. Even these workers need frequent rest breaks and cooling periods. Some body cooling devices are available as an aid to heat management, but frequent rest and cooling periods are still recommended. Unless a physician determines a worker is capable of such work, do not assign it on a routine basis.

In the case of transient heat fatigue, remove the person to a cooler environment. Gradually adjust workers to the hot environment. This adjustment period or acclimatization should take about two weeks.

Table 15-9
Physiological Responses to Protective Clothing

Core temperature change	0.76- 1.82°F
Heat storage	210.1-598.0 Btu
Mean skin temperature rise	3.53- 6.08°F
Sweat rate	2.4- 12.4 g/min
Working heart rate	108.7-145.7 beats/min

Table 15-10
Physiological Responses Measured at Two Air Temperatures

	WGBT=64°F	WGBT=79°F
Core temperature rise, °F	0.63-1.42	1.08-2.16
Evaporation rate, g/min	1.50-6.10	1.70-10.7
Heat storage, Btu	160.3-528.8	269.5-658.4
Skin temperature rise, °F	3.17-5.47	4.00-6.59
Sweat rate, g/min	2.5-9.5	2.70-14.7
Working pulse, beats/min	108-128	127- 147

For a person with heat cramps, have him or her rest in a cool environment and drink salted water containing 0.5% salt solution. Massaging cramping muscles helps. Plenty of fluid and salt replacement are indicated for someone suffering dehydration. Victims with heart problems or who have been placed on low sodium diets are at particular risk; consult their physician before giving them electrolyte fluids.

For heat exhaustion, have the victim lie down flat on his or her back in a cool environment. Give the victim some water to drink and loosen clothing. Contact a physician to determine whether the condition requires further treatment, even if spontaneous recovery seems to occur. Further oral or even intravenous fluids may be required for full recovery.

For victims of heat stroke, immediately and aggressively start cooling them. Beside removing the victim to a cool environment, use water and a fan if one is available, the air to evaporate the water. Evaporation causes a cooling effect. You need not douse the victim, but sprinkle enough water to dampen the clothes and skin, especially the trunk, face, and head. Take the patient immediately to the hospital. Pack the victim with ice when he or she is transported. Prevent direct contact of the ice with the skin by applying some towels or other materials between the skin and the ice.

Acclimatization

Acclimatization is a process of getting accustomed to heat gradually. The American Industrial Hygiene Association (AIHA) recommends between four to five days to become used to hot weather. Some physicians recommend two weeks while as much as one month has been necessary for some. The acclimatization period needed is directly proportional to the temperature and the workload:

$$t \propto TW \qquad [15.16]$$

where

t = time, hours
T = absolute temperature, degrees Rankine
W = workload, Kcal/hr

Research would have to provide an empirical constant that took into account unit conversion in order to make this an equality. The idea is that the higher the temperature and the heavier the workload the more time the worker needs to acclimatize. A healthy laborer doing heavy manual labor at 77°F is at risk, more so without acclimatization.

Reducing the Chances of Heat Stress

When the temperature heats up, allow exposed employees additional breaks. Reduce the strenuousness of the task as much as possible. According to AIHA, a worker performing light tasks 25 percent of the time and resting 75 percent, at 90°F, is still at risk. A worker performing heavy tasks 25 percent of the time and resting 75 percent, at 86°F, is equally at risk.

Figure 15-1 lists OSHA's recommendations for dealing with heat stress.

Supervisors may need to assign more people to a task in hot conditions in order to accomplish the job more quickly. Not everyone can perform at the same level in hot conditions, so allow for variations among employees.

Commercially available personal protective clothing for heat exposure include umbilical systems, vests with ice or frozen gel packs, and garments that provide evaporative cooling.

Salt tablets are not recommended, unless the person drinks a lot of water with each tablet. Whether or not salt tablets are used, keep in mind that the thirst mechanism of the body shuts off before an adequate amount of water is consumed during periods of extreme fluid loss. Workers who drink only

Figure 15-1
Ways to Avoid Heat Stress

- Let workers acclimatize to the heat.
- Install engineering controls: ventilation, spot cooling, shielding.
- Use flexible scheduling (to avoid the hottest parts of the day).
- Provide plenty of drinking water every 15 to 20 minutes.
- Train and educate employees at risk about heat stress.
- Allow the exposed workers to take longer breaks and to retreat to cooler areas.
- Provide appropriate protective clothing.

when they are thirsty will become dehydrated. Avoid caffeine and alcoholic beverages as these increase urine production, exacerbating dehydration.

Electrolyte replacement beverages provide the best source of fluids. To prevent heat stress, electrolyte drinks replace water, sodium, potassium, chloride, and glucose. Fluid replacement is essential for workers who may lose up to a gallon of sweat (just over eight pounds) per shift. It is recommended that they drink four to eight ounces of electrolyte solution per twenty minute period of work. If possible, heat exposed employees should record their body weights at the beginning and end of each shift. Investigate any substantial losses in body weight.

Spot cooling can be provided by directing airflow from portable fans or from air supply systems onto the workers. The air temperature must be less than 95°F or dry, to do any good. Local exhaust ventilation can be used to draw heat outside the work area. Increasing the amount of general ventilation may reduce the temperature in the work area. The most effective method is to deliver the air about eight or ten feet above the door.

When the heat source is radiant heat, heat flux from the source can be shielded. A reflective shield can be placed between the source and the worker. The emissivity of the surface of the source can be reduced by painting it silver. If steam leaks are the heat source, eliminate them.

Evaporative cooling and refrigeration can be used to reduce the temperature of the air supply. This is called air conditioning. In Japan, some factories run mountain brooks through the plant for evaporative cooling. So I am told, anyway. Unfortunately, not every plant has a mountain brook nearby to divert and not every manufacturing process can tolerate the humidity level created. In the southern United States, workers would drop like flies if it were not for that marvelous invention, the air conditioner, which has the advantage of controlling humidity as well.

You might consider relocating hot processes or, alternatively, the exposed workers themselves. Also, insulation can be achieved by either building an insulated enclosure around the process or, alternatively, by placing the workers in an insulated enclosure, such as a control room. Use insulation liberally. Insulation on steam and hot process lines conserves heat and generally saves money. It also reduces the heat contribution to surrounding work areas. You would be amazed how much insulation a single workers' compensation claim or a survivors benefits check will buy.

COLD EXPOSURE

Cold exposure mostly has to do with climate exposure, though a few workers work in walk-in freezers. The easiest way to deal with cold exposure is to wait until the weather warms up until work is performed. However, this is not always an option, so the safety manager needs to know how to protect his or her employees accordingly.

Cold Illnesses and Injuries

Cold exposure causes the body to lose heat. As the heat loss increases, the body loses its ability to retain and produce heat.

Hypothermia is one of the main health risks of cold exposure. It is a life threatening condition where the core temperature of the body drops below 95°F. The condition involves a paradoxical response to sudden cold: the shutting off of blood flow to the surface of the body. Wind speed, moisture, high altitudes, alcohol, tobacco, and the victim's physical condition all increase heat loss and are synergistic with cold exposure to multiply the damage. The hypothermia victim may be shivering, experiencing numbness, have a glassy stare, feeling weak, acting irrationally, acting apathetic, or losing consciousness.

Acute hypothermia, or *immersion hypothermia,* is caused by sudden immersion in water colder than 70°F. The victim may appear cold and dead, but attempt to resuscitate and gradually warm the body nonetheless. Do not give up until the victim is rewarmed or has been declared dead by medical personnel. Handle the victim gently in this near-death condition. The heart is very irritable and may fibrillate if the body is roughly handled.

As the core temperature decreases, the victim will experience uncontrollable shivering. This episode is followed by loss of coordination and manual dexterity. Next, speech becomes difficult. Unless this trend is reversed, death ensues.

Frostbite is another major health hazard of cold exposure. This condition consists of the formation of ice crystals in the tissue cells of the skin and deeper tissues of the body. Frostbite usually develops when the air temperature is below 10°F, but may occur at a temperature nearer the freezing point (32°F), when the synergistic elements mentioned above are present. Initially, frostbite causes little discomfort and may go unnoticed by the victim because the cold has an anesthetic effect on the tissues. The frostbite develops in three stages. The first stage is a reddening of the skin. Blisters are formed in the second. Finally, death of some of the skin cells and the underlying tissues occurs. Clots often form in the blood vessels in the affected area. Mild cases of frostbite often result in chilblain; more severe cases result in a dangerous gangrene. Free circulation of the blood inhibits the onset of frostbite. The parts of the body most often affected by frostbite are the hands, feet, ears, cheeks, chin, and nose. The most vulnerable parts of the feet are the heels and toes. Frostbitten skin appears waxy or discolored. The skin may be flushed, white, yellow, or blue.

Chilblain is a painful injury where the skin becomes tender, red, and swollen. The injured skin may also be numb. Another serious cold injury, *trenchfoot,* is caused by exposure to cold and moisture. Untreated trenchfoot may require amputation.

Cold Exposure and Work

Temperature, wind speed, and humidity are the pertinent factors in cold exposure. The combined effects of these factors are called *dry shade cooling. Wind-chill factor,* so familiar from television weather shows, is the mathematical expression of dry shade cooling. On television, wind-chill factor is reported in degrees Fahrenheit, but in the safety profession it is expressed as the number of kilogram calories the atmosphere can absorb from the body in an hour. Table 15-11 summarizes the implications of wind-chill factor.

Table 15-11
Implications of Wind-chill Factor

Wind-chill Factor, kCal/hr	Effect
0	Calm conditions in darkness at 91.4°F.
100	Warm weather in arctic mid-summer.
200-300	Pleasant conditions.
400-500	Snow surface becomes tacky and soft.

Table 15-11 *(continued)*

600-900	Woolen underwear, socks, mittens, ski boots, ski headband, and thin cotton windbreaker suits are comfortable.
1000-1100	Travel is unpleasant on foggy and overcast days.
1400-1500	Human flesh may freeze depending on activity, amount of solar radiation, character of skin, and blood circulation. The normal wind-chill experienced in December, January, and February.
1600-1800	Winds approaching blizzard force are required to produce this wind-chill.
1900-2200	Travel and survival are extremely tenuous.
2300-2500	Condition reached in midwinter darkness. Exposed areas of face freeze in less than one minute. Travel extremely dangerous.
2500+	Exposed areas of face freeze in less than thirty seconds.

Both the metabolic rate of the body and solar radiation offset the effects of wind-chill. Metabolic heat varies from person-to-person and depends on age, fitness, degree of obesity, and gender. The heat radiated from the sun varies according to cloud conditions and the latitude. Clothing that traps metabolic heat and absorbs solar radiation can go a long way in offsetting wind-chill. Tables 15-12 and 15-13 list the benefit of metabolic heat and solar radiation, respectively, in terms of kCal/hr.

Managing Cold Exposure

Earlier, we talked about avoiding outdoor cold exposure by waiting until the weather warmed up, but not every employer has that option. The stan-

Table 15-12
Metabolic Heat Generation

Activity	Body Heat, kCal/hr
Sleeping	40
Sitting, resting	50
Driving a vehicle	65
Standing	75
Walking slowly, level surface	100
Walking normally, level surface	150
Walking fast, level surface	200
Sustained heavy work	250
Brief heavy work	300

Table 15-13
Solar Heat Radiation

Cloud Cover	Sky Condition	Radiation, kCal/hr
0/10	Clear, midday	200
2/10	Light clouds or clear early morning or late afternoon	160
5/10	Moderate cloudiness	100
8/10	Heavy cloudiness	40
10/10	Solid overcast	0

dard management strategy for cold exposure is a three-fold approach. The first step is to make the management team aware of injuries that may result from cold exposure. Next, workers are instructed in how to prevent injury by using proper clothing, taking warming breaks, and using the buddy system. Finally, employees are reminded often to avoid dangerous conditions.

This simplistic strategy sometimes needs a more practical boost. First, increase the number of workers in the exposed crew. This gets the job done faster and lets the folks get back inside where it is warm. Another version of using more employees is to rotate them so that no one stays exposed for very long. Use warming sheds. These can be built of inexpensive plywood so that a small space heater provides a lot of warmth, allowing cold workers to recover a normal core temperature. An unheated shed may reduce the wind-chill enough to prevent frostbite or other cold stress injury. Eliminate overtime work until it warms up. Alter work/rest cycles. Reduce the time exposed to cold and increase the time resting in warmth.

The U.S. Army field manual for cold weather is a model for cold weather protection. Workers exposed to cold should maintain peak physical fitness. Periods of inactivity in cold exposure should be minimized. The exposed workers should eat adequate amounts of food, including frequent nutritious snacks. Sweating dehydrates the body, so drink plenty of water. The Army recommends sixteen ounces with each meal, sixteen ounces before going to sleep, and sixteen ounces every hour during work. More water is needed by workers performing strenuous activities. Workers should avoid alcohol, caffeine, and tobacco. Workers should keep their hands and feet dry. Lip balm may be used to prevent chapped lips and moisturizing lotion may be used for the skin.

Preventive Clothing

Workers exposed to cold should dress in layers, just as their mothers probably dressed them before sending them out as youngsters to sled down their favorite hill. Layering keeps the clothing adjacent to the skin dry. Also, it is easy to remove layers as necessary if the temperature warms up or if physical exertion requires ventilation of the body to dissipate some heat.

Inner layers should be composed of light material. The American Red Cross recommends thin, snug-fitting long underwear and a long-sleeved shirt. Underwear is the basic insulation and pulls moisture away from the skin. Natural fibers such as wool, cotton, or silk are quite warm for light activity. Synthetic fabrics that absorb less moisture and carry water droplets away from the skin are better suited for heavier tasks. Capilene is a lightweight synthetic fabric that makes a suitable underwear. Another is polypropylene.

One or more insulating layers are worn over the underwear. A wool sweater or down jacket are examples of such clothing. Wool pants are preferred over jeans or corduroys. Synthetic insulators recommended by the American Red Cross are Thinsulate™, Qualofil™, and pile. Qualofil™ is exceptionally warm, wet or dry. Pile is a plush, nonpiling polyester fabric. Down is a good lightweight insulator, but it is useless when wet.

The outer or shell layer should be windproof. Synthetics that are windproof include Supplex™, Silmond™, Captiva™, or ripstop nylon. Hypalon™ is a versatile, synthetic rubber that is applied to lightweight nylon, making it completely water-repellent, tolerant of saltwater, and abrasion resistant. Supplex™, a lightweight nylon that is cotton-soft but strong, is windproof yet breathable, has some ability to repel water, and dries quickly. A waterproof barrier may be necessary in potentially wet environments. Waterproof, windproof synthetic fabrics that breathe are Gore-Tex™, Thintech™, Ultrex™, and Super Microft™. The shell layer should be checked for seals at the waist, neck, wrists, and ankles.

If the feet get wet, change socks. Clean clothing is preferred to dirty clothing. Many workers who otherwise may wear a set of work clothes for two or even three days, should be encouraged to wear clean clothing if they are to be cold exposed. Dirty clothing packs down closer to the skin, does not insulate as well, and does not allow sweat to evaporate as well. Clothing that has some sort of ventilation allows sweat to evaporate and serves better in cold weather.

Waterproof boots and gloves are also recommended. Mittens provide the best insulation for the hands, but the least dexterity. Glove liners are recommended when cold exposed tasks require some dexterity. Always wear a hat in cold environments as it is vital for staying warm. Stocking caps are best because they are better at retaining heat in the body. If hard hats are worn, issue winter liners for them.

Cold Exposed Workers

Cold exposed workers must take frequent breaks from the cold. They may take shelter or be relieved by another worker. In wind chills from -30°F to -40°F, workers should return to warmth every ten minutes. In these situations, it might be prudent to have a number of people relieving each other so the work may go on fairly uninterrupted. The workers should stay out of the cold until they are well-warmed. In spite of scheduled breaks, workers should return to a heated area if they are shivering, perspiring greatly, or feel numbness or pain in the extremities. The buddy system is recommended because workers do not always recognize these systems in themselves. Workers should watch each other for changes in skin appearance, loss of interest in work, confusion, slurred speech, or difficulties on moving. The first is a symptom of frostbite, the rest of hypothermia.

Heavy-duty, strap-on grips with spikes that slip over shoes are used to prevent slips on ice.

If a worker is overexposed to cold, move him or her to a warm place. Remove wet clothing, and dry the victim if necessary. Warm the victim by wrapping him or her in warm, dry clothing or a blanket. Heating pads and other sources of heat may be applied, but not directly to the body. Place a dry towel or blanket between the heat source and the victim as a barrier. Do not immerse the victim in warm water or warm quickly if he or she suffers from hypothermia. Doing so might trigger heart problems. Give the victim warm liquids to drink, if he or she is still conscious. Do not try to pour warm liquids down the throat of an unconscious person. Alcohol causes loss of heat and should not be given. Send the victim to a hospital with emergency medical services if breathing slows or stops, or if the pulse slows or becomes irregular. Treatment for hypothermia at a hospital involves slowly raising the body temperature by various means.

The recommended first-aid treatment for frostbite is the immediate application of warmth to the injured parts. If possible, the affected areas should be soaked in warm water, but no warmer than 105°F. Without a thermometer, the water should be neutral to touch, neither cool nor too warm. Do not allow the frostbitten area to touch the container. Soak until the frostbitten part looks red and feels warm. Loosely bandage. If fingers or toes are frostbitten, place gauze between them. Handle the frostbitten area gently. *Do not rub!* Rubbing or vigorous massage should be avoided since it can cause further harm to the damaged tissues. Do not break any blisters that may have formed. Heparin or some other agent that prevents clotting of the blood is administered in severe cases by paramedics, registered nurses, or physicians.

Bodily injury due to slipping and falling on ice is a hazard we have not covered yet. Figure 15-2 lists eight steps proposed by Smith for managing this potential hazard.

Ice melting agents include calcium chloride or sodium chloride (salts), sand, cinder, or a combination of these. Give priority to removing snow from the path people will take from the parking lot to the entrances to the plant. Then clear the fire exits, hydrants, fire department connections, post-indicator valves, and evacuation rally points.

Figure 15-3 adapted for Smith's article defines employee responsibilities. Figure 15-4 are recommended electric space heater requirements.

Figure 15-2
Managing Icy Conditions

1. Select and disperse an ice melting agent.
2. Develop a snow and ice removal plan.
3. Draft a late opening and early closure policy.
4. Inform workers of their accountabilities concerning safety and injury prevention.
5. Prepare fire protection and heating and cooling systems.
6. Write a policy for portable electric space heaters.
7. Take into account facility design details and miscellaneous issues.
8. Announce the winterization plan to employees.

Special Nonionizing Radiation Considerations

Electromagnetic radiation is the transport of energy through space. In the previous chapter, we discussed electromagnetic radiation in the form of visible light and related forms of energy. Earlier in this chapter, we have discussed electromagnetic radiation as thermal energy. Now we turn our attention to the remaining forms of electromagnetic radiation, less that from radioactive sources.

Figure 15-3
Employee Responsibilities in Icy Weather

1. Wear boots with treads.
2. Wear strap-on cleats.
3. Don't wear tennis/walking/ running shoes.
4. Don't wear leather soles.
5. Don't wear high heels.

Figure 15-4
Requirements for Portable Electric Space Heaters

1. Electric space heaters must be in good working order.
2. Keep combustible materials away from the side or back of a heater and more than three feet away from the front.
3. Employees should report malfunctioning heaters immediately.
4. Allow only UL-listed or other listed heaters to be used.
5. Use only low-watt-density heaters.
6. Prohibit abuse or modification of heaters.
7. Prohibit operation of heaters with damaged cords or plugs, or after an unrepaired malfunction.
8. Prohibit extension cords with portable electrical space heaters.
9. Employees should notify management rather than reset a tripped circuit breaker.
10. Require employees to turn heaters off when unattended.
11. Require portable heaters to be turned at the end of each shift.
12. Prohibit employees from bringing portable space heaters from home.

Generally, we speak of electromagnetic forces (EMF) when we discuss nonionizing energy. Evidence of harm to exposure from EMF fields is anything but certain. Epidemiologic studies have delivered mixed results to date. Some studies have uncovered a weak dose-response effect, but others have found no correlation whatsoever. We do not know if EMF fields are harmful or not. Prudent avoidance is the practice of reducing human exposure to EMF fields where it is easy and inexpensive to do so. While some scientists are having an intellectual tantrum about the concept of human avoidance, even calling it superstition, others are recommending it. As a practicing safety engineer, I recommend it for awhile longer—at least until some definitive studies can get us some answers.

Radio-Frequency (RF) Energy

The RF band of EMF lies between zero and 3¥10^{12} Hz in the frequency range with wavelengths exceeding 10,000 mm, or 100 m.

Physical Properties

Table 15-14 summarizes the bands of radiofrequency energy.

Sources

RF energy is received from any electrical device that emits energy in the frequency and wavelength ranges listed above.

Biological Effects

No conclusive and consistent evidence has shown that exposure to EMF fields produces cancer, adverse neurobehavioral effects, or reproductive and developmental effects. Some experiments have demonstrated frequency and power density effects on living tissue. The calcium efflux from brain tissue is enhanced by millimeter magnitude EMF wavelengths. Also, the same wavelengths have affected bacteria growth rate in experiments. However, considerable controversy exists over the ability of RF to induce biological effects at field strengths less than 10 mW/cm^2.

Table 15-14
Radiofrequency Bands

Band	Wavelength, m	Frequencies	Applications
Low frequency (LF)	10^4-10^3 m	30-300 KHz	Radionavigation, radio beacon
Medium frequency (MF)	10^3-10^2 m	0.3-3 MHz	Marine radio telephone, LORAN, AM radio
High frequency (HF)	10^2-10 m	3-30 MHz	Amateur radio, world-wide radio, diathermy, radio astronomy

Health Hazards

Some studies have shown weak links to EMF field exposure and leukemia, brain cancer, and other tumors. Two types of EMF that have particular significance in industry are ELF and microwaves.

Extremely Low Frequency (ELF) Energy

Exposure to modulated ELF has been reported to affect the biological rhythms of human beings. Preliminary studies in Europe and Japan of rail workers on electrified train systems insinuated an elevated risk for brain cancer or leukemia. Yet other investigations detected no such increase in risk. Wenzl estimated the exposure to rail maintenance workers on an electrified railroad.

These rails have overhead catenary wires that deliver 11,000 volts of single phase power at 25 Hertz (Hz) along a trolley wire to the train. The power returns through the rails to complete an electrical circuit. Higher catenary wires also carry 25 Hz power in 138,000 volt transmission lines. Yet another wire carries 100 Hz ungrounded power at 6,900 volts for signal power. The power company transmission lines carry from 33,000 to 230,000 volts three phase at 60 Hz.

Table 15-15 summarizes Wenzl's findings. Again, considerable controversy exists over the ability of ELF to induce biological effects at field strengths below 10 mW/cm^2.

Table 15-15
Magnetic Flux Density on an Electric Railway

Location	ELF MF median mG	25 Hz median mG	25 Hz mG range
Rural, 3 tracks	28.8	16.7	4-185
Urban industrial 4 tracks	24.9	22.3	9.1-34.3
Urban large station 11 tracks	8.15	6.54	1.5-69.8
Urban power station	43.2	40.3	3.7-178
Urban 6 tracks	22.6	16.7	6.2-72.1

Figure 15-5
Sources of Microwave Radiation

Acquisition and Tracking Radar (military)
Air Radar
Air Traffic Control Radar
Diathermy Equipment
Industrial Drying Equipment
Industrial Heating Equipment
Microwave Ovens
Natural Microwaves:
Satellite Communications Systems
UHF-TV Transmitters
Weather Radar

Microwaves

The range of microwave radiation covers wavelengths from one millimeter to one meter. The occupational sources of microwaves are listed in Figure 15-5. Natural microwaves are generated by cold fronts. The sun also emits extremely low levels of microwave

Microwave energy on the order of three centimeters or less in wavelength is absorbed in the outer skin. From three to ten centimeter wavelength microwaves penetrate the outer skin and travel from one millimeter to one centimeter into the underlying tissue. When the microwave wavelength exceeds ten centimeters, the energy can penetrate internal organs. Table 15-16 summarizes the microwave bands. Table 15-17 lists the microwave frequency band designations.

The more hazardous and irreversible damage due to microwaves is caused by the heating of tissue. When the energy level exceeds 100 mW/cm^2, microwaves may contribute to cataracts at frequencies between 800 and 10,000 MHz. Long term exposure may also lead to neurological effects similar to stress-induced illness. Teratological effects occur only if the core temperature of the mother is raised significantly.

Frequency controls the amount of microwave energy that is absorbed by biological tissue. Body size, configuration, and homogeneity also affect

Table 15-16
Microwave Bands

Band	Wavelength	Frequencies	Applications
Very High Frequency (VHF)	10-1 m	30-300 MHz	FM radio, television, air traffic control, radionavigation
Ultra High Frequency (UHF)	1-0.1 M	0.3-3 GHz	Television, citizens band, microwave ovens, microwave communications, telemetry, weather radar
Super High Frequency (SHF)	10-1 cm	3-30 GHz	Satellite communication, airborne weather radar, marine navigational radar, microwave communication
Extra High Frequency (EHF)	1-0.1 cm	30-300 GHz	Radio astronomy, cloud detection radar, space research, HCN emission

Table 15-17
Microwave Frequency Band Designation

Letter Designation	Frequency Band, MHz
L	1,100-1,700
LS	1,700-2,600
S	2,600-3,950
C	3,950-5,850
XN	5,850-8,200
X	8,200-12,400
Ku	12,400-18,000
K	18,000-26,500
Ka	26,500-40,000

damage. Specific absorption rate (SAR) is the rate of energy absorption per unit body mass, W/kg. The threshold SAR is believed to be 4 W/kg. Scientists give this a safety factor of ten, so the accepted whole body SAR is 0.4 W/kg. This value can be relaxed outside the wavelength range from thirty to one hundred MHz. Performance standards for new microwave ovens is set at one mW/cm^2, measured five centimeters from any oven surface. During the life of the oven, the standard is five mW/cm^2.

The vector components of microwaves, as any EMF, are the electric field (E) and the magnetic field (H). Microwave propagation consists of a plane wave moving in an unbounded isotropic medium where E and H are mutually perpendicular to the direction of wave propagation. This simple proportionality is valid only in free space or the far field of the radiating device. Power density in the far field is inversely proportional to the square of the distance from the source. To estimate power density, W, in the near field

$$W = \frac{16P}{\pi D^2} = \frac{4P}{A} \qquad [15.17]$$

where

P = average power output
D = diameter of the antenna
A = effective area of the antenna

If 10 mW/cm^2 or more is estimated from the previous equation, the assumption is that the same power density exists throughout the near field and the Friis free space transmission formula is used to calculate power density in the far field.

$$W = \frac{GP}{4\pi r^2} = \frac{AP}{\lambda^2 r^2} \qquad [15.18]$$

where

G = the far field antenna gain
z = distance from the antenna
λ = wavelength

Human Exposure Guidelines for Nonionizing Radiation

The ACGIH has established TLVs for nonionizing radiation in the frequency range from 30 kHz to 300 GHz. These TLVs are supposedly conditions under which nearly all workers may be repeatedly exposed without adverse health effects. They were selected by the ACGIH TLV Committee to limit the whole body SAR to 0.4 W/kg for any six-minute period for frequencies between three MHz and 300 GHZ. An energy flux plateau is set at 100 mW/cm² for 0.4 W/kg exposures between 30 kHz and 3 MHz. This is to prevent electrical shock and burns.

ACGIH measures the power density as

$$W = \frac{E^2}{3770} = 37.7H^2 \qquad [15.19]$$

where

E^2 = volts squared per meter squared
H^2 = amps squared per meter squared

ACGIH issues these guidelines:

- Avoid needless exposure to all RF radiation.

Table 15-18
ACGIH TLVs for RF/Microwave Exposures

Frequency Band	Power Density, mW/cm²	Electric Field Strength Squared, V²/m²	Magnetic Field Strength Squared, A²/m²
30 kHz-3 MHz	100	377,000	2.65
3-30 MHz	$900/f^2$	$3770(900/f^2)$	$900(37.7 \times f^2)$
30-100 MHz	1	3770	0.027
100-1000 MHz	f/100	3770(f/100)	f/37,700
1-300 GHz	10	37,700	0.265

f = frequency in MHz.

- Determine the fraction of the protection guide incurred within each frequency level and sum these amounts. The sum should not exceed unity.
- Average the power density over a six-minute period if pulsed or continuous wave fields are involved.
- The protection factor may be exceeded for partial body exposures at frequencies between 30 kHz and 1.0 GHz, but only if the output power of the source is 7 watts or less.
- The TLVs may be exceeded if exposure conditions produce a whole body SAR of less than 0.4 W/kg and spatial peak SAR values less than 8.0 W/kg averaged over any 1.0 gram of tissue.
- Measurements should not be made closer than 5 cm from an object.
- Exposures should be limited to a peak electric field intensity of 100 kV/m.

Table 15-18 lists the TLVs recommended by ACGIH.

References

The American National Red Cross. *Community First Aid and Safety.* St. Louis: Mosby-Yearbook, Inc., 1993. pp. 221-222.

Bedny, Gregory Z. and Mark H. Seglin. "The Use of Pulse Rate to Evaluate Physical Work Load in Russian Ergonomics." *AIHA Journal.* May 1997, pp. 375-379.

Bielmaier, Michael. "When Cold and Water Meet, Hypothermia Is Close behind." *Industrial Fire Chief.* November/December 1993, pp. 24-25.

Boyd, Vicky. "Dealing with Heat Stress." *Occupational Health & Safety.* July 1996, pp. 37-39.

Dennis, Clyde H. "Thermal Burns: A Plan for Protection." *Industrial Safety & Hygiene News.* August 1992, p. 41.

Industrial Hygiene Study Guide. 4th ed. New Jersey Section, American Industrial Hygiene Association, 1989.

The Industrial Environment—Its Evaluation and Control. U.S. Department of Health, Education, and Welfare. Public Health Service. Center for Disease Control. National Institute for Occupational Safety and Health, 1973.

Johnson, Sandra J. "Fluid Loss Can Lead to Heat Stress, though Replenishment Offsets Effects." *Occupational Health & Safety.* June 1993, pp. 62-65.

Lewis, Darcy. "EMFs: Are We Paranoid or Just Playing It Safe?" *Safety + Health.* May 1997, pp. 60-63.

Malley, C. Brian. "Cold Stress Revisited." *Professional Safety.* January 1992, pp. 21-23.

Murray, Robert. "Prevention Is the Key to Avoiding Heat-Related Injuries." *Industrial Hygiene News.* September 1986, pp. 56-57.

O' Connor, John S. and Kim Querrey. "Heat Stress and Chemical Workers: Minimizing the Risk." *Professional Safety.* December 1993, pp. 35-38.

Reneau, Paul D., Phillip A. Bishop, and Candi D. Ashley. "Comparison of a Military Chemical Suit and an Industrial Usage Vapor Barrier Suit across Two Thermal Environments." *AIHA Journal.* September 1997, pp. 646-649.

Smith, Joseph A. "Preparing for Winter: Proactive Measures to Prevent Injury and Property Damage." *Professional Safety.* August 1997, pp. 28-32.

"Sun Care Tips Promote Safety." *Compliance Management.* June 1997, p. 6.

Wenzl, Thurman B. "Estimating Magnetic Field Exposures of Rail Maintenance Workers." *AIHA Journal.* September 1997, pp. 667-671.

Willen, Janet. "Take the Chill out of Winter Work." *Safety + Health.* January 1997, pp. 26-30.

Winter Safety. Virginia Beach, VA: Coastal Video Communications Corp., 1994.

16

Physical Injury Prevention Management

Machine Guarding

A guard must be provided on any machine part, function, or process that may cause injury to the operator. Safety standards are found at 29 CFR 1910.211-222, 241-247, and 250-257. Essentially, these regulations require machine hazards to be either eliminated or controlled where the operation of the machine or accidental contact with it can injure either the operator or those in the vicinity. Machines causes personnel accidents in three basic ways:

1. A person makes contact with the machine at any point of contact.
2. The machine or related equipment contacts the person.
3. The person is struck by work in progress.

Figure 16-1 summarizes the hazardous motions and actions of machines that may be harmful.

Machine safeguards are classified as either *guards* or *safety devices.* Four kinds of guards and five types of safety devices are available for worker protection.

Figure 16-1
Hazardous Mechanical Motions and Actions

Motions	Actions
Reciprocation	Bending
Rotation	Cutting
Traversing	Punching
	Shearing

The four kinds of guards are fixed, interlocked, adjustable, or self-adjusting. Each of these prevent contact with the hazardous motion or actions of the machine.

A cage around a pulley belt is an example of a *fixed guard.* The fixed guard is a permanent part of the machine and is not dependent upon moving parts to perform its function. Expanded metal, Plexiglas, sheet metal, screen, wire cloth, bars, or other materials substantial enough to withstand impact by a person are used to make guards. The fixed guard is preferable to other types of guards due to its simplicity.

Interlocked guards prevent the machine from operating when the guard is removed, such as an open access door. These guards may use electrical, hydraulic, mechanical, or pneumatic power or a combination of these to operate.

> **Warning!!**
> Replacement of an interlocking guard must not restart the machine. The restart must be manual.

Adjustable guards allow for different heights of workers, or longer or shorter arm reaches. The operator must physically adjust the guard to fit the stock size or his or her physique.

Self-adjusting guards do not require manual adjustment, but adjust openings dependent on the movement of stock. As the stock is fed into the danger area, the guard is pushed away. This provides an opening large enough for the stock to enter, but not the operator's body parts. After the stock is removed, the guard returns to the rest position.

Machine Safety Devices

The five types of safety devices used on machines are presence sensors, pullbacks, restraints, safety controls, and gates. Devices may stop a machine if a body part is inadvertently placed in the danger area. It may restrain or withdraw the operator's hands from the danger area during operation. Or it may require the concurrent use of both hands to operate controls, and thus keep them out of harm's way. Again, it may provide a barrier tied into the operating cycle of the machine that prevents the operator or anyone else's hand from being placed into the danger area.

Presence sensors may operate on the principal of photoelectrics, radiofrequency, or electromechanics. *Photoelectrical sensors* shut down the machine when any part of the operator's body breaks a light beam. *Radiofrequency* devices do the same thing with capacitance. When the capacitance is broken the machine stops. Either type can only be used on machines that can be shut down before the worker can reach the hazard area. Machines that continue to be dangerous for several minutes after killing power are not good candidates for these devices.

Electromechanical safety devices use solenoids, servos, and motors to provide protection. These devices consist of a probe or contact bar that descends to a predetermined distance when the operator initiates the machine action. If an obstruction prevents the probe from lowering completely, the machine will not cycle.

Pullbacks have attachments to the operator's wrists that pull his or her hands out of the way when the hazardous operation starts. *Restraints* are similar in that they attach to the operator's body, but work by limiting his or her field of activity at all times.

Three safety controls are trip controls, two-hand controls, and two-hand trips. *Trip controls* allow a machine to be shut down immediately by pressing a button or throwing a switch. Typically, by convention, a large round red palm button is used. Another type of trip control is the *pressure-sensitive body bar.* If the operator touches it, the machine trips off. A *tripwire cable* can be used in place of the pressure-sensitive bar. If speed is not required, a *trip rod* may be used.

Two-hand controls require the operator to have both hands on control levels or push control buttons to operate the hazardous process. This keeps the hands away from harm. If the machine operates without the operator's assistance, a *two-hand trip* can be used. The machine will not operate until the operator closes these switches, sometimes called *dead man* switches. If he or she lets go of either of the two-hand trip switches, as if dead, the operation shuts down immediately.

Gates are interlocked guards that can be opened for access to the machine internals. When the gate is open, the machine is off.

Yet another way to protect workers from machine hazards is by locating the operation remotely from the operators. Machines must be oriented so that the dangerous areas are inaccessible or else do not present a hazard to workers during normal operation. Another option is to locate the danger area high enough to be out of reach of employees. Walls and fences have also be used to isolate the danger areas. Sometimes a dangerous stock feeding process can be located far enough from operators to prevent them from reaching into it. The location method of hazard control is not always feasible, however.

Often, the hazardous operation involves feeding material to or ejecting material from the machine. *Automatic feed or ejection* removes the operator from the process. Used in conjunction with other safeguards, this method can protect other workers, too. *Semi-automatic feeding or ejecting* may be sufficient to remove the operator from the dangerous portion of the event. *Robots* are another option for removing the operator from harm's way.

Awareness barriers warn the operator of danger, but does not provide protection. *Protective shields* are used when the hazard is chemical or electrical in nature. *Hand-feeding* tools or *holding fixtures* are two more options when none of the preceding methods will work.

> Give machine guarding a high priority when purchasing a new machine or overhauling an old one.

Consider these questions for employee safety around machines:

- Do machine safeguards meet OSHA requirements?
- Is the machine designed to prevent the operator from having body parts in the danger zone during operation?
- Are all points of operation, nip points, and rotating parts guarded?
- Can all sources of energy be locked out during maintenance?

Lockout/Tagout

The key to a good lockout/tagout program is knowledge. All parties involved must know and understand the different types of harmful energy. Each authorized person must be able to identify the multitude of energy sources in a workplace. In the least, he or she must be able to identify those sources on the machine being worked on. Next, he or she must understand the loss causation theory based on the energy transfer concept. Authorized persons must be able to apply the strategies for energy control in a workplace environment and be able to prevent the unwanted transfer of energy. The goal of the program is to apply the necessary measures in the workplace environment to educate employees in control and prevention of energy releases as potential accident or incident causes.

Electrical

Electrical energy is found in three forms: alternating current, direct current, and static electricity. Alternating current, or AC, electricity follows the path of least resistance to ground. In AC currents, the flow of electrons continuously reverses at a particular frequency (60 times per second in the U.S.) Direct current, or DC, electricity produces severe burns. In DC currents, electrons continuously flow through a conductor in the same direction all the time. High voltage DC will produce serious shock. Static electricity is generally caused by friction between two dissimilar materials with no ground present. Severe sparks or shock can be produced by static charges.

Hydraulics

Hydraulics is a form of mechanical energy utilizing pressure stored in a liquid. The study of flow of liquids is the science of hydraulics. The flows of liquids can be either *viscous* or *nonviscous. Viscosity* is the resistance to flow by a liquid. No real liquid is nonviscous, but two liquids will have high viscosity or low viscosity relative to each other.

Three principles are important in hydraulics: the law of conservation of mass, the momentum equation, and the energy equation.

In Newtonian mechanics, mass can be neither created nor destroyed. Therefore, the mass taken out of a system has to equal the mass put into a system or the system must have a net change of mass exactly equal to the difference. Mathematically, this is expressed as:

$$\dot{m}_{out} - \dot{m}_{in} = \frac{(\Delta M)}{(\Delta t)} = \frac{dM}{dt} \qquad [16.1]$$

This equation is express for the system. In hydraulics, the mass flow rate can be written:

$$\dot{m} = \rho VA \qquad [16.2]$$

where

ρ = density of liquid, lb_m/ft^3
V = velocity, ft/sec
A = cross-sectional area, ft^2

The momentum equation is derived from Newton's Second Law based on pressure and shear forces in the liquid. Essentially, it states that the force on a liquid particle equals the rate of change of momentum:

$$F = \frac{\Delta(mv)}{\Delta t} = \frac{dmv}{dt} \qquad [16.3]$$

where *mn* is the product of mass and velocity, or momentum.

The energy equation is derived from the momentum equation and its most useful form is called the Bernoulli equation. The energy in a liquid stream is constant, where the energy consists of work and kinetic energy.

$$p + \tfrac{1}{2}\rho V^2 = constant = P_0 \qquad [16.4]$$

where

p = static pressure, lb_f/ft^2
ρ = density, lb_m/ft^3
V = velocity, ft/sec
P_0 = stagnation pressure, lb_f/ft^2

Stagnation pressure is the pressure in the liquid when it is stationary. This is the pressure that can release and harm a maintenance worker.

Pneumatics

Pneumatic power is a form of mechanical energy that uses the force of compressed air to do work. Pressure can be positive or negative. Positive pressure is produced by air compressors. The outputs and work of negative pressure systems are produced by vacuum pumps. Actually, all pressure is positive compared to absolute zero. What we refer to as negative pressure, or vacuum, is merely a pressure that is less than atmospheric.

While pressurized air and vacuum are mechanically advantageous, they are also hazardous. Pressurized vessels, containers, and conduits may rupture, spraying the workplace with expanding gas and missiles that are fragments of the ruptured equipment. In the case of industrial machinery, the possibility that a pressurized air receiver may discharge and operate the machine or a dangerous subsystem of the machine. Therefore, it is important that all pneumatic pressure or vacuum is released and equalized to atmospheric pressure before conducting work on a machine. Alternatively, the receiver must somehow be locked out of the operating system such that the pressure or vacuum cannot be released. Less preferable, the device may be tagged out to warn persons not to release the pressure or vacuum.

The behavior of gas systems is governed by several physical laws well known among engineers and scientists.

Boyle's Law: The absolute pressure of a fixed mass of gas varies inversely as the volume, provided the temperature remains constant.

Charles' Law: The volume of a given mass of gas is directly proportional to its absolute temperature, provided the pressure is held constant.

Pascal's Law: A pressure applied to a confined fluid at rest is transmitted with equal intensity throughout the fluid at right angles to containing surfaces.

Mathematically, Boyle's Law is

$$P_1V_1 = P_2V_2 \qquad [16.5]$$

where

P = pressure
V = volume

Charles' Law is expressed

$$\frac{V_1}{V_2} = \frac{T_1}{T_2} \quad [16.6]$$

or, if volume is held constant and pressure varies:

$$\frac{P_1}{P_2} = \frac{T_1}{T_2} \quad [16.7]$$

The combined gas law is a form of these equations that combines Charles' and Boyle's laws.

$$\frac{P_1V_1}{T_1} = \frac{P_2V_2}{T_2} \quad [16.8]$$

Finally, the equation of state for an ideal gas:

$$PV = nRT \quad [16.9]$$

where

n = mass expressed in molecular weight
R = ideal gas constant

Some values of R:

1.987 BTU/lb-mole °R
8.31×10^7 erg/g-mole K
0.08207 l-atm/g-mole K
10.73 psia-ft^3/lb-mole °R

Potential Energy in Machine Components

Potential energy is the energy stored due to position, such as a liquid reservoir overhead or a flywheel about to fall.

Springs

Springs contain potential energy until released. The work of deflection, W, is

$$W = {}^{F\Delta}\!/_{2} \quad [16.10]$$

where

F = force, lb_f
Δ = deflection, inches

This work can be transmitted to the human body where damage may be done. The amount of force and deflection depends on the particular spring in use. The four types of spring are 1) a deflected bar, 2) a semi-elliptic laminated spring, or leaf spring such as used in the automotive industry, 3) a spiral coil, and 4) a helical coil.

The force, F, in a deflected bar in in.-lb. is

$$F = \frac{bh^2S_b}{6l} \qquad [16.11]$$

where

b = width of the bar, inches
h = thickness of the bar, inches
S_b = safe unit stress in bending
l = length of the bar

The safe bending stress is 50,000 for spring brass and 80,000 for spring steel. The deflection, D, in inches is

$$\Delta = \frac{4Fl^3}{bh^3E_t} \qquad [16.12]$$

where

E_t = modulus of elasticity in tension.

This modulus is 9,000,000 for spring brass and 30,000,000 for spring steel. A lot of force is available when a deflected bar is released—enough force to break an arm.

In the case of a semi-elliptic laminated spring, the force is

$$F = \frac{nbh^2S_b}{6(l+m)} \qquad [16.13]$$

where

n = number of spring leafs
b = width of each leaf
h = thickness of each leaf
l = overall length of all leaves
m = deflected distance between the free and attached end of the leaf bundle

The deflection in this case is

$$\Delta = \frac{6Fl^2(l+m)}{nbh^3E_t} \qquad [16.14]$$

A spiral coil spring is dangerous because it can unravel and cause harm to an unsuspecting maintenance worker. The force available when such a spring is coiled is

$$F = \frac{bh^2S_b}{6r} \qquad [16.15]$$

where

b = width of the coil material
h = thickness of the coil material
r = mean radius of the coil

The deflection is calculated

$$\Delta = r\phi = \frac{12Flr^2}{bh^3E_t} \qquad [16.16]$$

where ϕ is the angle made between the end of the coil at its loaded and at rest positions and the center of the coil.

The spring we usually think about is the helical coil. This spring may be compressed when loaded to hold two objects apart. Or it may be stretched when loaded to hold two objects together. The force of a loaded helical coil is

$$F = \frac{\pi d^3S_s}{16r} \qquad [16.17]$$

where

d = diameter of the coil material
S_s = safe unit stress in shear
r = radius of the coil

The safe shear stress is 30,000 for spring brass and 80,000 for spring steel. The deflection of this spring is

$$\Delta = \frac{4\pi nr^2S_b}{dE_s} \qquad [16.18]$$

where the modulus of elasticity in shear, E_s, is 6,000,000 for spring brass and 12,000,000 for spring steel.

Springs are often used in multiples. The spring rate, R lb/in, is additive. For springs mounted parallel to each other

$$\sum R = R_1 + R_2 + \mathrm{L} + R_n \qquad [16.19]$$

Springs mounted serially are added this way

$$\sum R = 1 \Big/ \left(\frac{1}{R_1} + \frac{1}{R_2} + \mathrm{L} + \frac{1}{R_n} \right) \qquad [16.20]$$

Counter-Weights

Counterweights are a force waiting to let go. Maintenance workers need to be aware of counterweights and somehow block them from falling before commencing work. The impact, K, of a falling weight dropped from a height h ft or struck with velocity v fps is

$$K = 1 + \left(\sqrt{d^2 + dv^2} \right) \Big/ d \qquad [16.21]$$

where

$$v = \sqrt{2gh} \qquad [16.22]$$

Flywheels

Another device that stores energy is the flywheel. Make sure that the flywheel is in a rest position and locked before proceeding with maintenance.

Other Forms of Energy

Radiation is a form of energy from either ionizing or non-ionizing sources. Ionizing radiation takes the form of X-rays, alpha particles, beta particles, or gamma rays. This type of radiation comes from radioactive isotopes or equipment that produce x-rays. Non-ionizing radiation takes the form of ultraviolet, infrared, microwave, visible light, as well as low frequency elec-

Figure 16-2
Lockout/Tagout Provisions

1. Preparing for shutdown.
2. Shutting down the machine, operation, or process.
3. Isolating the machine, operation, or process.
4. Applying lockout/tagout devices.
5. Releasing stored or residual energy.
6. Verifying isolation of the machine, operation, or process from energy sources or stored energy.

tromagnetic forces (EMF). Arc welders produce ultraviolet radiation. Infrared is produced by lasers.

Thermal energy is in the form of high or low temperature. Heat may be generated by the abrasion of two surfaces. Heat exhaustion or stroke may be the consequence.

Chemical energy can be harmful by inhalation, ingestion, or skin absorption. More importantly, for energy release, is the production of heat in some chemical reactions. Toxic and non-toxic chemical can have an effect.

A good lockout/tagout procedure includes provisions for the activities summarized in Figure 16-2.

Complete employee protection must be provided through energy isolating devices and the use of locks and tags. Provide initial training and annual retraining to all *authorized* and *affected* employees. Certify that such training has been given. Perform annual inspections to assure that the energy control procedures are implemented properly.

PORTABLE TOOLS

Hand tools and portable tools are typically an overlooked part of the safety process.

Hardened steel striking tools such as hammers develop spalling, chipping, and deformation on the perimeter of the striking face. This goes for struck tools, such as chisels and punches, as well. Cracking, chipping, or deformation of the striking or struck face or the perimeter of the face should disqualify the tool from service. Any tool that is impacted by another may sustain such degradation. Even a small defect such as a nick or chip on the hard areas of a chisel or punch may result in a spall flying off as a missile during an impact load. Glancing blows generally are hardest on impact tools.

Inspect tool handles regularly. Wooden handles may splinter, crack, or loosen. Replace such handles or remove them from service.

Wrenches are probably the most abused tools. They are used for pry bars, levers, and hammers. Have ratchet wrenches cleaned and maintained regularly. Take damaged wrenches out of service.

It is poor safety practice to allow workers to modify tools. If the tool they have is not the right one for the job, get the right one. If it is an urgent situation, make the modification, inspect the tool for damage beyond the modification, and dispose of it immediately after use. Even if used effectively for the job it was modified for, how do you know it will sustain additional force? You do not. Get rid of it.

Electrical Safety

The most violated of all the OSHA standards are those dealing with electrical safety. The perennial electrical citations in order of most violated to least are listed in Figure 16-3.

The OSHA requirements for electrical safety are not some onerous rules that were dreamed up by some bureaucrat one day when he had nothing better to do. All the OSHA requirements were copied from the National Electrical Code (NEC) as it existed in 1981. Not all of the NEC was extracted, however, so even more electrical standards apply to your facility than what you will find in the OSHA standards. OSHA took only those requirements that apply to worker safety and were least likely to change with each revision of the NEC.

The four most common electrical hazards are shocks, burns, explosions, and fires. *Shocks* occur when electrical current flows through sensitive parts of the body. Indirectly, shocks cause harm by making people fall from ladders, scaffolds, and high places, or making them fall into a greater, or at least another serious hazard. Electrical shock causes direct damage when the flow of electricity through a vital organ leads to death or serious damage

Figure 16-3
List of Most Violated Electrical Requirements
(from 29 CFR 1910.x)

1. Guarding of live parts (303(g)(2)).
2. Identification (303(f)).
3. Uses permitted for flexible cord (303(g)(2)).
4. Prohibited uses of flexible cord (305(g)(1)(iii)).
5. Pull at joints and terminals must be prevented (305(g)(2)(iii)).
6. Effective grounding (304(f)(4)); (303(b)); (304(e)(1)(i)); (304(f)(6)); (305(a)(1)(i)); and (305(h)).
7. Grounding of fixed equipment (304(f)(5)(iv)).
8. Grounding of equipment connected by cord and plug (304(f)(v)); (303(a)); (303(b)(1)); (303(b)(2)); (304(f)(iv)); (304(f)(6)(i)); (305(g)(1)(i)); and (305(g)(2)(iii)).
9. Alternating-current circuits and systems to be grounded (304(f)(1)).
10. Location of overcurrent devices (304(e)(1)(iv)).
11. Splices in flexible cords (305(g)(2)(ii)).
12. Electrical connections (303(c)).
13. Marking of equipment (303(e)).
14. Working clearances about electrical equipment (303(g)(1)).

to the organ. Many electrocutions are caused by exposure to less than 600 volts, which are typical of workplace and residential voltages. Exposure to 115 volts for as little as 3 to 4 seconds can start heart fibrillation. The onset of fibrillation starts more quickly if the current and voltage are higher. At 100 ma, fibrillation begins in 1.5 to 3.9 seconds. Figure 16-4 lists the occupational categories of people who face a higher than normal risk of hazardous electrical contact.

Excessively hot conductors, as sometimes occurs in faulty wiring, lead to *burns*. An arc blast caused by a short circuit may also lead to serious burns.

Explosions may happen when electricity becomes a source of ignition for an explosive mixture in air. Excessively high conductor temperatures or arcing at switch contacts are ignition sources, for example. *Fires* are typi-

Figure 16-4
Employees Who Face Higher than Normal Risk of Hazardous Electrical Contact due to Occupation

Electricians
Welders
Material Handling Equipment Operators
Riggers and Roustabouts
Painters
Mechanics
Electrical and Electronic Technicians
Industrial Machine Operators
Stationary Engineers
Electrical and Electronic Equipment Assemblers
Line Supervisors of Blue Collar Workers
Electrical and Electronic Engineers

cally caused by deterioration of insulation on old wiring, defective or misused electrical equipment, loose connections, or misused electrical devices.

General Considerations

OSHA electrical safety rules apply to anyone who faces the risk of electrical shock in your facility. The standards apply to both qualified and unqualified persons working on or near exposed energized parts.

Qualified electricians must deenergize circuits or parts and lock or tag them out to work on them. If not deenergized and isolated, qualified electricians must use safe work practices, isolating equipment, and PPE. Safe work practices, then, are procedures for minimizing risk when working on live circuits. OSHA organizes these practices into groups, summarized in Figure 16-5.

Some general considerations are in order when using safe work practices. First, some safety planning is always prudent prior to working on energized parts—each time, every time. The worker and his or her supervisor should mentally walk through the task and sub-tasks, asking these three questions: 1) What do I need to do to be safe? 2) What do I need to wear to

Figure 16-5
Safe Work Practices on Energized Parts

1. Lockout/Tagout
   a. Deenergizing equipment
   b. Application of locks and tags
   c. Verification of deenergized conditions
   d. Reenergizing equipment
2. Overhead Power Lines
   a. Qualified persons
   b. Unqualified persons
   c. Vehicular and mechanical equipment
3. Illumination
4. Confined or enclosed work spaces
5. Conductive materials and equipment
6. Portable ladders
7. Conductive apparel
8. Housekeeping duties
9. Interlocks

be safe? 3) What tools or devices do I need to use to be safe? This dictates that the employee and his or her supervisor are both qualified persons, even if the supervisor is not a hands-on qualified person. This also dictates that such electrical equipment is guarded from incidental or accidental contact by unqualified persons. That is why not only those who work on electrical circuits, but also those who work near or in the area must receive some basic electrical safety training.

In general, part of the electrical safety planning process before undertaking the hazardous task is to review electrical prints and drawings as well as the labeling, marking, and identification features of equipment, such as name plates. It does not hurt to say often: KEEP NAME PLATES CLEAN AND READABLE.

Specific Considerations

The first safe work practice to consider is always isolation of the equipment for work: lockout/tagout. Obviously this is not always possible for electricians who often have to work on energized circuits, so the following work is exempt from the lockout/tagout standard.

1. When your qualified persons are engaged in construction activities (their activities, not your company's—determine whether construction rules apply).
2. When your qualified persons are working under the EXCLUSIVE control of a electrical utility company for the purpose of installing power generation, transmission, and distribution equipment, including related communication or metering equipment.
3. When work on live parts is required, Subpart S safety-related work practices apply.

A partial lockout or tagout, or none at all, is sometimes needed and it is why electricians are exempt from the OSHA lockout/tagout standard. That standard requires complete deenergization and complete isolation. Electricians would be unable to perform their work in many instances if the complete deenergization policy were followed. The isolation of equipment rules in the OSHA standard on electrical safety (1910.333(b)(2)) allows the electrician to isolate what can be isolated, and to leave energized that branch of the circuit that requires his or her attention. Procedures for the electrician must be in writing and must cover how to deenergize, apply locks or tags, verify the deenergized conditions, and reenergize equipment for work.

Most violations concerning the guarding of live parts (1910.303(g)(2)) are simple and inexpensive to correct. Failure to guard live parts is indicative of poor maintenance. Live parts of electrical equipment are required to be guarded against accidental contact. Approved enclosures or other means may be used to guard the parts. We are talking here of circuitry carrying from 50 to 600 volts. The purpose of this requirement is to protect any person who may be in the vicinity. If the equipment is accessible only by qualified persons, then the requirements at NEC 110-16 apply.

Put all live parts and conductors in approved enclosures. Use covers, screens, or partitions that can only be removed by the use of tools. Equipment that is elevated eight feet or more above the working floor is considered to be guarded by isolation, so long as material handling equipment operates lower than eight feet. Continuous vigilance is necessary to make sure that covers are replaced when removed for maintenance. Constant vigilance is required for changing conditions leading to new violations of the electrical safety standard. Network changes frequently lead to new hazards and poor maintenance practices allow old hazards to reappear.

High on the list of violations is the one found at 29 CFR 1910.303(f), which requires identification so that the purpose of switches and circuit breakers is obvious. Identification assists personnel in determining which device controls power to a particular circuit. NEC 110-22 is similar, requiring the legible marking of each disconnecting means, unless its location and arrangement make its purpose obvious. An example of the latter would be a sole power switch on the side of a machine.

Figure 16-6
Electrical Safety-related Work Practices

- Use PPE, guarding, or insulation to protect from electrical shock.
- Use insulated gloves with hand tools that may contact live circuits.
- Survey every job before beginning work to determine potential for contact with live circuits.
- Post warning signs where energized circuits exist.
- Advise employees of locations of power lines, associated hazards, and protective measures to be taken.
- Ensure employees are qualified to work on energized circuits.
- Keep passageways clear.
- Prohibit use of worn or frayed cords.
- Fasten extension cord loops together, hang from nails or suspend with wire to keep out of traffic. Remember, extension cords are temporary tools, not permanent circuits.
- Permit only qualified personnel to bypass interlocks.
- Train employees to use portable equipment to prevent damage.
- Check attachment plugs prior to connection.

Figure 16-6 *(continued)*

- Require dry hands when plugging in cords.
- If a connection provides a conductive path, handle energized plugs with protective equipment.
- Permit only qualified persons to test electric circuits or equipment.
- Visually inspect testing equipment and accessories for damage before each use.
- Ensure testing equipment is rated for its intended use.
- Take appropriate precautions when electrical equipment will be used in the vicinity of flammables.
- When work is to be performed near overhead lines, deenergize and ground the lines and take other protective measures.
- Ensure unqualified persons remain at least ten feet away from voltages at or below 50KV.
- Ensure qualified people who must go inside this clearance distance are using insulating equipment.
- Maintain vehicles at the proper clearance.
- Provide adequate illumination.
- Train employees to use protective barriers, shields, on insulation when entering confined spaces with exposed live parts.
- Employees in contact with conductive materials must prevent contact with energized circuits.
- Use non-conductive portable ladders.
- Do not permit the wearing of conductive apparel, if the wearer may be contacting live parts, unless rendered non-conductive.
- Provide safeguards for employees performing housekeeping duties in areas with exposed live circuits.

The electrical system for the workplace will originate at a service entrance from which feeders carry current to branch circuits, which carry current to outlets. Sometimes branch circuits are disconnected for one reason or another. If the appropriate disconnecting means is not obvious in such cases, mistakes could be made and vital time may be lost.

Figure 16-7
OSHA's List of High Risk Personnel

Occupation
Supervisors of blue collar workers
Electrical and electronic supervisors
Electrical and electronic equipment assemblers
Electrical and electronic technicians
Electricians
Industrial machine operators
Material handling equipment operators
Mechanics and repairers
Painters Riggers and roustabouts
Stationary engineers
Welders

Trace out all existing circuits from service entrance to utilization equipment. Clearly mark each disconnecting device to indicate what circuit of what equipment it disconnects.

Figure 16-6 lists general electrical safety-related work practices.

Training

Employees who face a risk of electric shock that is not reduced to a safe level by installation techniques must be trained. Figure 16-7 list persons OSHA deems to facing a higher than normal risk of electrocution.

All but electricians and welders may be exempt from training. The condition for this exemption is that their work or the work of those they supervise must not bring them or the employees they supervise close enough to exposed parts of electric circuits for a hazard to exist. Only those circuits operating at 50 V or more to ground are covered.

The content of training must cover the contents of the OSHA electrical standard. Safety-related work practices must be covered so that employees are familiar with them as they pertain to their job assignments. General electrical safety must be covered with all employees as necessary for their safety. Persons who work on or near exposed energized parts shall receive the following additional training:

1. Skills and techniques necessary to distinguish exposed live parts from other parts;
2. Skills and techniques necessary to determine the nominal voltage of exposed live parts; and
3. Clearance distances required by OSHA and their corresponding voltages.

Electrical PPE

PPE is required to protect employees from exposed live parts when installation practices and safety-related work practices do not reduce the risk sufficiently. PPE is to be inspected or tested periodically or both.

Provide nonconductive headgear where a danger exists of head injury from electrical shock, burns, or missile hazards from electrical explosions. Hard hats that are insulated from electrical contact are suitable. See ANSI Standard Z89.1, *Safety Requirements for Industrial Head Protection.*

The danger of electrical arcs and flashes demands the use of eye and face protection. See ANSI Standard Z87.1, *Practice for Occupational and Educational Eye and Face Protection.*

Other PPE include:

- Insulating rubber gloves
- Glove protectors
- Sleeves
- Line hoses
- Blankets
- Hoods
- Mats

Slip, Trip, and Fall

"Have a nice trip, see you next fall," is a crude and often cruel remark made to someone who falls in the workplace. Often a slip, trip, and fall results in no more than a few strained muscles or a bruise that lasts for a few days. Little sympathy is offered in those quarters where such an accident is attributed to clumsiness or carelessness. In such a situation it is easy not to care. A splash of machine coolant or a drop of cutting oil on the floor of a machining plant, however, may lead to a slip, trip, and fall accident where OSHA strictly demands that management care.

Slip, trip, and fall safety measures apply to several aspects of work life. Fall protection is required in the construction standards, for instance, and these may apply to industrial situations, particularly where maintenance employees are engaged in construction activities: demolition, renovation, painting, erection of new or additional structures, building repairs, building maintenance, and electrical installation. OSHA also requires safe walking surfaces in general.

Water on a walking surface creates a hydroplaning effect that causes skidding, slipping, and falling with accompanying injuries, often severe. Water is the most prevalent hazard in the workplace because it is found in so many areas. It is used to hose down and clean in some plants. Outside loading docks are exposed to rain. Food processing areas use water. Chemical manufacturing and mineral beneficiation processes are also big users of water. Many plants also have areas for washing company vehicles.

OSHA claims that a conservative estimate is that 1,500 workers die annually due to slips, trips, and falls, making this hazard the number one cause of deaths in the workplace. Slips and falls are also the leading cause of disabling injuries in the workplace, according to the National Safety Council: over one-half million persons disabled. Also, according to the National Safety Council, over 11 million work days are lost annually due to slips and falls. The Census of Fatal Occupational Injuries found that 38 percent of industrial fall fatalities occur at construction sites. The Census found in one year that 51 percent of fall fatalities were craft workers, 31 percent were operations personnel, and 20 percent were executives.

Four common slip, trip, and fall conditions are found in many facilities at any given time. Walking surfaces are worn smooth in areas with high pedestrian or light rolling traffic. Surfaces exposed to heavy rolling equipment, such as forklifts, may be worn smooth or, alternatively, may have broken up walking surfaces. Areas of low ventilation may obscure visibility by allowing airborne chemicals or particulate matter to linger as a cloud. Finally, areas with little or no lighting may also cause slips, trips, and falls

Figure 16-8
Slip and Fall Traps

Carpets: not tacked down or without rubber pad underneath.
REPAIR.

Hidden Steps: around corners or going into a higher light level.
USE WARNINGS.

Icy Areas: treacherous.
USE SAND OR SALT. WALK SLOWLY.

Loose Flooring: tiles, bricks, pavement, floorboards.
REPLACE OR POST WARNINGS.

Smooth Surfaces: waxed floors, worn floors.
USE CAUTION OR REPLACE.

Wet Areas: oil, grease, liquids.
CLEAN UP IMMEDIATELY.

Adapted from Channing L. Bete Company, Inc.

due to poor visibility. Figure 16-8 lists some slips and fall traps.

When a person falls, before any fall-arrest equipment does its work, the beginning is free fall. The arrest system activates after a certain distance. More distance, called the *deceleration distance,* is needed to bring the person to a complete stop. Force is needed to stop a fall and this force enters the body through the straps of the safety belt or body harness. The sudden jolt is the transfer of the force to the body and can damage internal organs or the spine. Deceleration systems absorb the fall arrest force. Body harnesses distribute the force to areas of the body that are protected by bones.

Slip and Fall Prevention

OSHA requires a 0.50 coefficient of friction (COF) for walking surfaces in workplaces. Researchers have always agreed that this is a sufficient level of friction to permit safe walking. More recent research demonstrates that slip and fall prevention is much more complex than a high COF. Latter day studies show that the degree of change of COF is the major factor leading to a slip and fall. As a person walks, his or her inner ear and Eustachian tube track the position of the body, its balance. Falls rarely occur when the degree of friction provided by the foot remains relatively constant. People can even walk on slippery surfaces, such as ice, as long as the COF is constant. However, when the amount of friction underfoot changes suddenly, especially unexpectedly, slips occur, leading to falls. The force required to move a stationary object is called the static COF or SCOF. The Americans with Disabilities Act (ADA) recommends 0.6 SCOF for level floors and 0.8 for ramps.

Kendzior provides an excellent summary of slip resistance as defined by various committees of the American Society of Testing and Materials (ASTM). The ASTM D-21 committee on floor polishes uses the James Machine as a test device to measure SCOF under dry conditions. As long as the SCOF of the applied polish measures 0.50 or more, the product is classified as "slip resistant." This is the characteristic (SCOF) and value (0.50) OSHA has adopted. The ASTM C-21 committee on floor tile, on the other hand, uses a Horizontal Pull Drag Meter to measure SCOF under wet or dry conditions. The same value, 0.50, is considered to be the borderline of slipperiness.

Dynamic COF, or DCOF, is the measure of the COF of an object in motion. Many professionals support the use of DCOF, which is measured by a Brungraber Mark II or English XL meter. The ASTM F-13 committee on footwear has issued standards of operation of these meters.

Metal-bonded anti-slip coatings for floors and other walking surfaces are formed of abrasive grit particles encapsulated in metal that is bonded to a metal substrate in the molten state. Surfaces that are so treated offer the highest coefficient of friction of any metal surface even after much wear and abrasion. Aluminum oxide grit encapsulated with steel, on steel substrates, is ideal for all types of steel plate surfaces in walking or standing areas. Carbon steel plates can be zinc coated to provide long-term corrosion resistance. Aluminum oxide grit can also be encapsulated with aluminum, on aluminum, carbon steel, or stainless steel substrates. Steel grit can also be encapsulated with steel, on carbon steel substrates, with zinc coating for long-term corrosion protection. An epoxy coating system may also be used to prevent slips and falls. Epoxy coatings create distinctive peaks and valleys when applied to a surface. Water and chemical contaminants flow into the valleys, while the exposed peaks offer high traction.

Other preventive measures include restructuring work areas to reduce exposure to slip, trip, and fall. Can leaky machines be moved away from high traffic areas? Good housekeeping is important. Pick those spills or leaks up right away. Nonslip or slip-resistant floor mats are useful where flooring cannot be modified. Nonslip shoes is a last resort, preferably. Also, do not underestimate the importance of employee training in the prevention of slips, trips, and falls.

A comprehensive floor safety strategy consists of general and specific measures. Generally, wet floor signs and walk-off mats should be used in preventing slips, trips, and falls. Wet processes require well maintained drainage systems in the floor. False floors, platforms, and mats can also be used to maintain dry walking conditions. Test to determine the safest (least slippery) floor cleaners and floor treatment products and specify these for ongoing use. Write maintenance procedures for cleaning up slippery floors and for cleaning floors with the least slippery method. Provide handy spill cleanup equipment and train the appropriate personnel in how to clean up without creating even greater hazards. Do not assume that common sense will prevail. Another specific measure is appropriate footwear. Many work shoes offer nonslip soles as a feature.

Figure 16-9
Slip, Trip, and Fall Prevention

Identify areas:

- of high moisture
- where outside drafts occur
- that are exposed to weather
- of condensation
- with metal floor or stair treads
- where oil or grease spills occur
- with poorly lit stairs or passages
- of high traffic between high and low traction floors

List substances that workers walk on or touch: mud, grass, oils, greases, production chemicals, steam, paints and coatings, caustic or hazardous chemicals.

Prevention Checklist

Bell provides a checklist for planning slip, trip, and fall prevention, which is summarized in Figure 16-9.

Fisher recommends keeping thorough records of all maintenance procedures, products used, names of workers performing the tasks, as well as the date and time the workers performed the tasks, along with the dates and results of floor tests and inspections. These records are a sign that you have an ongoing floor safety program.

Fall Protection

OSHA recently developed a new fall protection standard for construction, Subpart M to 29 CFR 1926. Employees engaged in construction work that exposes them to a fall of six feet or more must have fall protection consisting of at least one of:

- guard rail
- safety net
- personal fall arrest systems

Please Note: Body belts are not acceptable after January 1, 1998.

The slip, trip, and fall hazard of construction continues to be a major cause of occupational deaths. Stanley cites a case where a bridge erector installed a horizontal lifeline system on the top flanges of plate girders as fall protection during the girder placement process. The lifeline is reinstalled for the deck forming operation. To backup the lifeline, the contractor emphasizes slip, trip, and fall during the initial safety training. Employees are provided with full body harnesses and two shock-absorbing lanyards and trained how to inspect and don the harness, attach the lanyards, and tie off to an adequate anchor point. One hundred percent enforcement of safety measures is the key to keeping workers alive where they may fall to their deaths otherwise.

Floor openings are void spaces that measure at least twelve inches in the smallest dimension. Hatchways, stair openings, ladder openings, pits, and manholes are examples of floor openings. *Floor holes* measure less than twelve inches but more than one inch at the smallest dimension. Floor holes may lead to tripping or falling materials (to a surface below).

Personal fall arrest systems are designed to catch workers after they have fallen. For working on elevated platforms without railings and toeboards, a body harness is attached to an anchorage with either a self-retracting lifeline or lanyard. Deceleration devices are used to take the force from the sudden stop at the end of the fall. Positioning devices help prevent falls by supporting a worker in a fixed position. Window cleaners use positioning systems. Power company and telecommunications linemen use belts and pole straps. Restraint lines are another example of positioning systems. Other systems are used to protect workers who must climb. A ladder safety device that attaches to a belt on the worker and a slide on the surface behind the ladder is an example of the latter.

Equipment supplied to employees for fall protection must meet OSHA standards for strength—they must be designed to stop falls quickly with limited force on the body. If unsure about which equipment to use, consult the distributor or manufacturer. Don't guess—you are literally playing with someone's life. Require employees to use the equipment exactly per manufacturer's instructions. If the job interferes with proper use, modify the job, not the equipment. Never use a lineman's pole straps as lanyards unless they have been designed for fall-arrest. Make sure the deceleration device is supposed to be used with the body harness system you have. They are not necessarily interchangeable. Snap hooks are another type of device that is not universally interchangeable.

Eliminate free fall where feasible. Do not allow more than two feet of free fall under any circumstances. Two methods can be used to limit free fall. The more common method is to use shorter lanyards between lifelines and body harnesses. The second method is to reduce slack in the lanyard by raising the tie-off point to the lifeline. This tie-off point must always be level with or higher than the connection to the body harness.

The anchorage must be strong enough to withstand the fall-arrest force. Most companies use the four inch rule.

Rule of Thumb
Use only four inch or greater diameter unheated steel lines with welded fittings as anchor points.

Before hooking up, the anchorage must be inspected for damage. No obstacles should be located under the anchorage. Allow for the free fall distance, deceleration distance, and any distance the lifeline stretches. Manufacturer's labels will state equipment stretch and deceleration distances. The further attachment to the lifeline is from the line's anchor, the more the line can stretch.

Rule of Thumb
Maximum deceleration distance allowable is 42 inches (3.5 feet).

Foot Protection

Foot protection is about more than slip, trip, and fall but we will address it as an issue now. Besides being a good slip-resistant sole for the type of surfaces in your workplace, foot protection needs to address three chief hazards: impact, compression, and puncture protection. *Impact protection* is needed during materials handling, such as heavy packages, parts, or tooling. *Compression protection* is required when the work involves manual handling of carts, bulk rolls, and heavy conduit. *Puncture protection* is needed for work around sharp objects, such as fasteners, wire, and scrap metal. *Dermal protection* is required where dangerous chemicals in a liquid form may spill into the workplace, or where feet would be constantly wet by water, oil products, or sludges. No one shoe or boot will satisfy all needs, so in a workplace where more than one hazard is present, a combination of footwear may be required.

Like abatement of other hazards, the first question is "What are the hazards?" You have to know what the worker needs to be protected against: impact, penetration, compression, chemical contact, heat, or electrical power contact. Next, you need to understand the tasks that workers will be expected to perform while exposed to the hazard. What are the environmental influences? For instance, what will the temperature be? Will the workers have to work in mud? Are steep slopes involved? What are the wind and precipitation conditions expected to be? Is lighting adequate?

CRANES AND HOISTS

Many industrial facilities use cranes and hoists for lifting heavy weights. The most common types are jib booms, monorails, gantry cranes, and bridge cranes. Electricity is the most common energy source for operating indoor cranes, followed by compressed air. A *jib crane* or *jib boom* consists of a boom suspended from a single building support. *Monorails* are usually an I-beam suspended between two or more building supports. A *gantry crane* is suspended between two mobile A-shaped structures that ride along a set of tracks similar to railroad tracks embedded in the floor. A *bridge crane* is suspended between two fixed set of I-beams set in the roof structure. Typically, each of these types of cranes can handle an increasingly heavier load in the order listed. A *hoist* is the mechanism that allows the load to be raised and lowered, consisting of a motor, a drum, wire rope, and a hook.

Most industrial facilities schedule monthly and annual crane and hoist inspections as part of an ongoing attempt to comply with federal and/or state OSHA standards. The intent is good, but the equipment is being used daily. Without fail, breakdowns will occur between maintenance inspections. Daily inspections by operators are needed to avoid disaster between maintenance inspections. A brief, practical inspection allow operators to discover small maintenance problems before large problems overwhelm them.

General

A lift plan should be devised for each operating site. A check sheet can be used to review ground conditions, congestion, complexity of the lift, power lines, and lifting radius needed. Operating rules should be posted near the crane.

Personnel and Training

The OSHA standard does not specify operator qualification and training. However, professional consensus standards by ANSI, ASME, and CMAA outline well-defined operator qualifications. Pre-employment and periodic physical examinations are highly recommended.

Lifting fixtures and jigs are unique to the load. The manufacturers of these devices typically provide operations and maintenance manuals or other printed guidelines for their safe use. Those operators who are assigned to work with lifting fixtures or jigs need additional safety training, based on manufacturer recommendations. As for all lifting equipment, include fixtures and jigs on the daily operator's inspection checklist.

Regularly scheduled training is suggested. All operators should complete a training course provided or recommended by the crane manufacturer. Second in order of preference is the use of commercial training films.

Inspection and Maintenance

The inspection and maintenance of cranes and hoists on a regular basis is critical to the safety process. On a daily basis the operator must perform visual and operational inspection tasks as indicated in Figure 16-10, which is adapted from a checklist put together by Creative Media Development.

A checklist can be maintained on the crane itself, mounted behind a sheet of hard plastic or preserved in a soft plastic document protector. If equipment is found functional and safe when inspected, the checklist can be completed with a crayon or similar marker and cleaned off before recording the subsequent daily inspection. If equipment needs repair, it should be tagged out until repaired.

The daily equipment inspection can be performed in no particular order, so long as the operator is systematic and consistent.

The Crane Manufacturers' Association of America recommends certain inspection and maintenance activities on a weekly, monthly, and semi-annual basis. Figures 16-11, 16-12, and 16-13, adapted from the CMAA 5M checklist, summarize this maintenance, respectively. If any of these items are found upon inspection to be in less than satisfactory condition, the necessary adjustments, repairs, replacement of defective parts, lubrication, cleaning, or other maintenance shall be taken immediately.

The wire lifting rope must not be kinked, twisted, crushed, or heat damaged. The rope must be properly seating on the drum and sheave grooves with the absence of overlapping or slack. The rope on the drum must not lay across groove ridges. Do not use the rope if any strand has more than five broken wires or any rope lay has more than ten.

Slings and Hoists

When inspecting slings, pay attention to the wire rope, end attachments, and sling hooks. Wire rope is particularly strong and widely used in industry and maritime where heavy loads are routinely lifted. The rated capacity of a wire rope depends on several factors. The heuristics for removing wire rope from service vary, but most rules are based on the number of broken wires per unit length. Randomly distributed broken wires do little damage to overall strength of the rope. However, if ten or more wires are broken in one rope lay, reject the sling as unsafe. Likewise, if you find five or more breaks in one strand in one rope lay, reject the sling as unsafe. Also reject the sling if

Figure 16-10
Daily Crane and Sling Safety Inspection

If equipment needs repair, lock out or tag out.

VISUAL INSPECTION OF BRIDGE CRANES:

Bridge runway	Hoist Line
Bridge	Hoist Chain
Trolley	Running Block
Hoist	Hook
Hoist Drum	Controller

VISUAL INSPECTION OF JIB BOOMS:

Mounting Brackets	Carrier Stops
Mounting Hinges	Hoist
Support Rod	Hoist Cable/Chain
Boom Arm	Hook
Carrier	Controls

OPERATIONAL INSPECTION OF BRIDGE CRANE OR JIB BOOM:

Controls
Upper Limit Switch
Brakes

VISUAL INSPECTION OF SLINGS:

Abrasion	Corrosion
Metal Loss	Chemical Damage
Kinks	Loose Fittings
Deformed Links	Rough Fittings
Crushed Fibers	Twisted Hooks
Discoloration	Loose Connections
Broken Wires	

you find wear or scraping of outside individual wires affecting at least one-third of the original diameter.

Also, check the wire rope for these faults: kinking, crushing, bird caging, distortion, heat damage, and corrosion. Excessive heat may damage any

Figure 16-11
Weekly Crane Inspection and Maintenance

BRIDGE:

- Brake and hydraulics

CAB:

- Master switches
- Warning device
- Mainline disconnect

HOISTS:

- Holding brake
- Wire rope
- Upper limit switch
- Hook and latch

TROLLEY:

- Brake

MISCELLANEOUS:

- General condition and housekeeping
- Grabs and attachments
- Push button station

Figure 16-12
Monthly Crane Inspection and Maintenance

BRIDGE:

- Control operation
- Line shaft bearings
- Lights
- Wheels
- Runway collectors
- Wheel gearing
- Reducers
- Wheel bearings
- Couplings
- Guards and covers

CAB:

- Fire extinguisher

Figure 16-12 *(continued)*

HOISTS:

Electrical control brake	Reducer
Control operation	Couplings
Mechanical load brake	Bottom block

TROLLEY:

Control operation	Wheels
Trolley collectors	Wheel gearing
Reducer	Wheel bearings
Couplings	

MISCELLANEOUS:

Cable reels	Wind anchors
Warning signs	

Figure 16-13
Semi-Annual Crane Inspection and Maintenance

BRIDGE:

Motor	Alignment and tracking
Control panels	Trolley rails and stops
Resistors	Bumpers
Trolley conductors	Rail sweeps
Girder connections	

HOISTS:

Motor	Upper sheaves
Control panels	Rope drum
Resistors	

TROLLEY:

Motor	Bumpers
Control panels	Rail sweeps
Resistors	

MISCELLANEOUS:

Runway rails - span	Runway rails - wear
Runway rails - joints	Main conductors

sling. Therefore, a history of heat exposure is enough to retire a sling. How can such exposure occur? From cutting torches, molten metals, and any process or device that has a heat source in excess of 180°F. Chemicals can also cause serious damage to slings. Chemical damage to synthetic slings may appear as brittle or discolored fibers. The most common symptom of chemical attack on alloy chain slings or wire rope is a decrease in diameter. Therefore, inspection of slings should always involve measurement of sling diameter where chemicals are present.

Reject the end attachments if cracked, deformed, or worn. Similarly, inspect any corrosion areas for potential impact on the strength of the attachment. If more than one wire is broken within one lay length from the end attachment, reject the sling.

Sling hooks should be rejected if the throat opening has increased by 15 percent or more. If the hook is twisted more than ten degrees off center, reject it. When a hook is stretched, do not replace the hook latch with a longer one and go on operating. Replace the hook. If not stretched and the latch is broken or missing, by all means replace the latch, but with a proper sized one.

Teach operators to avoid unsafe hoisting practices and enforce this religiously. Whether the hoist is electrically or pneumatically driven, operators should follow the guidelines of Figure 16-14 to avoid injury.

Specific Safety Practices

Before making a lift with a crane or hoist, compare the rated capacity for each component device with the load to be lifted. Do not proceed if the load is too heavy. Remove loose items, such as tools, from the crane and the load. Position the hoist system directly over the load. Mentally plan the safest route to take to get the load from its rest position to the desired position.

When using slings, the safe working load (SWL) is sometimes exceeded unwittingly. Each sling has one or two metal clips that tell the SWL of the single sling with a sling angle greater than 45°. If the sling angle does not exceed 45°, the load on each leg could exceed the SWL. The sling angle is the angle between the sling leg and the load, not the angle between the legs. See Figure 16-15.

While making the lift, use correct hand signals to your partner. Move slowly and check and recheck clearance. When possible, use brakes well ahead of time. Keep the rate of travel as steady as possible. Avoid shock load, such as pulling a load off an edge. The operator should remain in a position from which he or she can see the load and the path of travel at all times. No riders allowed.

The operator should position the crane directly over the load's desired position. The rate of the load's descent should be steady and even. After the

Figure 16-14
Hoist Operator Instructions

1. Familiarize yourself with hoist operating controls procedures and warnings.
2. Make sure hook travels in the same direction as on the controls.
3. Test hoist limit switches before picking up load.
4. Maintain firm footing during operation.
5. Load slings and other attachments must be properly sized and seat well in the saddle of the hook.
6. If a hook latch is used, it must be closed and may not support any part of the load.
7. Inspect the lift area to make sure the load is free to move and will clear all obstructions.
8. Take up slack carefully by lifting a few inches. Check load balance and action before continuing.
9. Avoid swinging the load or load hook.
10. Keep personnel from standing or walking beneath the load.
11. Warn unwary personnel that a load is approaching them.
12. Protect wire rope and hoist chain from damaging contaminants, including weld scatter.
13. Promptly report any malfunction, unusual performance, or damage to the hoist.
14. Replace damaged or worn parts as soon as discovered and keep detailed records of maintenance.
15. Lift no more than the rated load. If each component device of the system has different ratings, use the lowest rating for load limits.
16. Do not use the load limiting device to measure the load.

Figure 16-14 *(continued)*

17. Do not use a damaged hoist or one that is working improperly.
18. Do not use the load rope or load chain as a sling or wrap rope or chain around the load.
19. Make sure all supporting ropes and chains have equal loading.
20. Do not apply the load to the tip of the hook.
21. Center the load under the hook before operating.
22. Keep your attention fixed on operating the hoist once you start.
23. Do not use limit switches as routine operating stops.
24. Do not use hoist to lift, transport, or support people.
25. Do not lift loads over the heads of people.
26. Attend suspended loads at all times. If you must leave, lower the load to the floor.
27. Prevent sharp contact between two hoists or between hoists and obstructions.
28. Do not allow rope, chain, or hook to be used as a ground for welding.
29. Do not allow rope, chain, or hook to be touched by a live welding electrode.
30. Do not remove or obscure warning signs.
31. Do not adjust or maintain hoist unless qualified.
32. Do not attempt to repair or lengthen load rope or chain.

lift is complete, the operator should either secure the crane in the normal stored position. Another secured position should be used only if a reason exists not to return to the original secured position. Controls need to be disengaged. Report any damage or breakage resulting from the use of the crane.

The most valuable safe use concept is consistency in operation. Like all people, crane operators develop habits. The safety manager's job is to assist the crane operator's supervisor in making sure the operator develops good habits.

Figure 16-15
Safe Working Load with Slings

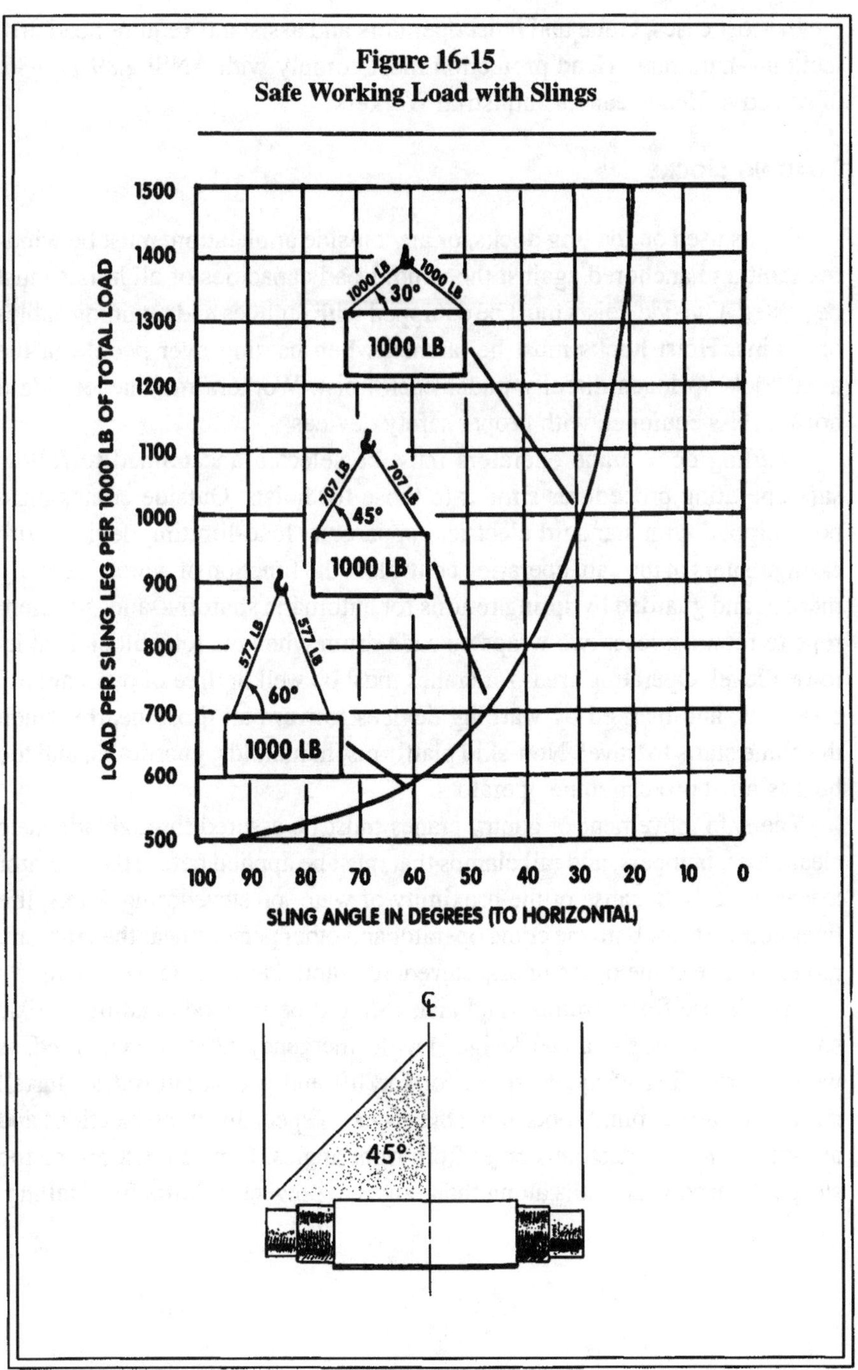

In most cases, crane and hoist operators and assistants require head protection—hard hats. Head protection must comply with ANSI Z89.1-1986, "Protective Headwear for Industrial Workers."

Loading Docks

Hoists used on loading docks, or any outside application, must be wind-resistant and anchored against the wind. Load capacities of all hoists must be posted. Outside hoists must be equipped with taglines and retaining cables or chains. Hoist hooks must be latched when passing over people in the area. Pick up loads directly under each hoist. Workers may never ride a hoist unless equipped with proper safety devices.

Loading dock crane operators must be selected and trained to follow safe operating procedures similar to those for hoists. Outside cranes must be equipped with standard electrical apparatus, load-limiting devices, fire extinguishers in the cab, operating controls—the function of which is clearly marked and guarded by spring returns for automatic shutoff—and sufficient rope to retain at least two wraps on each drum when the load block is at its lowest level. Operating areas for cranes must be well lit, free of overhanging obstacles, and marked by warning devices to warn all those nearby when the crane starts to travel. Non-skid platforms, handholds, guardrails, and toe boards must protect crane operators.

The safe movement of gantry cranes must be assured through adequate clearances, bumpers, and rail clamps that must be applied before the operator leaves the cab. Because of the proximity of water on stevedoring docks, life lines and vests for both the crane operator and other persons near the crane are required. The crane operator and stevedores should also be able to swim.

Warehouse floors, ramps, and aisles should be in good condition. Aisle spaces and widths should be designed with emergency evacuation in mind, as well as providing adequate space for forklift and pedestrian traffic. Install convex mirrors at blind spots along the aisles—especially at intersections and at places where pedestrians enter from right angles. Ramps must not be too steep and need guard rails along their edges to prevent forklifts from falling.

References

Bell, Jonathan. "Planning Can Prevent Slip and Fall Accidents." *Occupational Health & Safety.* December 1996, pp. 33-35.

Crane Manufacturers' Association of America. *Crane Inspection Schedule and Maintenance Report.* CMAA 5M 6/29/84.

Creative Media Development, Inc. *Daily Crane and Sling Safety Checklist.* 1993.

Daugherty, Jack. *Assessment of Chemical Exposures: Calculation Methods for Environmental Professionals.* Boca Raton, FL: Lewis Publishers, 1998.

David, David A., Jr. "Proper Fall Protection Training." *Occupational Health & Safety.* June 1996, pp. 54-59.

Eckhardt, Robert. "Crane Safety." *Occupational Health & Safety.* Pp. 40-44.

Fisher, Kenneth. "Dodging the Negligence Bullet." *Occupational Health & Safety.* December 1996, p. 32.

Head, George L. "Safety Practices and Loss Prevention Methods." *Professional Safety.* January 1992, p. 37.

Indoor Cranes: Safe Lifting Operations. Virginia Beach, VA: Coastal Video Corporation, 1993.

IKG Borden, A Harsco Company. *IKG Mebac® Slip Resistant Metal Surfaces.* 15M-10/95-PM.

Jeffries, Neal P. and Richard L. Shell. *Engineering Fundamentals.* Dearborn, MI: Society of Manufacturing Engineers, 1975.

Kendzior, Russell J. "Defining Slip Resistance." *Occupational Health & Safety.* April 1997, pp. 55-56.

LaBar, Gregg. "Are You Picking the Right Foot Protection?" *Occupational Hazards.* August 1997, pp. 43-44.

Lonsdale, C.P. "Don't Overlook Hand Tool Inspections." *Occupational Health & Safety.* October 1997, p. 155.

Mahan, Kirk. "Slips, Trips, and Falls." *Compliance Magazine.* April 1997, pp. 10-13.

National Screw Machine Products Association. *Safety & Health News.* "New OSHA Rule on Personal Protective Equipment. S&H Bulletin 228-SM. NSMPA, June 7, 1994.

OSHA Electrical Hazard Fact Sheets. Nos. 1-16. Reprinted January 1992.

OSHA Training Electrical Course 203/309A. OSHA Training Institute, July 15, 1997.

Personal Fall Protection. Virginia Beach, VA: Coastal Video Communications Corporation, 1992.

Smith, S.L. "Sharing Tips on Trips and Falls." *Occupational Hazards.* May 1997, pp. 61-64.

Stallcup, James G. *OSHA Electrical Regulations Simplified.* Fort Worth, TX: Grayboy Publishing, 1992.

Stanley, Rod. "Fall Protection in Bridge Construction." *Occupational Health & Safety.* December 1996, pp. 28-29.

Steer Clear of Slips, Trips & Falls. South Deerfield, MA: Channing L. Bete Company, Inc., 1993.

Thomas, W.A., H.A. Spalding, and Zarko Pavlovich. *The Engineer's Vest Pocket Handbook.* Baltimore: Ottenheimer Publishers, Inc., 1960.

Vacuum and Pressure Systems. Benton Harbor, MI: The Gast Company, 1976.

17

OCCUPATIONAL OPERATION OF VEHICLES

INDUSTRIAL VEHICLES

Industrial vehicles include industrial power lifts (forklifts), hydraulic cranes (cherry pickers), scissors lifts, low-lift trucks, front-end loaders, pallet-movers, and powered hand trucks.

Use and Selection of Vehicles

Cushion tire or solid tire industrial lift trucks, called forklifts, are designed for use on smooth dry surfaces such as a concrete floor in a warehouse or plant. Pneumatic or air-filled tire trucks are used on improved surfaces inside or out. Rough terrain requires special air-filled tires. Do not attempt to use standard pneumatic tires or solid tires or you will need a heavy-duty tractor to pull it back onto an improved surface.

- Check the vehicle's stability before lifting loads.
- Stay within the vehicle's rated capacity for lifting loads.
- Rig properly before lifting a load. Rigging practices vary from machine type to machine type.
- Lift only one load at a time.
- Watch the load. Assign lookouts and/or signalmen as necessary.
- In the case of cherry pickers, use the boom for hoisting only. Do not use the boom to move the cherry picker sideways or as a ram.

Industrial hand trucks or low-lift trucks are often used to reduce back injuries from handling materials. Both manual and powered low-lift trucks can cause other injuries if used improperly. Wheels can roll off the edges of ramps or loading docks. Operators can push the truck into an obstacle. Hands can be pinched between the truck and other objects. Some operators have been run over and crushed. The danger is greater with powered trucks than for manual trucks, of course, though the possible types of accidents are the same.

Personnel and Training

Forklift rallies create safety awareness and provide an opportunity to recognize and reward operators who promote safety on the job. Speed has no place at the rally, but deliberate driving and demonstration of skills do. Some skills needed for forklift operation are itemized in Figures 17-1, 17-2, and 17-3.

Employees need to be trained on how to work with and near cherry pickers, too.

Skills for powered hand truck operators are listed in Figure 17-4.

Inspection and Maintenance

Industrial lift trucks have a nameplate that lists its type, capacity, load center, and truck weight. Operators should conduct a daily maintenance inspection before using the truck. Have repairs made before using it.

Check the truck's fork retainer pins and locks. Inspect the upright, but keep hands out. Make only minor adjustments. Leave major adjustments to factory authorized mechanics.

Check the overhead guard and backrest extension. Ensure the horn, backup alarm, and rotary warning lights are working properly. Check tire pressure, but face the treads not the side and use a long-handled tire gauge. If the tire is low, call the factory-authorized mechanic. Make no repairs yourself.

Figure 17-1
Forklift Operator's Inspection

1. Proper use of tilt.	10. Check the warning light.
2. Proper use of raise/lower.	11. Check the rear view mirror.
3. Sound the horn.	12. Check battery retainer.
4. Check for oil leaks.	13. Check the discharge indicator.
5. Check mast chains.	14. Check the back-up alarm.
6. Check the brakes.	15. Check the hoses/hose reel.
7. Check tires and wheels.	16. Check the overhead guard's light.
8. Check hour meter.	17. Know the capacity of the forklift.
9. Check the scissors reach.	

Figure 17-2
Operator's Knowledge of Facility Safeguards

Dock Area

- Wheel chocking
- Dock plate
- Condition of trailer floor
- Don't jump off dock
- Keep clear of others
- Be aware of signs
- Correct height of empty pallets

Battery Charging

- Protective equipment
- Acid neutralizing
- MSDS
- No smoking
- Plug/unplug procedures
- Clean-up procedures
- Eye wash station
- Commercial battery rules

Fire Safety

- Location of extinguishers
- How to use extinguisher
- Type of extinguisher to use

Personal Safety

- Eye protection during operation

Figure 17-3
Operations Skills

1. Pull forward toward designated section of rack without striking anyone.
2. Properly place forks under pallet.
3. Properly raise or tilt load.
4. No part of the container may strike any part of the racking while removing pallet from rack.
5. Lower the pallet before moving/backing out.
6. Drive at a safe rate of speed.
7. Slow down or stop at cross aisles.
8. Sound horn at appropriate times.
9. Pull into the proper area of racking to place pallet.
10. Avoid striking any racking while lifting load or going into rack.
11. Back out and lower forks before moving.
12. Always check behind the forklift before backing up.

Figure 17-3 *(continued)*

13. Wear protective equipment as required.
14. Remove obstacles on the floor.
15. Set the load on the floor before getting off the forklift.
16. Put on a hardhat, if required, before getting off the forklift.

Figure 17-4
Skills for Powered Hand Trucks

1. Allow new operators some time to get to know the truck before using it.
2. Teach them the proper operation of forward, reverse, brakes, and lift controls.
3. Let them practice maneuvering in an open area before driving the truck in a crowded area.
4. Teach the operator to judge the clearance between his or her truck and the objects around it.

Change liquefied petroleum gas (LPG) tanks in well-ventilated areas only. Unconfined outdoor spaces are great. Turn the ignition off. Check the system for leaks. Inspect for nearby sources of ignition, such as open flames. Store tanks in accordance with local fire regulations. When refilling LPG tanks remember that LPG is heavier than air and will settle on your clothes and the ground. Check all connections when replacing the tank on the truck. Check all connections for damage or leaks. If the truck does not start after changing a tank, get the factory-authorized mechanic.

Refuel gasoline and diesel fuel trucks in designated areas only. Use clean, properly marked fuel cans. Clean up spills immediately. Use vented fuel caps.

Trained maintenance people must check electric trucks on a regular schedule. Batteries contain acid and can generate hydrogen, a highly flammable gas.

Maintenance of cherry pickers must be performed in accordance with manufacturer's instructions. Inspect cherry pickers—especially brake and safety restraining systems—daily and at periodic intervals, as recommended by the manufacturer.

Operator Safety Practices

Operators must check the truck weight and capacity before using the truck on any surface. Compare the data with elevator capacity before entering the elevator. Verify that floors can handle the weight. Stay away from soft ground unless you have a rough terrain truck.

Slack chains mean rail or carriage hang up on forklifts. Raise the upright before moving. When stacking with a forklift, watch the chain lifts. If they go slack, stop; raise the load, then lower again.

Forklifts

Do not operate a forklift that has a maintenance problem. Remove the key and put an "Out of Service" tag on it.

Figure 17-5 lists general safety rules for forklifts.

Figure 17-5
Forklift Safety Rules

- Watch for pedestrians.
- Wear safety equipment when required.
- Do not block safety or emergency equipment.
- Drugs and alcohol are prohibited.
- Always buckle up.
- Be alert for no smoking areas.
- Allow no riders.
- Lift material, not people.
- Never allow anyone to walk under raised forks.
- Never try to repair chains or any other part of the upright, carriage, or attachments.
- Do not use the upright for a ladder.
- Keep hands, feet, arms, legs out of the chains
- Keep hands, feet, and legs out of the upright, carriage, or attachments.

Figure 17-6
Tip Over Prevention on Grades, Ramps, or Inclines

Empty: travel reverse up and forward down

Loaded: travel forward up and reverse down

Travel in a straight line and never turn on a grade.

Forklifts are known to tip over on grades if the operator does not understand where the center of gravity is located and how to operate in spite of it. Figure 17-6 gives general safety precautions to prevent tip over.

For cherry pickers, clean the cab frequently during the work shift. The operator should remove clutter and debris every time he or she enters the cab. The operator should also take a minute or two to remove mud or grease from shoes. Control pedals are difficult to operate with slippery shoes or debris blocking the way.

Observe safety requirements posted on placards placed on industrial vehicles.

The industrial vehicle operator should enter and exit carefully, using both hands to get aboard and being certain of footing. Never get on or off a moving vehicle.

Cherry picker operators must secure the turntable before traveling. Use the swing lock or house lock if the vehicle has one, or lower the boom to its horizontal travel position.

Operators of any industrial vehicle, but especially cherry pickers and scissors lifts, must maintain ten or more feet clearance from power lines. Clearance must also be maintained from physical obstructions such as walls, ceilings, and other overhead and side obstructions. When maneuvering in tight situations, it is best to post a lookout to prevent collisions and bumping.

Industrial vehicles should back up carefully. A signalman is advisable to verify that blind areas are clear and to direct the operator. Check load capacities of floors, driveways, docks, and bridges before using an industrial vehicle on them for the first time. Forklifts and cherry pickers can be quite heavy.

Avoid dangerous inclines. Industrial vehicles are typically top heavy and tilt over easily. In the case of cherry pickers, use outriggers fully extended to stabilize the vehicle before lifting heavy objects.

Do not allow riders on any industrial vehicle.

Hand Trucks

For two-wheel hand trucks, make sure the chisel is pushed all the way under the load before taking off with the load. Strap bulky or dangerous cargo to the frame. Put heavier objects on the bottom of stacks. Put heavier objects to the back when two or more objects are in one layer.

Operators of two-wheel trucks need to protect their hands. Grip the truck behind the widest part of the frame. Never put you foot on the wheel to brake. Keep the truck ahead of you going down a ramp. Pull the truck behind you going up a ramp. Except for going uphill, face the direction you are traveling in. Store two-wheel trucks on the chisel, with the handle leaning against a wall or another truck.

When operating four-wheel trucks, protect hands and feet. Prevent the truck from tipping over. When loading, arrange items so they will not fall. If the truck has no drawbars, push it, keeping hands behind the cart. If the truck has a drawbar, pull it so you can see better. Standing to one side, pull it in such a way that the truck does not run over your ankles. Stay within the aisle boundary lines. Store the truck by locking or blocking its wheels.

Stabilize the load with shrink-wrapping or strapping before moving a pallet mover. Block or bind round or irregular shaped objects. Stack solid rectangular objects neatly, interlocking them when possible. Inspect pallets regularly for damage. Remove defective pallets from service. Stack unused pallets flat in an out-of-the way place. Store large amounts of pallets outside away from the building, since they are combustible.

Powered Hand Trucks

Walkies (or walking trucks) and walkie riders are two types of powered hand trucks. Walkies are electric-powered pallet movers that are controlled by an operator who walks alongside of it. Walkie riders are driven by an operator who stands on the truck. A dead-man control is a safety device that shuts off the power if the operator lets go of the handle. A belly-button stop switch keeps the operator from getting caught between the handle and a stationary object, such as a wall. Brakes on a walkie are applied by moving the steering handle up and down.

With a powered hand truck, approach the pallet with the forks squared to the load. Lower the forks and align them with the openings on the pallet. Space forks evenly on either side of the center stringer for balance. Drive the truck slowly in reverse until the forks are completely under the pallet. Avoid striking the stringers or deckboards with the forks. Raise the load before traveling.

Stop the truck if it runs but will not lift the load. Check the battery. Remember that the battery can generate explosive gases. The operator should not operate the truck with wet or greasy hands. Face the truck in the direction of travel, with forks in a raised position. Normally the truck will be pulled. Stand to one side of the control handle. Keep feet clear of the moving truck. Do not lead with the forks in blind areas or doorways. Load first into elevators and tight spots. Do not exit a walkie rider until it stops moving.

Fleet Vehicles

Motor vehicle deaths are the single largest cause of accidental death on the job. The two principal factors of motor vehicle accidents are driver error and vehicle failure. An effective motor vehicle safety program will pay dividends and will include driver selection, training, supervision, safety meetings, preventive maintenance, and fleet accident data.

To the extent available, vehicles should have certain safety features that are still not necessarily standard equipment. Air bags and seat belts are a necessity. Although the federal government now allows airbags to be disengaged in situations where their use might be more hazardous than non-use, insist on air bags unless you are certain the vehicle is safer without one. Because the designers intended air bags to be used with shoulder and lap restraints, insist on the proper use of seat belts.

Four-wheel disc brakes should be inspected regularly for wear limits. Power-assisted brakes are preferred. I recommend the anti-locking brake systems (ABS). Some people misuse these systems by pumping them like standard brakes, but if used properly they are remarkable for fast stops without skidding and sliding.

Power-assisted rack and pinion steering is also recommended for more road sensitivity and a smaller turning radius.

A tilt steering wheel allows the driver to make adjustments for his or her comfort depending on driving conditions.

Whole body vibration for long hours in highway driving causes fatigue. Drivers should understand their vehicle's suspension system.

Ensure tires are properly inflated.

Drivers need good quality, optical-grade-tinted glasses with polarized lenses to reduce glare. Avoid sunglasses that block peripheral vision. Keep the dashboard clear of papers and junk as these items make reflections on the glass that can interfere with vision.

Integration of OSHA, DOT, and EPA

Over-the-road driving is regulated by the U.S. Department of Transportation (DOT) if goods are involved and the Occupational Safety and Health Administration (OSHA) with respect to the employees involved. The OSHA regulations are found at 20 CFR 1910, Subpart N, Paragraphs 177-178 (29 CFR 1910.177-178). The U.S. Environmental Protection Agency (EPA) also has regulations that govern the transportation of hazardous waste. All these regulations overlap and you need to be familiar with all of them.

Personnel and Training

Part of an effective motor vehicle safety program is the selection of drivers. Certainly if a person does not have a current valid state operator's license you do not want them driving company vehicles.

Company drivers should be a least twenty-one years old, and a twenty-five year old age minimum is recommended. The candidate should be able to read and speak English fluently in order to converse with the general public, understand road signs, and speak to police officers in case of an accident. A candidate should demonstrate the ability to effectively operate all company vehicles. If DOT regulations apply, the candidate must undergo physical examinations by a physician, take written and road tests administered by the state, and pass a drug screening. The candidate for DOT regulated over-the-road vehicles must have an acceptable driving history and the same criteria is recommended for all driver candidates. Past violations, if any are on the candidate's records, should be thoroughly investigated.

No matter how small your over-the-road fleet, a training program is required. DOT has strict training rules that must be repeated every other year. If rules changes are made in the interim, even more training is required.

Maintain meticulous written records of training activities. Drivers must have certified operator training for the type vehicle they will be driving. A current state operator's license, valid for the desired vehicle, is evidence of this training. It is recommended that you hold refresher training from time to time to cover road rules, DOT regulations, and OSHA and EPA regulations to the extent that they apply.

Assign a person to supervise the operations of all commercial vehicles. This person should assign loads, route drivers, and plan loads and delivery. He or she should keep written records of the operations of all commercial vehicles. He or she should verify periodically that drivers maintain their logs properly. The fleet operations supervisor must maintain a complete file of accident records.

Safe Driving Policy

Companies that have fleet vehicles need to have a written policy that requires employees to operate the vehicles safely, wear restraints, and obey traffic rules.

Inspection and Maintenance

Inspect the fuel gauge regularly and refuel with plenty to spare. Inspect and keep tires properly inflated. A tire gauge should be carried in the glove compartment, as many filling stations concentrate more on convenience groceries than car maintenance. Remember, underinflation of tires causes them to wear faster and provides less control over steering. Clean the windshield frequently. Check the windshield wiper fluid level too. Clean the headlight lens.

The fleet operations supervisor should keep records for daily preventive maintenance checks.

Operator Safety Practices

Safe driving is more than safe practices, it is an attitude. A safe driver does not bully others into changing lanes. A safe driver does not follow too closely the vehicle in front of him or her. He or she does not shout at or make gestures towards other drivers. He or she does not drive on the shoulder of the road. Figure 17-7 lists some distractions a safe driver should avoid.

Figure 17-7
Dangerous Distractions while Driving

Eating and/or drinking
Reading maps, directions, newspapers
Using a cellular telephone
Smoking
Using rearview mirror to apply makeup or brush hair
Reading billboards
People watching

Safe drivers do not drink and drive. Nor do they use or abuse any other substance that may impair their judgment, slow their reflexes, reduce their coordination, or blur their vision.

Use shoulder restraints and lap belts. Sit straight up in the seat while buckling up. Adjust the lap belt so that it is snug yet comfortable. Adjust the shoulder strap so that it fits tightly.

Air bags are designed to be used with passive restraints for maximum protection.

Drive with both hands on the steering wheel at the ten and two o'clock positions so that hand positions can be changed rapidly if needed. Keep scanning the road to develop a picture of the what is going on. Look down the road far enough to anticipate the next several seconds. Normally, you should adjust your speed with traffic. This reduces relative speed differences and gives you more reaction time. Also, maintain a space around your car, as best you can, to give yourself more reaction time, stopping distance, and visibility.

Slow down before you enter a curve to avoid decrease of traction.

Communicate with other drivers. Use the horn, headlights, brake lights, turning signals, and emergency flashers to let others know where you are and what you are doing. Tap your brakes two or three times as a warning signal before braking quickly. Use turn indicators even when no one else is around.

Visibility of one hundred feet or less calls for speeds of thirty miles per hour or less. Turn low beams on whenever visibility is less than optimal.

Use polarized sunglasses and the sun visor on bright, sunny days. At dusk, remove the sunglasses and use full headlights. Use of parking lights while driving at dusk leads to accidents.

The eyes should be kept moving at night. Night vision is slightly better slightly off center. Avoid looking directly into oncoming headlights and other bright lights. Use bright beams, but not when another car is approaching head on. Most rearview mirrors have a day/night feature to reduce headlight glare from cars approaching from behind at night.

Driving in the Rain

Reduce speed. Use low beams. Use the defroster. If the brakes get wet, pump them several times to dry them out. The first few minutes of a rain are the most dangerous, as stormwater mixes with oil on the street. Avoid wet leaves on the road. Avoid driving through pools of standing water.

Driving in the Snow

Drive slow and steadily. Look one full block or more ahead. Avoid slush. Do not change lanes if traveling more than twenty-five miles per hour. Never pass another car in the snow. Approach a hill well behind the vehicle in front of you. Build speed up before the hill. Do not accelerate while on the slope of the hill. Allow the momentum of your vehicle to carry you up, keep a steady speed.

Driving on the Ice

Ice is wetter and more slippery near the freezing point than at colder temperatures. Ice develops in shady areas, under bridges, and on overpasses before other areas.

If your vehicle skids, do not panic. Do not use the brakes. Take your foot off the accelerator and steer into the skid, which is the direction the rear of your vehicle is going. Begin straightening your wheels as the vehicle starts to come out of the skid. Use your brakes to stop after the vehicle is straight. Pump the brakes if you do not have anti-lock brakes, but apply steady pressure to anti-lock brakes.

Accident Scene Procedures

The fleet operations supervisor should maintain fleet accident data. Complete records are important, so all drivers must know and understand accident scene procedures.

Reference

Industrial Low-lift Trucks. Virginia Beach, VA: Coastal Video Communication Corporation, 1993.

Lightfoot, Ethan and Jo Bennett. "Cherry-Picker Safety." *Occupational Health & Safety.* Date unknown, p. 88.

Meola, John J. "Better, Safer Driving." *Occupational Health & Safety.* October 1997, pp. 142-144.

Motor Vehicle Awareness. Virginia Beach, VA: Coastal Video Communications Corporation, 1992.

The New Professionals: Rules for Safe Industrial Truck Operation. Battle Creek, MI: Clark Equipment Company, 1983.

Norris, Eileen. "So Long, Forklift Cowboy—Hello, Safe Operator." *Safety & Health.* July 1993, pp. 58-61.

OSHA Manual. The Merritt Company, 1988.

Safe Winter Driving. Virginia Beach, VA: Coastal Video Communications Corporation, 1994.

Swartz, George. "Forklift Safety Training: Tips for Improving Your Program." *Professional Safety.* January 1993, pp. 16-21.

18

Biological Hazard Management

Identification of Biological Hazards

In 1969, the U.S. Surgeon General made the statement, "We have won the war on infectious diseases." For several years afterward, the general public really thought we had eradicated serious infections leading to such diseases as plague, smallpox, diphtheria, and polio. Further, we gloried in vastly reducing the effects of such infectious childhood diseases as measles, rubella, chicken pox, and mumps.

The decline in infectious diseases in the mid-twentieth century came about with improvements in basic hygiene, public health, and disease prevention measures. For example, many diseases now have vaccines that were unavailable just three generations ago. In 1968, around 150,000 cases of mumps were recorded. Now that the mumps vaccine is available, about 4,000 cases are reported annually.

However, from around 1980 to the present, new and more vigorous infectious agents have shown up. Potent diseases such as AIDS (Acquired Immune Deficiency Syndrome), Legionnaires disease, Lyme disease, the drug resistant strain of tuberculosis, and Ebola are the apparent plagues of the twenty-first century. In 1995, more than 17 million people died from infectious diseases! Internationally acclaimed epidemiologists are currently expecting the outbreak of a hybrid strain of flu virus that could be as virulent as the flu that killed millions in the second decade of the twentieth century. In hindsight, we certainly have not won the war against infectious diseases.

Occupational Exposure to Health Hazards

With respect to the workplace, OSHA issued voluntary guidelines on dealing with Hepatitis B virus in 1983. Then, in 1992, OSHA issued a final rule on occupational exposure to bloodborne pathogens in general industry. This standard does not apply to the health care industry.

Biohazards are etiological agents, viable microorganisms, and toxins secreted by these microorganisms that cause human disease. *Etiologic agents* include chemical toxins. *Infectious agents* are biological organisms that are capable of establishing an infection process in living organisms. *Pathogenic organisms* are capable of infections leading to disease.

Biological Hazards

Biological hazards include bacteria, viruses, fungi, rickettsia, chlamydia, protozoa, and other living organisms that can cause acute and chronic infections by entering the body either directly or through breaks in the skin. Occupations that involve handling plants or animals or their products or food and food processing may expose workers to biological hazards. Laboratory and medical personnel also can be exposed to biological hazards. Any occupations that result in contact with bodily fluids pose a risk to workers from biological hazards. In occupations where animals are involved, biological hazards are dealt with by preventing and controlling diseases in the animal population, as well as properly caring for and handling infected animals. Also, effective personal hygiene, particularly proper attention to minor cuts and scratches (especially on the hands and forearms), helps keep worker risks to a minimum. In occupations where there is potential exposure to biological hazards, workers should practice proper personal hygiene, particularly hand washing.

Bioaerosols are airborne particles of biological origin, also known as biogenic particles, airborne organisms, microbes, microorganisms, microbiological agents, microbial agents, and viable pathogenic aerosols. Among bioaerosols are fungi, bacteria, viruses, algae, amoebae, pollen grains, and dead particles produced by all of these. Bioaerosols may also include plant parts, insect parts, animal parts, and wastes such as saliva, urine, feces, and dander.

Obligate parasites are biological organisms that can only survive in living cells. *Facultative saprophytes* can also survive on dead organic materials. Biological hazards need a *reservoir* to survive. A fungi may grow in a stagnant water reservoir in an air conditioning duct. HIV accumulates in human body fluids. A source of nourishment is also needed. This may be dirt and oil in a duct system or living cells in a human body. *Amplification* refers to the growth of the organism. *Dissemination* refers to the aerosolization and distribution of water containing the microorganisms.

Places where injured and ill patients are treated or given aid should have proper ventilation, proper personal protective equipment (such as gloves and respirators), adequate infectious waste disposal systems, and appropriate controls, including isolation in instances of particularly contagious diseases (such as tuberculosis).

The interval of time between when an infection occurs and the subsequent development of disease is independent of the time to infection. Time to infection is a function of the exponential probability density function: pdf f(t).

$$f(t) = \lambda_1 \bullet \exp(-\lambda_1 \bullet t) \qquad [18.1]$$

where

λ_1 = rate of infection

Another way of expressing λ_1 is the probability of infection per unit time. The subscript denotes the unit time as one month.

After one month, the time to disease is modeled by the exponential pdf g(t):

$$g(t) = \lambda_2 \bullet \exp(-\lambda_2 \bullet t) \qquad [18.2]$$

Nicas sets λ_2 at 0.004274/month so that after one year, the risk of developing disease is 0.05. At one year, the exponential pdf h(t) is

$$h(t) = \lambda_3 \bullet \exp(-\lambda_3 \bullet t) \qquad [18.3]$$

where λ_3 = 0.00009195/month to give a 0.05236 risk of developing disease after forty-nine years.

Nicas has paired the λ_2 and λ_3 infection rates such that a 0.10 risk of disease development exists for any 50 year lifetime. He bases this on the statistical principle that the probability that an event of interest will occur at some point during a series of independent trials is the complement of the probability that the event does not occur in any of the trials. Let's say that the probability of an event occurring in a trial is p_1. The probability that the event occurs in trial two of two is p_2. The probability that the event occurs in neither one of the two trials is $(1-p_1)(1-p_2)$. The probability that the event does occur in one of the two trials is $[1-(1-p_1)(1-p_2)]$.

The time to disease, then, over a one-year interval starting at time t_0 is an exponential pdf s(t):

$$s(t) = \frac{\lambda_1 \lambda_2}{\lambda_1 - \lambda_2} \left[\exp(-\lambda_2 \bullet t) - \exp(-\lambda_1 \bullet t) \right] \qquad [18.4]$$

Airborne Diseases and Transmission

Bacteria are one cell organisms. If the environment is right, they can reproduce themselves. Why is it risky to eat undercooked meat? Because bacteria that are toxic to human systems may have survived the cooking process.

Viruses are little pieces of rogue chemistry. They are little more than bits of protein running amok, yet scientists consider them to be living creatures. Viruses cannot survive by themselves, however. They need a host cell in which to live and reproduce themselves. Hepatitis-B (HBV) and AIDS are caused by viruses.

Infection is a condition where disease-causing organisms have entered the body and set up temporary housekeeping. Infection ends with either the restoration of health of the victim or the death of the victim. In either case, the bacteria or virus have lost a host. Therefore, in major epidemics, infectious agents will not kill the entire population. Rather these agents will, over a period of time, adapt themselves so that their hosts can better tolerate their presence. If the infectious agent ever killed the total host population, it would die off too. Nevertheless, even with this eventual lessening of vigor, the infectious agents will continue to do a lot of damage to their human hosts.

The *aflatoxigenic* strain of *Aspergillus flavus* and *Aspergillus parasiticus* is ubiquitous and a worldwide occupational concern, especially among agricultural and grain processing workers. These workers are exposed to aflatoxin by means of inhaling grain dust. Pulmonary interstitial fibrosis is one health hazard of airborne aflatoxins. The respiratory system metabolizes aflatoxin to a carcinogenic form. The study of grain processing facilities by Ghosh, *et al,* found the aflatoxin in the respirable as opposed to the total dust portion.

Fungi, mold, and mold spores come from outside. These microorganisms may grow in settled water within air handling systems. Their size ranges from 3 to 200 µm, but most are in the 10 to 20 µm range. Mold blooms results in the release of billions of tiny molds that remain airborne for a very long time. Mold spores cause hypersensitivity to molds. *Aspergillus,* mentioned above, *Penicillium,* and *Cladosporium* are important indoor molds. Most molds are saprophytic fungi, or fungi of decay. Mold detritus, especially the carbohydrates mannans and glucans, arouse direct antibody-independent interactions with bronchial tissue cells. As a result, cellular and humoral inflammatory mechanisms are activated. Airborne glucans have been blamed for the symptoms in sick building syndrome. Glucans have been used as a general marker for mold exposure.

Molecules that initiate a humoral immunoglobulin (Ig) response are called *antigens. Allergens* are antigens that cause allergic reactions, usually by combining with an IgE antibody.

Figure 18-1 lists some typical microorganisms that may be found in a workplace. This list is not all inclusive.

Figure 18-1
Typical Workplace Microorganisms

Bacterial Cells

Actinomycetes
Bacillus subtilis
Brucella suis
Escherichia coli
Flavobacterium
Pasteurella pestis
Pasteurella turalensis
Pseudomonas aeruginosa
Pseudomonas syringae
Saccharomyces cerivisae
Serratia marcescens
Staphylococcus epidermis

Bacterial Spores

Bacillus subtilis
Bacillus subtilis var *niger*

Fungi

Aspergillus
Cephalosporium
Cladosporium
Cryptococcus
Fusarium
Mucor
Harposporium
Ostracoderma
Rhodotorula
Trichoderma

Mold Spores

Penicillum chrysogenum

Nematodes

Rhabditis

Protozoa

Acanthamoeba
Vorticella

Legionnaire's Disease is caused by strains of bacteria referred to as *Legionnella bacillus.* The disease causes pneumonia, especially in the elderly, immunosuppressed individuals, and in those already suffering from some respiratory ailment. *Pontiac fever* may be an allergic reaction to Legionnaire's disease and not a true infection in its own right. Nonetheless, it is a non-pneumonia form of Legionnaire's disease.

High levels of airborne noninfectious microorganisms are known to cause respiratory symptoms and disease where workers handle biological materials: agricultural work, sawmills, municipal waste, and fuel chips.

Endotoxin is a *lippopolysaccharide* (LPS) consisting of the biologically active lipid part (Lipid A) and a hydrophilic polysaccharide. Endotoxins are responsible for acute and chronic respiratory health effects.

Figure 18-2, from De Chacon and Van Houten, lists the etiologic agents found in certain building components.

Bloodborne Diseases and Transmission

A *pathogen* is a disease-causing organism, such as a bacteria or virus. A *bloodborne pathogen,* then, is an organism that is carried in the blood and certain other fluids in the body.

The list of bloodborne pathogens includes AIDS, HBV, syphilis, malaria, brucellosis, babesiosis, leptrospirosis, and viral hemorrhagic fever. In the U.S., 300,000 new HBV infections are reported each year. One of every 300 people in the U.S. are infected with HBV.

HIV (human immunodeficiency virus), the virus that causes AIDS, was identified in 1981. About 40 million people worldwide will be HIV infected by the year 2000. Of all U.S. males aged 24 to 39 in 1995, one in 139 whites, one in 60 Hispanics, and one in 39 blacks were HIV positive. AIDS is the number one killer of people aged 25 to 44, though that trend is changing at the end of the 1990s. More young people die from AIDS than are killed with guns. About 200 medical employees die each year from HBV. Many more are made sick.

How do we protect our human resources from these diseases? The first line of defense is awareness. A vaccine is available that is 90 percent effective for preventing HBV infection. No vaccine is yet available for HIV infection. Primarily, employees should avoid contact with infected blood and other body fluids and follow the OSHA bloodborne pathogens requirements.

Figure 18-2
Etiologic Agents in Building Components

Etiologic Agent	Location in Building
Actinomycetes, thermophilic	Dust in HVAC ductwork HVAC water spray
Amoebae, nonpathogenic	Microbial slime in reservoir
Clostridium Perfringens	Dust in HVAC ductwork
Cryptococcus	Feces of pigeons
Flavobacterium spp. or their Endotoxins	Aerosol from humidifier reservoir
Penicillin	Fan coils
Unknown	Stagnant humidifier water

Additional suspects: acinetobacter subtilis, bacillus subtilis, blastomyces, candida albicans, coccidioides, dermatophytes, histoplasma, and Legionella pneumophilia.

Body fluids that should be considered potentially infectious are blood (including menstrual blood), semen, vaginal secretions, saliva in dental procedures, and any body fluid visibly contaminated with blood. Other potential sources of infection are the fluids found around the brain, backbone, joints, chest, heart, abdomen, and fetus. Body fluids that are not considered infectious for HBV and HIV are feces, nasal secretions, saliva, sputum, sweat, tears, urine, and vomit.

HIV and Hepatitis

You cannot catch HIV through the air or by casual contact with infected persons. Nor can you catch HIV by sharing bathrooms and eating areas with infected people. *Vectors* (insects, mosquitoes, rats) do not transmit HIV from one person to another. As stated above, you cannot get HIV from non-infectious body fluids. Except for the presence of infected blood during an injury, the high risk personal factors for HIV infection are highly unlikely in the workplace. The high risk personal factors all involve sexual practices or sharing intravenous drug needles.

When HIV enters the body, the immune system begins working to kill the virus by producing HIV-antibodies. This battle can be waged for years. A few HIV are killed, but a few reproduce. Eventually HIV wins the war against the HIV-antibodies. Unfortunately, the immune system is destroyed in the process. That is when an opportunistic disease—AIDS—occurs.

The symptoms of AIDS are fatigue, fever, chills, night sweats, and swollen lymph glands in the neck, armpits, and groin. The swollen glands are a sign that the immune system is under attack. Pink to purple spots on the skin are another sign. In the mouth, white spots and sores are common. AIDS patients often have diarrhea without reason. They typically gain or lose weight for no reason. They develop a dry cough and short breath.

Hepatitis-B comes in a self-limited acute form and a chronic HBV infection. Hepatitis means inflammation of the liver. One-third of the victims of the acute form have no symptoms whatsoever. These patients do not realize they are sick until their blood is tested. One-third of acute hepatitis patients merely develop mild flu-like symptoms. Again, blood screening confirms their ailment. The final third of patients become extremely ill. They develop jaundice, dark urine, extreme fatigue, and nausea. One in one thousand becomes very sick and 85 percent of this latter group die from the illness.

From 6 to 10 percent of newly infected adults become long term carriers of HBV. Twenty-five percent of carriers have chronic persistent hepatitis, which is a mild, non-progressive liver disease. Another 25 percent develop chronic active hepatitis, which leads to cirrhosis of the liver. Chronic active hepatitis can also lead to primary hepacellular carcinoma, liver cancer, and death.

What is the risk of being occupationally infected with HBV or HIV? HBV infected needle puncture wounds occur from 6 to 30 percent of the time. HIV infected needle puncture wounds occur about 0.5 percent of the time. However, you could be infected on the first occasion of a puncture. How long do the viruses live outside the body? In a dry environment, HIV lives a few hours and HBV lives about seven days. So, old, dry blood is not necessarily safe unless you know it has been around for more than seven days.

Ingestion Potential

Enteric diseases are caused by *enteric* bacteria: *Escherichia coli, shigella,* and *salmonella. Enteric* refers to the intestines so these bacteria are generally ingested in order to cause infection.

Bacteria are single-cell plants requiring warm, moist, nutritious environments in which to breed. Extremely cold temperatures stop, or at least slow, growth of bacteria, but extremely hot, about 180°F, is required to kill bacteria. Bacteria flourish between 45° and 140°F.

Pathogenic bacteria cause human illnesses in two ways. Some bacteria are naturally infectious or poisonous to their host. Some bacteria are not virulent of themselves, but discharge a toxic metabolite called an *endotoxin* that poisons their victim.

Round or spherical bacteria are called *cocci.* Rod-shaped bacterium are called *bacilli.* A comma or spiral configuration is called a spirilla.

The gram-negative family called *enterobacteriaceae* is particularly dangerous in the workplace. They cause acute diarrheal disease, bacteremia, and urinary tract infections. E. coli is the predominant enterobacteria found in feces and normally lives in harmony with the human body. However, E. coli is opportunistic and will invade the bloodstream and contaminate injured sites in the body. Also, since the late 1960s, several strains of E. coli have mutated and possess virulent properties that the normal flora does not. Complications include urinary tract infection, meningitis, cystitis, pyelonephritis, bacteriuria, abscesses, and pneumonia.

Shigella is a non-motile gram-negative bacillus causing severe intestinal disturbances in humans and animals. Shigellosis indicates poor sanitation and personal hygiene practices.

Salmonella is a gram-negative bacillus that is also common in the human intestinal tract, but which is pathogenic. Salmonellae typically enter the host through ingestion of contaminated food and water. Species of Salmonella include *Salmonella enteritidis, Salmonella typhi, Salmonella paratyphi,* and *Salmonella choerae-suis,* causing gastroenteritis, typhoid fever, paratyphoid fever, and septicemia, respectively.

Bloodborne Pathogen Management

How do we know if someone has been infected? You cannot tell merely by looking. Anyone can be infected—and not even know it. So, treat all potentially infectious body fluids as if they are infectious. General methods of compliance with the bloodborne pathogen standard are given below:

- Communicate hazards to employees.
- Communicate protective and preventive procedures.

- Observe universal precautions.
- Use engineering controls where economically feasible.
- Use good work practices and housekeeping.
- Provide appropriate PPE.
- Develop and implement sound decontamination procedures.
- Manage wastes according to regulations.
- Maintain exposure records and documentation of follow-up activities.
- Make the Hepatitis B vaccination available to all employees with occupational exposure to bloodborne pathogens.

Exposure Control Strategies

Occupational exposure of bloodborne pathogens is regulated by the 1992 OSHA standard. Any reasonably anticipated contact skin, eye, or mucus membrane with potentially infectious material during the performance of employee's duties is covered. Who might be covered? First-aid responders, janitors, nurses, security officers, tattoo artists, dentists, firemen.

OSHA requires a written exposure control plan and adherence to universal precautions.

General Precautions

Employers must offer HBV vaccine to any potentially exposed employees. A post exposure evaluation and follow-up is also required. The hazards of bloodborne pathogens must be communicated to employees who may be exposed. Training must be conducted and records must be kept.

The training given to employees who may be exposed to bloodborne pathogens must be at no cost to them and must be conducted during working hours. Initial training was required when the standard took effect in 1992. Annual refresher training is also required. Training is required whenever a new exposure occurs. The training must be performance oriented and matched to the education, literacy, and language of the employee. Allowance must be made for an interactive question-and-answer session during the training. That implies that the trainer must be knowledgeable.

The required training includes making a copy of the OSHA standard accessible to the employees, but also making available an explanation suited to their level of education and abilities. The epidemiology of bloodborne pathogens must be discussed and understood, as well as the symptoms of exposure. The employees must understand the modes of transmission. They must also understand and be able to implement the company exposure control plan and know where to get a copy of the plan. They must be able to explain events that could lead to exposure and what personal protective equipment is available to prevent exposure. The HBV vaccine must be explained and the company must offer vaccinations to those who want them. Employees must know what procedures to follow once they discover they have been exposed. Finally, the training must cover signs and labels. Obviously, showing employees a talking head video will not suffice as training. A doctor, nurse, industrial hygienist, or other health care provider needs to be available for the training.

Housekeeping

Housekeeping and waste disposal are essential to the exposure control plan. Commercial disinfectants that are tuberculocidal are effective against both HBV and HIV. Household bleach diluted with water 1:10 is also effective. Dilutions as weak as 1:100 are effective, but stronger is recommended.

Work Practices

Treat all contact with blood and other bodily fluids as potentially infectious. Practice good hygiene. Wash your hands immediately after handling blood or body fluids.

Vaccinations

Offer exposed employees vaccinations, where such are available. Preferably, vaccinations are given before the fact. That way, at least the probability of protection is increased. Some exceptions may exist, so check with your physician. Hepatitis B vaccinations are more effective if received before exposure. They are nearly 90 percent effective after the fact.

PPE

The OSHA standard on Bloodborne Pathogens (29 CFR 1910.1030) does not contain specific requirements for protective clothing but it is a general requirement. NFPA 1999, *Standard on Protective Clothing for Emergency Medical Operations,* gives more detailed requirements for emergency medical responders. This situation fits most industrial situations involving biological hazards.

Gloves

Surgical thin latex gloves or similar should be worn when handling patients who are bleeding. Elastomeric gloves that cover the full hand and wrist are recommended as they provide a good barrier to biological hazards while offering essential tactility and dexterity. Thicker rubber utility gloves can be worn for cleanup and disposal. In a pinch, wear any blood-impervious glove to protect yourself. Remove gloves in such a way that you do not contaminate yourself. Some people are allergic to latex gloves, but nonallergenic gloves are available.

Airways

Plastic airways are recommended for giving resuscitation rather than direct mouth-to-mouth. Be sure to handle airways in such a way that you do not contaminate yourself.

Facewear

Masks, goggles, safety glasses, or respirators may be worn as conditions require. All mucous membranes on the face must be protected. This may require more than one facewear item. Tests of N95 respirators for bacteria of the size and shape of *Mycobacterium tuberculosis* showed filtration efficiencies of 99.5 percent or better. The respirators are effective protection when a good face seal exists.

Garments

Full body clothing may be used to protect the torso, arms, legs, and head. Partial body clothing such as jackets, smocks, aprons, sleeves, or shoe protectors may be used as necessary.

Airborne Pathogen Management

The management of airborne pathogens is similar to the management of air contaminated with particulate matter.

The least toxic effect of airborne pathogens is unpleasant odor. Controlling odor is not always an easy task. In addition to the fact that odor thresholds vary over six orders of magnitude (from parts per billion to parts per thousand), individual response to odor varies widely.

The source of the biological odor may be outside air used as fresh air makeup. Decaying indoor microbes can also produce odors. Moist filters contaminated with organic dust are one site where odoriferous microbes accumulate. Another site is window unit ventilators that are contaminated with dust, organic debris, and moisture.

Look for these conditions where biological odors are generated.

1. A moisture reservoir
2. A source of nutrients
3. Amplification (growth)
4. Dissemination pathway

Burton provides us with the most frequent causes of microbiological contamination:

- Contaminated cooling coils, humidifiers, and air washers
- High air velocity through wet coils
- Inadequate preventive maintenance
- Improperly designed air-handling equipment
- Inaccessible air-handling equipment
- Porous man-made fiber insulation inside air-handling and fan-coil ventilation equipment
- Contaminated sound liners in supply ducts
- Stagnant water in drain pipes
- Relative humidity greater than 70%
- Recirculation and accumulation of human-shed pathogens
- Dissemination of pathogens on human-origin aerosols (coughing, sneezing)
- Water-contaminated organic fiber furnishings (cotton, wool)
- Cool-water room humidifiers
- Water runoff from window into ventilators
- Flooding
- Outdoor air intakes located near external bioaerosol sources (cooling towers)

To minimize moisture buildup where biological pathogens may grow, follow these suggestions:

- Clean up spills at sinks, coffee areas, and break areas.

- Use exhaust fans in restrooms.
- Do not use room humidifiers.
- Repair roof leaks immediately.
- Remove standing water on roofs, in basements, in crawl spaces, and on internal surfaces.
- Provide adequate drainage for flat roofs.
- Do not allow condensed water to accumulate in air-handling systems, on perimeter windows, or on basement walls.

Reduce microbial contamination with the following preventive measures:

- Do not allow moisture to accumulate in occupied spaces.
- Do not allow moisture to collect in HVAC components.
- Remove stagnant water and slime from mechanical equipment.
- Use steam for humidifying.
- Do not use water sprays in HVAC systems.
- Maintain the relative humidity less than 70%.
- Use filters with a 50-70% collection efficiency.
- Discard microbial damaged furnishings and equipment.
- Remove room humidifiers.
- Conduct frequent preventive maintenance on building systems.
- Place pigeon screens on intakes and exhausts.

REFERENCES

Burton, D. Jeff. *IAQ and HVAC Workbook.* 2d Ed. Bountiful, UT: IVE, Inc., 1995.

Burton, D. Jeff. "Proper Action Can Help in Alleviating Most Indoor-Air Bioaerosol Hazards." *Occupational Health & Safety.* November 1990, p. 38.

De Chacon, Jeffrey R. and Norman J. Van Houten. "Indoor Air Quality." *The National Environmental Journal.* November/December 1991, pp. 16-18.

Eduard, Wijnand and Dick Heederik. "Methods for Quantitative Assessment of Airborne Levels of Noninfectious Microorganisms in Highly Contaminated Work Environments." *AIHA Journal.* February 1998, pp. 113-127.

Ghosh, S.K., Manisha R. Desai, G.L. Pandya, and K. Venkaiah. "Airborne Aflatoxin in the Grain Processing Industries in India." *AIHA Journal.* August 1997, pp. 583-586.

Nicas, Mark. "A Risk/Cost Analysis of Alternative Screening Intervals for Occupational Tuberculosis Infection." *AIHA Journal.* February 1998, pp. 104-112.

Qian, Yinge, Klaus Willeke, Sergey A. Grinshpun, Jean Donnelly, and Christopher C. Coffey. "Performance of N95 Respirators: Filtration Efficiency for Airborne Microbial and Inert Particles." *AIHA Journal.* February 1998, pp. 128-132.

Stull, Jeffrey O. "Selecting Protective Clothing for Emergency Medical Operations." *Occupational Hazards.* November 1993, pp. 45-47.

Windler, Kathleen K., Dennis F. Sigwart, and Janet Della-Giustina. "Enteric Disease." *Professional Safety.* July 1994, pp. 33-36.

19

Process Safety Management

OSHA's process safety management (PSM) standard sets forth requirements for the prevention of, or at least the minimization of, consequences subsequent to a catastrophic release of toxic, reactive, flammable, or explosive materials.

Is your plant covered by the OSHA PSM? You must implement PSM measures if your facility:

- Manufactures explosives or pyrotechnics
- Processes at least one highly hazardous chemical defined by OSHA in an amount that exceeds the listed threshold for that chemical
- Processes flammable chemicals in quantities greater than ten thousand pounds

History

What place names bring an image to mind that literally shouts PROCESS SAFETY MANAGEMENT? Bhopal. How about Three Mile Island and Chernobyl. After that, Institute, West Virginia. These are not the only process disasters that ever occurred, of course, but they are the most recognizable in the public imagination. Flixborough, England was a disaster that everyone cited before Bhopal. Pasadena, Texas was another that was close to home. All of these accidents were avoidable. Some of the catastrophes are summarized in Figure 19-1.

The 1990 amendments to the Clean Air Act required EPA and OSHA to implement regulations that address the public risk of highly hazardous chemicals.

Figure 19-1
Recent Process Disasters

Deaths	Injuries	Process	Company	Place	Date
6	0	Phosphorus	Tenneco, Inc.	Charleston, SC	June 1991
8	128	Nitroparaffin	IMC Fertilizer-Angus Chemical Co.	Sterlington, LA	May 1991
1	0	Ethylene Oxide	Union Carbide	Seadrift, TX	March 1991
2	0	Chemical	BASF Corp.	Cincinnati, OH	July 1990
17	0	Petrochemical	Arco Chemical Co.	Channelview, TX	July 1990
23	132	Plastic	Phillips Petroleum Co.	Houston, TX	October 1989
2,000	100,000	Methyl isocyanate	Union Carbide	Bhopal, India	1984
0	0	Methyl isocyanate	Union Carbide	Institute, WV	1986

Purpose of the Standard

The OSHA Process Safety Management Standard applies to all businesses that produce, store, transport, or use highly hazardous chemicals at greater than threshold levels. The standard applies to you if you deal with any one of the toxic chemicals OSHA lists in the standard at or above the threshold quantity. It also applies to you if you handle 10,000 pounds or more flammable fluids.

The purpose of the standard is to provide assurance that the risks of fire, explosion, and toxic release are prevented.

EPA Chemical Accidental Release Prevention

The U.S. Environmental Protection Agency (EPA) also has a rule that concerns the catastrophic release of chemicals to the environment. The chemical accidental release prevention (CARP) standard, issued under Section 112(r) of the Clean Air Act, expands risk management responsibilities for companies that have process safety management. EPA requires the identification of a worst-case scenario. The EPA required risk management plan, which must be made available to the public, includes:

- Summary of hazard assessments
- Worst-case scenarios
- Description of major hazards
- Summary of emergency response programs
- Description of the management system in use
- Certification of accuracy and completeness of the plan
- Five-year history of significant releases from the facility

Similarities and differences between the two standards are compared in Figure 19-2.

Overview of OSHA Requirements

As you might expect, the PSM standard requires that you develop a written plan, implement it, and allow employees to participate. The essential requirements of PSM are:

1. An initial process hazard analysis
2. A process hazard team
3. A system that effectively processes the team's findings
4. An update of the process hazard analysis every fifth year
5. Retention of process hazard analysis

The first process hazard analysis was allowed to be conducted in phases but the date for 100 percent completion, May 26, 1997, has passed. For information, the original dates for phasing in the hazard analysis were:

- May 26, 1994 - PHA 25% complete
- May 26, 1995 - PHA 50% complete
- May 26, 1996 - PHA 75% complete
- May 26, 1997 - PHA 100% complete

Figure 19-2
Chemical Process Safety: EPA and OSHA

EPA's risk management program and OSHA's process safety management standard are similar.

Nine sections are found in both EPA regulation and OSHA standard:

1. Process hazard analysis
2. Process safety information
3. Standard operating procedures
4. Worker training
5. Prestartup safety review
6. Maintenance (EPA) or mechanical integrity (OSHA)
7. Management of change
8. Safety audits
9. Accident investigation

EPA regulation only:

- Hazard assessment
- Emergency response
- Registration, submission, and auditing of risk management plan
- Defining the program management system and naming person responsible

OSHA standard only:

- Worker consultation
- Hot work permits
- Contractor safety
- Trade secrets

Role of HAZOP

Process hazard analysis is the foundation of the Process Safety Standard. Process hazards must be systematically identified and analyzed. Methods for doing this were discussed in Chapter 5. These include failure mode and effects analysis (FMEA), fault tree analysis, hazard and operability studies (HazOP), what-if analysis, and checklists.

Critical Compliance Management Issues

Good effective operating procedures are essential to process safety management. Written safety procedures must be developed for each process. An information package must be compiled that describes the hazards of each process operation. Part one should include the explanatory section. Information on equipment design makes up part two. In the third part, process documentation is compiled. When bringing a new contractor on board, you need to evaluate his safety record before hiring him and hold him accountable for following your safety procedures after you hire him. Ensure that your employees participate in all process safety management activities. Below are some other issues in process safety management.

Training

The standard requires formal process safety training for employees who operate covered processes. Rules are specified for initial training, refresher training every three years, and training documentation on each employee.

Each employee who will be assigned to operate a covered process must receive an overview of the process and training on the operating procedures and safe work practices. Safety and health hazards must be communicated during the training. Emergency operations, including shutdown procedures, must also be covered.

The risk of a process incident increases immediately after a responsible employee's break.

Mechanical Integrity

Corrective maintenance alone is an inadequate safety strategy for hazardous processes. Testing and inspections must be conducted on critical equipment and vessels with regular frequency. Equipment inadequacies must be corrected before startup or as soon as possible after discovery. Regular preventive maintenance (PM) is another requirement. However, avoid over commitment on frequency of inspections, testing, and PM. If you do not do PM because you are backlogged, you have no PM program. Document when and by whom inspections, testing, and PM is done. Also document findings of inspections and testing and problems uncovered during PM. Finally, make sure your paper trail includes the disposition of these problems by logging when, where, and who took corrective action. Who verified the corrective action and when?

Keep recommended spare parts on hand. Review operating and maintenance manuals for other recommended maintenance materials.

Management of Change

The success or failure of process safety management depends on how change is managed. Before making changes to chemicals, equipment, technology, procedures, or facilities—not counting replacements in kind—potential impacts on process safety must be considered in a systematic and rational manner. Management must sign off on the change before it occurs.

What type of change is covered by the standard? The rules apply only to changes that affect a covered process. In other words, any change whatsoever, except for replacement in kind. The first question to ask then is this: does the proposed change trigger management of change requirements? Establish a procedure that causes all process changes to be considered in light of management of change rules. Consider the technical basis for the change, impact on safety and health, and whether modifications to operating procedures will be required. Also, how much time will it take to implement the change? What authorizations are required for the change?

Incident Investigation

Employers must investigate each incident that ends in a catastrophic release, or could reasonably have been a catastrophic release if the situation had been slightly different. Re-examine the standing procedure for investigating such incidents to ensure that it is consistent with the specific requirements of the process safety management standard. Put together a knowledgeable team of experienced personnel to investigate incidents. Investigation must be initiated within 48 hours of the incident. A detailed report should document and analyze the factors that contributed to the accident. Suggestions on eliminating problems and preventing recurrence should also be included.

Emergency Planning and Response

This section of the standard incorporates existing emergency plan requirements found at 29 CFR 1910.38(a) and emergency response requirements found at 29 CFR 1910.120. Employee training concerning emergency alarms must include process emergencies as well as natural disasters that could impact the process. Also, cover evacuation procedures and routes, assembly areas, head count procedures, emergency response duties of the fire brigade,

HAZMAT team, first aid/CPR team, and process shutdown responsibilities. Training should also cover the availability and use of respirators. The compliance officer will want to know how often emergency drills are conducted and documented. He or she would also like to see documentation that site emergency drills were coordinated with community emergency drills.

Compliance Audits

Every third year you must conduct an audit of how well your plant complies with this procedure. Since it is a specific requirement of the standard, an inspecting compliance officer may ask to see the documents. Therefore the usual legal games of having an attorney initiate and supervise the audit will not suffice to protect the information under attorney-client privilege. If you use that strategy you will still be out of compliance since you will have no documentation to present to the compliance officer.

Specific Process Issues

Some technical issues deserve our special, though brief, attention as they occur in several types of industries.

Carbon Adsorption Systems

In 1997, EPA issued an alert to owners of carbon adsorption deodorizing systems. These systems may pose a fire hazard when used to deodorize certain substances. If proper procedures are followed, these fires can be prevented. Particularly hazardous is the deodorization of crude sulfate turpentine in the pulp and paper industry.

A fire and explosion occurred in 1995 at a chemical terminal facility. The incident involved three tanks of crude sulfate turpentine that were connected to drums of activated carbon for deodorizing. The resulting fire and explosion damaged other storage tanks and toxic gases were released, forcing a large-scale evacuation of area residents. Previous crude sulfate turpentine deodorizing fires had not had the disastrous effects of the 1995 fire and explosion. Serious effects are not expected unless the fires get beyond the activated carbon containers and spread to tanks containing flammable or combustible substances.

Activated carbon is used to adsorb vapors to control releases to the atmosphere. The adsorption of certain chemicals onto the surface of the activated carbon particle releases heat, sometimes in large amounts. Hot spots can develop in the carbon bed. Organic sulfur compounds, such as mercaptans, plus ketones, aldehydes, and some organic acids release heat when adsorbed onto activated carbon. High vapor concentrations of organic materials can also lead to the hot spots mentioned. A fire hazard is created when flammable vapors are involved. This occurs when the hot spot temperature reaches the autoignition temperature of the vapor and oxygen is present to support combustion. This hazard may increase at night due to a vacuum created in the system when temperatures drop. The vacuum may pull air into the system. The mixture of air and a high concentration of vapors generated during the day can have disastrous consequences.

Follow the manufacturer's instructions for design and operation of activated carbon adsorption systems. A qualified engineer or technician should supervise the design, construction, and operation of carbon adsorption systems. Determine the composition of the vapors that contact carbon and respect the manufacturer's warnings about potentially hazardous reactions. If the vapor may contain sulfur, ketones, aldehydes, or organic acids, of if an organic vapor will have high concentrations, appraise the potential for development of hot spots on the carbon.

Test the action of the vapors on carbon for potential heat release before putting the system into service. If a potential for fire is indicated by test or evaluation of compounds, design the system so that air does not enter the system over the carbon bed. A vacuum breaker on the storage tank will prevent air from entering. Separate the carbon containers from containers of flammable or combustible substances. Install heat control measures where potential for high concentrations of organic vapors exist. Either dilute the inlet air, or time weight the inlet concentration, or pre-wet the carbon.

Conduct frequent visual examinations of the carbon containers for hot spots. Ensure that safety devices are still installed and operating.

Pressure Vessels

Improperly operated or maintained pressure vessels have failed catastrophically, leading to the death and injuries of workers and other persons. Even when the contents were not hazardous in themselves, serious damage has ensued from ruptured pressure vessels. A 1996 accident killed three workers and several other persons when a high-pressure vessel containing air and water ruptured. The design working pressure of the vessel was 1,740 pounds per square inch (psi). At the time of the rupture its operating pressure was between 2,000 and 3,000 psi. The vessel was old and had developed a pinhole leak, which was repaired, but not in adherence to recognized codes for pressure vessels. One month later, the weld failed catastrophically. Pieces of the vessel were launched as missiles through the roof of the building. Some of the shrapnel weighed from 1,000 to 5,000 pounds. A few pieces landed a half-mile from the plant. Travelers on a nearby highway and passengers on a commuter railway narrowly escaped injury. The plant itself was extensively damaged. Telephone and electrical services were interrupted for several hours in that part of the state.

Operation of pressure vessels above their maximum allowable working pressures can lead to such accidents. A worse situation is operating above the test pressure of a vessel. The improper sizing or pressure setting of relief valves or rupture disks may lead to failure. Faulty maintenance and failure to test regularly may lead to improper operation of a relief device. A vessel may fail due to fatigue from repeated pressurization, general thinning from corrosion or erosion, localized corrosion, stress corrosion, cracking, embrittlement, holes, or leaks. Failure to inspect frequently has allowed problems to set in and go unnoticed until too late.

Welding or annealing a pressure vessel is critical, as improper repair may lead to embrittlement and further weaken the vessel instead of strengthening it. Repair welds must be made while the vessel is shut down and depressurized. Water or other liquids in the vessel at the time of welding or annealing may quench the steel and embrittle it.

Exothermic reactions or polymerizations can lead to overpressuring and failure. Exposure to fire may also cause a vessel to fail.

Requirements for pressure vessels vary widely from state to state. Many states have a boiler law, but some do not. Boiler laws typically specify adherence to either the American Society of Mechanical Engineers (ASME) codes or to the National Boiler Inspection Code (NBIC).

Facilities should survey their vessels, review pertinent history and data to find hazards, and prevent vessel rupture or catastrophic failure. Vessels designed to operate above 15 psi should be designed, fabricated, and constructed according to the ASME code or other suitable code. The vessel code should be labeled or stamped somewhere on the vessel. The operating pressure and size of the vessel should also be on the nameplate. The design must consider concerns about temperature and characteristics of vessel contents, such as toxicity, corrosivity, reactivity, or flammability. Anytime vessel contents are changed from those it was designed for a risk assessment must be performed to determine if the vessel is still safe for the new materials.

In cases where the vessel cannot be constructed in full compliance with the appropriate code, a copy of drawings, calculations, service conditions, welding procedures, welder qualifications, welder performance tests, and a professional engineer's certification of these documents must be forwarded to the NBIC. NBIC will examine the submission and make a recommendation to the state as to whether the pressure vessel is approved for service. If an unmarked vessel is found, or salvaged one is to be brought into the state, a similar submission to NBIC, plus the vessel's repair history will be necessary to get a recommendation to the state for service. In states that do not have a pressure vessel law, the same submission should be made to a pressure vessel-consulting engineer or to an authorized inspector for safety review.

Vessel Maintenance and Repair

Vessels should be maintained, inspected, and repaired according to NIBC and/or API (American Petroleum Institute) 510. Ratings and settings of relief devices should be labeled. The relevant state authority should conduct an annual inspection. The safety devices must also be tested regularly. Monthly is recommended. The vessel itself should also be inspected externally by the owner on a monthly basis. Internal inspections should be conducted at least annually. Internal inspection requires confined space entry, which means the vessel must be taken out of service, cleaned, and prepared for the entry. The purpose of these inspections is to discover general thinning of walls due to corrosion or erosion, to find localized corrosion, stress corrosion cracking, embrittlement, pits, holes, leaks, or any other defects that could weaken the vessel. Records of thinning must be kept in the vessel's repair history folder. Follow-up corrective action is imperative before placing the vessel back into service. As a vessel approaches the end of its useful life (a date dependent on the design thinning rate) the interval between inspections must be shortened in order to take the vessel out of service permanently before it becomes dangerous.

When a vessel is to be repaired, the plan of repair, including welding techniques and safety tests, must be reviewed and approved by a certified or authorized inspector. A qualified welder must do the welding. Welding performance qualification tests should be insisted on and approved by an inspector. When the repair is complete, the vessel must be tested. If necessary, the vessel needs to be down rated, with appropriate changes in operating conditions and relief device settings. These changes must be stamped or posted on the vessel. Large temperature differences across the vessel wall must be avoided to prevent embrittlement or metal stressing.

Nondestructive testing of vessels may include such techniques as radiographic inspection, ultrasonic inspection, liquid penetrant testing, magnetic particle testing, eddy current testing, visual checks, and leak testing.

If exothermic reactions are carried out in a pressure vessel, the vessel must have an emergency relief system designed to handle runaway reactions. A runaway reaction ruptured a vessel at the ICMESA plant near Milan, Italy in 1976, scattering dioxin compounds over several square miles of countryside.

Storage Tanks

Aboveground, atmospheric storage tanks have failed catastrophically when flammable vapors in the tank exploded and either broke the shell-to-bottom or side seam. The tanks thus rip open and tanks or large pieces of tank are propelled as missiles through the air, potentially causing harm to workers and the community. Tank materials are spilled, potentially creating additional problems.

Properly designed and maintained storage tanks break along the shell-to-top seam, limiting any fires to the damaged tank. The contents are not spilled.

In 1995, while workers were welding the outside of a tank, combustible vapors inside two large, 30-foot diameter by 30-foot high storage tanks exploded. The shell-to-bottom seams split, hurling the tanks skyward. They landed more than 50 feet away—a remarkable distance considering their weight. The flammable liquid they had contained was released and also ignited. A massive fire caused five deaths and several serious injuries.

In 1994, a worker was grinding on a tank that contained a petroleum-based sludge. The tank was propelled upward. Seventeen workers were injured and the explosion forced the tank contents over a containment berm into a nearby river.

In another incident, workers were welding the outside of an empty tank in 1992 when residual vapor in the tank exploded and propelled the tank like a missile into a nearby river. Three workers were killed and one was injured.

Tank design, inspection practices, and maintenance practices have been directly linked to these catastrophic tank failures.

An accident where the shell-to-bottom seam ruptures is more common in older tanks. Steel storage tanks built before 1950 do not always conform to current erection standards, particularly for explosion and fire venting. Atmospheric tanks that hold combustible or flammable liquids must be designed to fail at the shell-to-top or -roof seam to prevent a tank explosion from propelling the tank upward or splitting down the side. The American Petroleum Institute has published API-650, "Welded Steel Tanks for Oil Storage." Additional codes and standards published by API, ASME, NFPA, ANSI, American Society of Nondestructive Testing (ASNT), American Welding Society (AWS), and UL all address tank design, construction, venting, and safe welding.

Controlling Vapor Explosions

Poorly maintained tanks, tanks that are rarely inspected, or tanks that are repaired without attention to design risk catastrophic failure should tank vapor explode. The shell-to-bottom seam can weaken through corrosion. The practice of placing gravel or spill absorbents around the base of a tank may contribute to bottom corrosion. Over the years, the bottom of the tank becomes lower than ground level due to buildup around it. Moisture trapped along the bottom of the tank also causes corrosion and weakens the bottom and the shell-to-bottom seam. Modifications to the roof, or attachments that are added to the roof over the years may make the shell-to-roof seam stronger relative to the shell-to-bottom seam. This situation may go unnoticed and is very dangerous.

Generation of combustible vapors can happen for the darnedest reasons. Where pure flammable liquids are stored, everyone tends to be vigilant. However, the storage of any liquid or sludge with a combustible component may produce a flammable vapor directly or by means of a chemical reaction. These are the situations that surprise tank owners in dramatic ways. Comments such as "I never knew that stuff could burn" are not uncommon. Sludge tanks, slop tanks, and oil/water mixture tanks that are open to the atmosphere are particularly vulnerable because combustible vapors can accumulate outside as well as inside the tank. Finally, the most insidious situation is the empty tank. An empty tank is usually full of explosive vapors If we had every life back that was lost when an empty tank exploded we could populate a small town.

In every case when these disasters occur, the potential for disaster goes unrecognized. Equipping tanks containing combustible vapors, or potentially containing these vapors, with flame arresters prevents external fires from reaching the vapor space inside the tank. Vapor control devices limit the hazardous vapor emissions from the tank. Forewarned is forearmed. Pay attention to your storage tanks.

Combustion occurs as combustible vapors escape from containment and mix with air in the presence of an ignition source. All possible ignition sources must be isolated from potential combustible vapors. Ignition sources include welding equipment, maintenance equipment that may cause sparking such as grinding or wire brushing, static electricity, lightning, hot work (welding, cutting, brazing), and electrical equipment that does not conform to the National Electrical Code (NFPA-70) for hazardous locations.

Danger of explosion is increased when tanks are not spaced with NFPA-30 requirements. To mitigate potential consequences to workers, as well as to other tanks, proper secondary containment should be considered.

Recognizing Risks

What storage tanks are at risk? Atmospheric storage tanks that do not meet API-650 or equivalent code requirements and contain flammable liquids or liquids that may produce ignitable vapors are risky. So are tanks where corrosion occurs around the base. Steel tanks where the base is in direct contact with the ground and exposed to moisture accumulation are potentially hazardous. Tanks or piping that have potentially weakened or defective welds are risky. Storage of mixtures of water and flammable liquids where the water phase is at the bottom may cause the tank bottom corrosion rate to exceed the design rate. Verify that the designer considered this situation. Tanks containing combustible vapor but not equipped with emergency venting, flame arresters and vapor control devices are candidates for explosion. Finally, go over your tank farm with a fine tooth comb for possible ignition sources. The one you overlook will be the one that burns you, so to speak.

API-653 gives inspection guidelines and procedures for periodic inspections and testing. Following these guidelines is particularly critical for older tanks that may lack safety features. Written documentation for the tests and inspections by certified tank inspectors is highly recommended and required in many cases. Also, review procedures for pressure testing, welding inspections, corrosion testing, and metal fatigue tests to ensure your tank management process is providing an adequate degree of protection.

Welding should be conducted in accordance with API-650 or equivalent code requirements. The welder should be prequalified by test welding. After welding, a certified expert should conduct an inspection using radiographic or magnetic particle examination or other suitable method.

Hot work safety procedures should be strictly adhered to, as nearly all tank explosions have occurred as a result of hot work. A fire watch with augmented extinguishing capability should be posted during hot work. The atmosphere should be tested for explosivity prior to commencement of the hot work. Cover and seal all tank drains, vents, manways, and open flanges. Seal sewers to prevent vapor migration.

Review OSHA standards and NFPA codes for compliance with ignition source identification and control. All electrical equipment in the vicinity must conform to NFPA-70. Ground tanks to dissipate static charge. Allow workers to use only non-sparking tools and equipment in potentially flammable atmospheres. Train employees to avoid creating sufficient heat or causing sparks where flammable vapors could be ignited.

Lightning

Lightning strikes have caused disasters when storage or process vessels or equipment containing flammable substances were hit. Refineries, bulk storage and transfer plants, processing sites, and even manufacturing plants with small storage tanks have been devastated by lightning initiated fires and explosions.

In 1996, lightning struck a storage tank containing between three and four million gallons of gasoline. A portion of the tank lid shot up into the air and came down on its side inside the tank. The heat ignited the gasoline, which did not spill out. The massive fire that was created burned for 28 hours before firefighters could put it out. Meanwhile, the firefighters had to spray about twenty nearby tanks with water to prevent secondary explosions. Nevertheless, the fire fatigued four additional tanks. No workers or firefighters were injured in this incident, but nearly two hundred nearby residents were evacuated from their homes as a safety precaution.

Lightning struck a fiberglass storage tank in 1992, setting off a series of secondary explosions. Toxic vapors and thick smoke settled over the town. More than one thousand people were evacuated with minor injuries including nausea, skin irritation, and shortness of breath.

The roof of a storage tank of diesel fuel was struck by lightning in 1977. Roof fragments struck and ignited two gasoline tanks. The tanks and gasoline were destroyed in the resulting fire. The loss of property plus the cost of cleanup amounted to $8 million.

Lightning causes more deaths, injuries, and damages than all other climate-related disasters combined. This includes all hurricanes, tornadoes, and floods. NFPA estimates over 26,000 lightning related fires occurred between 1989 and 1992. Property damage from these fires amounted to billions of dollars. The Insurance Information Institute reports that about 5 percent of all insurance claims paid for property damage are lightning-related.

Lightning is static electricity with extremely high-energy potentials. Extremely high temperatures are generated when lightning hits. Lightning strikes are random, erratic, and not well understood. It is known that lightning tends to strike the tallest object on the ground in the path of its discharge. Parts of structures that project above their surroundings are most likely to be struck by lightning. Structural parts such as vents, roof edges, wind socks, or weather vanes are highly susceptible.

The lightning bolt typically follows a conductive path to ground. Lightning may enter a structure by direct strike, through a metal object pointed up and out from the main structure, through a nearby tree, horizontally through another tall object nearby, or by conduction through power lines.

The National Severe Storm Laboratory (NSSL) reports that the frequency of lightning decreases towards the northwest. The highest frequency of lightning occurs between Tampa and Orlando, Florida. High frequency also occurs along the Gulf of Mexico coast westward towards Texas. The western mountains experience a relatively high frequency of lightning. The Atlantic coast in the southeast and inland from the Gulf of Mexico (Mississippi, Alabama, Georgia, and north Florida) experience lightning at a fairly high frequency. The least cloud-to-ground lightning occurs along the Pacific West Coast.

Lightning Protection

Lightning protection provides a controlled path for the current to follow to ground. A low impedance path prevents the lightning current from taking alternative routes that could be more destructive. A lightning rod to a ground path offers low impedance. Most metals are good lightning rod materials. A continuous path must be provided from ground to the air terminal (the lightning rod). All metal parts must be interconnected or bonded so they have the same electrical potential. If any substantial potential difference develops along the path, side-flashes or sparks will occur, which may be devastating where flammable vapors accumulate.

For flammable liquid storage tanks consider air terminals, bonding, grounding, conductors, masts, overhead wires, or other types of protection recommended by the National Lightning Safety Institute (NLSI). The bonding should be thermal, not mechanical, where feasible. NLSI recommends frequent inspections with resistance measurements of mechanical connectors.

The configuration of the grounding system is vital. Design depends on soil conditions, building construction, and the presence of other underground conductors. Grounding systems consist of either driven ground rods, plates, or a counterpoise, which is a buried cable that encircles the site. Low impedance materials that can withstand a lightning strike must be used. The resistance of the grounding system should be five ohms or less per linear meter of material.

Testing, inspection, and continuity measurement are essential for the maintenance of a lightning grounding system. Do not allow grounding cable connections on tanks to get painted. Look for corrosion, dirt, or accumulation of dead insect bodies that can create a path for lightning other than to ground. Place ohmmeter leads from cable to tank and clean the connections if the reading is not very low.

Lightning strikes may also damage control systems and electrical circuitry up to two miles away. Data may be corrupted; false signals may be generated; sensitive electrodes may be damaged immediately or fail in the next few hours or days. These situations can cause process upsets or unexpected releases of material to the environment. To prevent this, install surge protection for sensitive electronics. Place the surge protection device where it is easily inspected and replaced. Replacement of surge control must be the only alternative considered when a lightning strike occurs.

REFERENCES

Auger, John E. "EPA's Chemical Risk Management Exceeds OSHA's Process Safety Mandate." *Occupational Health & Safety.* April 1994, pp. 65-67.

Brown, J.D. "Preventing Accidental Releases of Hazardous Substances." *Professional Safety.* August 1993, pp. 23-27.

EPA 550-F-97-002a. *Rupture Hazard of Pressure Vessels.* United States Environmental Protection Agency. Office of Solid Waste and Emergency Response. Chemical Emergency Preparedness and Prevention (CEPP) Office. May 1997.

EPA 550-F-97-002b. *Catastrophic Failure of Storage Tanks.* United States Environmental Protection Agency. Office of Solid Waste and Emergency Response. Chemical Emergency Preparedness and Prevention (CEPP) Office. May 1997.

EPA 550-F-97-002c. *Lightning Hazard to Facilities Handling Flammable Substances.* United States Environmental Protection Agency. Office of Solid Waste and Emergency Response. Chemical Emergency Preparedness and Prevention (CEPP) Office. May 1997.

EPA 550-F-97-002e. *Fire Hazard from Carbon Adsorption Deodorizing Systems.* United States Environmental Protection Agency. Office of Solid Waste and Emergency Response. Chemical Emergency Preparedness and Prevention (CEPP) Office. May 1997.

Kearney, Kevin E. "Process Safety Management: An Overview of 1910.119 from a Hazardous Waste Facility Perspective." *Professional Safety.* August 1993, pp. 16-22.

Lastowka, James A. "OSHA's PSM Standard: Key Lessons from the First Five Years (Part II)." *Occupational Hazards.* August 1997, pp. 59-64.

Mansdorf, Zack. "Analyzing Process Hazards." *Occupational Hazards.* September 1994, pp. 11-13.

Roughton, Jim. "Process Safety Management: An Implementation Overview." *Professional Safety.* August 1993, pp. 28-33.

Wortham, Sarah. "Process-Safety Management Comes of Age." *SAFETY+HEALTH.* February 1995, pp. 48-52.

20

FIRE, EXPLOSION, AND DISASTER MANAGEMENT

How a plant prepares itself to respond to an emergency dictates the probability of saving lives, avoiding other injuries, and preventing property damage. A disaster can happen at any given facility at any given time. When this happens, some real number of deaths and injuries will occur. (Actual deaths and injuries in real accidents are summarized in the previous chapter on process safety.) Some measurable amount of property damage will also occur. These things are inevitable. Perhaps if you're lucky this disaster will not happen on your watch, but it will happen. Rather than despair, throwing up our hands and saying, "what can be done?" we need to be prepared to mitigate and alleviate the circumstances of a disaster before it happens.

The essential first step to emergency management is to appoint an emergency response coordinator and at least one back-up coordinator. Many plant managers designate some overworked junior. That would be otherwise acceptable, except that the emergency response coordinator must have the absolute authority to expend whatever resources are available and needed to mitigate the consequences of the emergency and to cause the emergency situation to end as soon as possible. If this means utilization of community resources, the coordinator must have a blank checkbook—no questions asked. If you are the emergency coordinator and your management does not give you this much authority, resign at once. A federal marshal may have a pair of handcuffs with your name on it.

Insist that whoever management assigns to this responsibility also be given unlimited authority to combat and mitigate the disaster aggressively. The coordinator must be able to ensure that outside emergency services such as medical aid and local fire departments will be called when necessary without undue hesitation. The coordinator must also have the authority to shut down operations when necessary without being second-guessed by more senior executives. This person must direct all emergency activities from a central location, including the evacuation of nonresponding personnel.

As soon as this person is assigned, he or she must determine what emergencies may happen and supervise the development and implementation of procedures that address them.

Fire Safety

Flash point is the lowest bulk temperature at which a pool of liquid gives off sufficient vapor to produce a momentary flame in the presence of an ignition source.

The material having the lower flash point is the more hazardous with respect to fire and explosion. *Boiling point* is the temperature at which the vapor pressure of a liquid just exceeds atmospheric pressure. This means the atmospheric pressure can no longer keep the material in the liquid phase and it vaporizes rapidly. Thus, the boiling point is a relative indicator of a liquid's volatility. Low boiling liquids are more hazardous than high boiling liquids. Liquids having flash points of 100°F or higher are called *combustible liquids.* If the flash point of a liquid is less than 100°F, it is called a *flammable liquid.*

Caution

Other English speaking countries refer to a low flashing liquid as an inflammable liquid. Do not confuse this expression with the American term nonflammable.

Flammable and combustible liquids are further divided into classes as shown in Table 20-1.

When petroleum liquids, such as crude oil or fuel oil, burn, Figure 20-1 lists the contaminants of concern.

Hot Work Permit

Hot work is welding, brazing, soldering, cutting, grinding, thawing pipe, or applying roofing with a torch. Hot work is the deliberate introduction of a nonroutine source of ignition to the workplace. Fire and explosion damage resulting from hot work incidents tends to be big. Several refineries have experienced deadly explosions due to hot work performed in the wrong

Table 20-1
Classes of Flammable and Combustible Liquids

Class I-A Flammable Liquids:	flash point <73°F boiling point <100°F
Class I-B Flammable Liquids:	flash point <73°F boiling point ≥100°F
Class I-C Flammable Liquids:	flash point ≥73°F but <100°F
Class II Combustible Liquids:	flash point >100°F but <140°F
Class III-A Combustible Liquids:	flash point ≥140°F but <200°F
Class III-B Combustible Liquids:	flash point ≥200°F

Figure 20-1
Contaminants in Petroleum Fire Plumes

Contaminant/Mean Concentration (μg/m³)

Total Hydrocarbons (as n-Hexane)/1,217
Total Suspended Particulate/128,000
Acenaphthene/1.4
Acenaphthylene/43.1
Acetaldehyde/37.7
Acetone/22.0
Acrolein/20.3
Anthracene/2.60
Benzaldehyde/59.3
Benzene/520
Benzo(a)anthracene/2.24
Benzo(b&k)fluoranthenes/4.87
Benzo(a&b)fluorenes/1.08
Benzo(g,h,i)perylene/5.4
Benzo(a)pyrene/3.37
Carbon Dioxide/1,994,000
Carbon Monoxide/33,500
Chrysene/2.37
Crotonaldehyde/7.5
Decane/24
2,5-Dimethylbenzaldehyde/17
Dodecane/37.7
Ethylbenzene/12
Ethyltoluenes/28
Fluoranthene/9.03
Fluorene/4.58
Formaldehyde/177.3
Hexadecane/53.3
Indeno(1,2,3-cd)pyrene/2.42
Isoveraldehyde/13.9
Methyl i-Butyl Ketone/14
Methyl Ethyl Ketone/6.23
1-Methylfluorene/0.453
Naphthalene/84
Nonane/17
i-Octane/11
n-Octane/8.8
Phenanthrene/16.5
Pyrene/8.58
Sulfur Dioxide/7,500
Tetradecane/98.3
p-Tolualdehyde/14
Toluene/53
1,2,4-Trimethylbenzene/41
n-Undecane/30
Xylenes/31

place or at the wrong time. Cutting work on roof HVAC units has set more than one manufacturing facility on fire.

To get a handle on hot work, establish a chain-of-command. Management appoints a fire safety supervisor to be responsible for the day-to-day activities of a fire safety program and gives him or her authority to administer and manage the program. The fire safety supervisor establishes a permitting system and ensures that employees and contractors who perform hot work do the job correctly.

Certain areas in the plant may be suited for routine hot work and do not need a permit for such work. Anyone with a need to perform hot work outside of one of these designated areas must obtain permission by getting the fire safety supervisor to sign a permit. Can the hot work be avoided somehow? Can flame cutting be done with a saw? Can a weld be replaced by nuts and bolts?

If the hot work must proceed, can all combustibles be removed from the area? Can the hot work be moved to one of the routine hot work areas? Presumably these areas are either already or readily made fire safe.

If the hot work still must proceed in other than a hot work designated area, the precautions outlined in Figure 20-2 must be taken.

Figure 20-2
Fire Safety Precautions for Hot Work

1. Sweep floor clean for 35 feet
2. Clean up and remove grease and oil
3. Cover combustible material with tarpaulin, unless it is moved more than 35 ft away.
4. Remove flammable liquids more than 35 feet away.
5. Eliminate explosive atmospheres.
6. Halt hazardous or explosive processes until work is complete.
7. Cover all walls and floor openings.
8. Seal ductwork and duct openings with metal covers or tarpaulin.
9. Close doors and fire doors within the 35-foot radius.

Smoke

Three-quarters of structural fire deaths are attributed to smoke inhalation. Smoke management by design is essential for life preservation and property conservation in building fires, which can grow fast and generate considerable smoke. Automatic sprinklers, standpipe systems, elevator fire service modes, and pressurized stairwells are part of good smoke management strategy. Tall buildings can produce a significant chimney effect to pull a draft and smoke from the fire source.

The standard practice for smoke control systems is NFPA 92A. The chief means of controlling smoke movement is by generating pressure differences across partitions, floors, and other building components. For a 1,700°F 4MW fire in a sprinklered building, the required pressure differential is 0.05 inches of water gauge (iwg). Smoke flow follows overall air flow through a building. Smoke can spread through openings such as cracks, pipe penetrations, ducts, and open floors. Stairwells and elevator shafts also spread smoke, but this can be countered by pressurizing these spaces.

Large volume spaces are a little different. Smoke generated in large open spaces is buoyant and rises in a plume above the fire. Buoyant smoke either reaches the ceiling or forms an inversion layer. The smoke supply rate from the plume is approximately equal to the air entrainment rate into plume below the smoke layer interface. Equilibrium is reached in the layer by exhausting smoke at the same rate as is supplied. Makeup air is required to keep smoke exhaust fans operating effectively.

The smoke generation rate, M kg/s, is

$$M = 0.188P^{3/2}y \qquad [20.1]$$

where

P = perimeter of the fire, m

y = the distance from the floor to the smoke layer, m

The flame height, H_f in meters, is

$$H_f = 0.011(kQ)^{0.4} \qquad [20.2]$$

where

k = wall effect factor

Q = heat release from fire, W

The wall effect factor, k, equals 1 if no walls are nearby. The temperature of the plume gas is determined by calculating the temperature rise,

$$\Delta T = 0.22(kQ)^{2/3} H^{5/3} \qquad [20.3]$$

where H is the distance from the top of the fuel packet to the ceiling, m. Natural draft from the chimney effect, D, of tall buildings is determined by

$$D = 2.96 HB\rho\left(\frac{1}{T_o} - \frac{1}{T_i}\right) \qquad [20.4]$$

where

H = stack height, feet
B = barometric pressure, inches Hg
ρ = density of air, lb/ft^3
T_o = outside air temperature, R
T_i = inside air temperature, R

If air flow is used to control smoke by creating a velocity to force smoke back, the required velocity, V in ft/min, is calculated as

$$V = 5.68 E/W \qquad [20.5]$$

where

E = energy release into the corridor, Btu/hr
W = corridor width, feet

Explosion Safety

Contrary to popular opinion, explosion-proof equipment and electrical circuitry does not exist. If you find anything that is actually labeled "explosion-proof," call me—I'd like to see it. The *National Electrical Code* defines *classified* locations as any place where ignitable atmospheres may exist. The three location classes are based on the physical properties of the contaminant. *Class I* locations contain flammable gases and vapors. *Class II* locations contain combustible dusts. Ignitable fibers and flyings would cause the location to be a *Class III*.

Each of these location classes has two divisions. For Class I and II locations, Division 1 means the hazard exists under normal conditions. If the hazard is present in an area only under unusual circumstances, it is a Division 2 designation. Class III, Division 1 means the fibers or flyings are normally suspended in the air. If the material is in storage and not usually suspended, it is a Division 2 designation.

The location classes are further divided into groups. The assigned groups are A through D for Class I, and E through G for Class II.

The principal factor used to establish materials into Class I groups is the *Maximum Experimental Safe Gap* (MESG), which is the gap between two metal plates, placed edge to edge, below which burning material on one side of the plates will not propagate to the other side. Class I, Group A and B materials are not easily classified, as the range of their MESGs are between 0.003 and 0.0012 inches. Whether they are Group A or B depends more on their explosion pressure. Materials with an explosion pressure similar to hydrogen gas are classified as Group B. In practice, acetylene is considered the only Group A material. Group B includes butadiene, ethylene oxide, and hydrogen.

Group C materials have an MESG greater than that of ethyl ether, ranging between 0.012 and 0.029 inches. Materials such as acetaldehyde, allyl alcohol, n-butyl aldehyde, crotonaldehyde, diethyl ether, diethylamine, epichlorhydrin, ethylene, methyl ether, 2-nitropropane, and nitromethane belong to Class I, Group C. Group D materials such as acetone, acrylonitrile, benzene, butane, cyclohexane, ethyl alcohol, gasoline, heptane, hexane, isopropyl acetate, isopropyl ether, methane, methanol, methyl ethyl ketone, octane, pentane, pentanol, 1-pentane, propylene, styrene, toluene, and xylene have MESGs greater than 0.029 inches (the MESG for gasoline). Remember, the smaller the MESG is, the greater the ignition hazard is.

The factor used to group Class II materials is their resistivity, measured in ohm-cm.

Class II, Group E combustible dusts include materials such as aluminum, magnesium, and zirconium, regardless of their resistivity. Other combustible dusts that have a resistivity of less than 102 ohm-cm are also included as Group E.

Class II, Group F includes carbon black, charcoal dust, coal dust, and coke dust. These materials contain more than 8 percent volatile material, as determined by ASTM Method 3175-82. Also included are dusts with a resistivity between 102 and 108 ohm-cm, and which other materials sensitize so they are explosive. Finally, Group G atmospheres contain Class II combustible dusts that have a resistivity greater than 108 ohm-cm.

Class III is not divided into groups. Ignitable fibers and flyings include cotton dust, lint, and wood chips.

Chemical Explosives

Substances with a positive enthalpy of formation release energy during decomposition. These substances typically have a high degree of unsaturation. They also tend to have either a large proportion of nitrogen or a heavy local concentration in the molecular chain. Nitrogen to halogen bonds also contribute to reactivity. Table 20-2 lists some compounds with high positive enthalpy of formation.

Chemical explosions are often compared to TNT explosions. The combustion energy of the chemical vapor is converted to a commensurate TNT charge with a given yield factor. The two have different blast characteristics, however. Specifically, the peak static overpressure, peak dynamic pressure, and positive phase duration differ. In the near field, TNT overstates overpressures. In the far field, vapor cloud explosions do not decay as rapidly as

Table 20-2
Compounds with High Positive Heat of Formation

	ΔH_f	
Compound	**BTU/lb•mol**	**BTU/lb$_m$**
Acetylene	97,600	3,740
Allene	85,600	2,064
Benzotriazole	107,500	903
1,3-Butadiene	48,200	903
Cyanogen	132,400	2,537
Diazomethane	85,600	1,978
Hydrogen cyanide	55,900	2,064
Nitrogen trichloride	98,900	817

Figure 20-3
Severity of Vapor Cloud Explosions

These factors influence severity of a vapor cloud explosion:

- Degree of confinement
- Degree of turbulence
- Composition
- Degree of cloud mixing
- Strength of ignition source
- Cloud configuration
- Reactivity of fuel

TNT explosions. When unconfined quiescent clouds are burned, no overpressure is produced.

Figure 20-3 summarizes the factors that influence an explosion when a vapor cloud ignites.

Typically, the degree of confinement in process plants is high, making blast strength more damaging. Turbulence increases blast strength. Composition of a vapor cloud is affected by the release mechanism. Explosively dispersed clouds increase turbulence. Aerosols have a greater explosive potential than vapors or gases.

Figure 20-4 summarizes some molecular structures that produce high energy levels.

Figure 20-4
High Energy Molecular Structures

Acetylenic compounds	Hydroxylammonium salts
Alkyl nitrites	Metal acetylides
Azo compounds	N-nitro compounds
Diazo compounds	N-nitroso compounds
N-halogen compounds	Peroxides

Figure 20-5
Potential Hypergolic Ignition

Alkali metal	Water
Chlorate	Acid
Chlorite	Acid
Hypochlorite	Acid
Nitric acid (concentrated)	Amines

In Figure 20-5, compounds that may ignite when mixed are listed. This is called hypergolic ignition.

Figure 20-6 lists potentially explosive mixtures.

Potentially explosive ammonium salts are formed when any ammonium salt is mixed with either chlorate or nitrite radicals. The mixture of acetylene and copper forms a salt that is sensitive to shock and friction.

Molecular structures that are capable of forming an explosive peroxide in air are listed in Figure 20-7.

Deflagration Venting

Incorporate deflagration venting into exterior walls or roofs of rooms where Class I-A, I-B, or unstable liquids are dispensed or where Class I-A or unstable liquids are stored in quantities exceeding one gallon. Inside storage rooms do not require venting, nor do rooms where Class I-B liquids are dispensed from containers of less than 60 gallon capacity.

Figure 20-6
Potential Explosion Hazards

Hydrogen peroxide (aqueous)	Amines
Potassium permanganate	Concentrated sulfuric acid

Figure 20-7
Compounds Susceptible to Peroxidation in Air

Acetals	Lactams
Aldehydes	N-alkylamides
Allyl compounds	Olefins
Cumene and its derivatives	Styrene and its derivatives
Cyclic ethers	Ureas
Dienes	Vinyl acetylenes
Ethers	Vinyl compounds
Haloalkanes	

A common method of mitigating explosions is gas explosion venting. A weak area is deliberately installed in an outer wall of the plant to act as a blowout panel. This permits unburned gases and combustion products to flow to the outside in case of an explosion. Three mechanisms enhance explosion effects: turbulence generated in a shear layer, oscillatory combustion, and Taylor instabilities. Also, ensuing explosions outside the facility impede the flow out and trigger flame instabilities inside the facility.

The maximum overpressure of natural gas explosions in an enclosure such as a room or small building is governed by oscillatory combustion. An opening pressure of an explosion vent of ten kPa or less has no subsequent effect on a vented methane-air explosion. The position of the ignition source has some effect on the maximum overpressure generated in large-scale enclosures without obstacles. When an explosion occurs at the wall opposite the wall containing the explosion vent, an external explosion determines the maximum overpressure. The shape of the explosion vent in a large-scale enclosure has little effect on the maximum overpressure generated by an explosion. The presence of an obstacle will increase overpressure only if the distance between the ignition sources and the obstacle is large. External explosions determine the strength of blast waves generated by vented explosions. This becomes significant when the ignition source is so far from the vent opening that a large amount of turbulent combustion gas is pushed out of the vent.

Kane provides a diffusion formula for the safe venting of hydrogen.

$$D = \frac{1 \times 13^{-3} T^{1.75} \sqrt{\frac{1}{M_a} + \frac{1}{M_b}}}{P\left[\left(\sum V_a\right)^{\frac{1}{3}} + \left(\sum V_b\right)^{\frac{1}{3}}\right]^2} \qquad [20.6]$$

where

M_a = atomic mass of component a
M_b = atomic mass of component b
V_a = atomic volume of component a
V_b = atomic volume of component b
P = pressure
T = temperature

Notice that rising temperature increases diffusion and rising pressure decreases it. The diffusion coefficient is a factor in Fick's Law:

> The mass flux of a constituent per unit area is proportional to the concentration gradient.

Mathematically, this is

$$\frac{\dot{m}}{A} = -D\frac{\partial C_a}{\partial x} \qquad [20.7]$$

where

$\dot{m}$ = mass flux of constituent a
A = area
D = diffusion coefficient
C_a = concentration of constituent a
X = distance across which the constituent travels

For gases, concentration is most conveniently expressed in terms of partial pressures, so

$$C_a = P_a = \frac{\rho_a M_a}{R_0 T} \quad [20.8]$$

where

ρ_a = density of constituent a
P_a = partial pressure of constituent
M_a = molecular weight
R_0 = universal gas constant

Combining equations 20.7 and 20.8,

$$\frac{\dot{m}}{A} = -D_{ab}\frac{M_a}{R_0 T}\frac{dP_a}{dx} \quad [20.9]$$

Integration of 20.9 yields

$$\frac{\dot{m}}{A} = \frac{-D_{ab}M_a}{R_0 T}\frac{\left(P_{a_2} - P_{a_1}\right)}{\Delta x} \quad [20.10]$$

and

$$A = \frac{\dot{m}_a R_0 T \Delta x}{-DM_a\left(P_{a_2} - P_{a_1}\right)} \quad [20.11]$$

but, from Dalton's Law and the ideal gas equation

$$\frac{P_i}{P} = \frac{V_i}{V} = \frac{n_i}{N} \quad [20.12]$$

If P_a at time 2 is the lower explosive level, and atmospheric at time 1,

$$P_{a_2} - P_{a_1} = P - LEL \quad [20.13]$$

and

$$A = \frac{\dot{m}_a R_0 T \Delta x}{DM_a\left(LEL\right)} \quad [20.14]$$

Electrical Wiring

Electrical wiring is the most common cause of fires in buildings. On the national level, the National Fire Protection Association, a consensus standard organization, issues the *National Electrical Code* (NFPA 72) in an effort to prevent electrical fires by getting everyone to follow standard practices. Some regions of the country have codes, such as the Southern Building and Construction Code (SBCC), and many municipalities have their own code. Which of the various available codes should you use? Whichever one your fire marshal tells you to use. This is not merely an original building erection and construction issue. It is a process issue, as well as a building modification and renovation issue. Every time you move a production machine from one department to another, do it by the code.

Storage rooms for Class I liquids must be wired to Class I Division 2 specifications. General-purpose wiring is acceptable in Class II or III storage rooms. If Class I liquids are to be dispensed, or Class II or III liquids are to be dispensed when their bulk temperature is at or above their flash points, all wiring within three feet of the dispensing nozzle must be Class I Division 1.

Motors are the second largest cause of fires. Set up a maintenance plan to check all motors periodically. Also, ensure they are installed and wired according to manufacturer's instructions or code.

Ventilation

Storage rooms generally may have natural special ventilation, even if flammable liquids are contained within. If flammable liquids are being dispensed, then mechanical ventilation equivalent to one cubic foot per minute (cfm) per one square foot of floor area is required. Minimum ventilation is 150 cfm.

Material Storage

Explosions have three essential ingredients: fuel, air, and a source of ignition.

Ordinary Combustibles

Even ordinary combustibles become explosive given adequate air and a source of ignition. Air can enter a tank or pipe during startup operations or be drawn in during production. New, unused tanks contain one hundred percent air. New tanks should be purged before use in order to prevent an LEL condition during filling. Tanks that have contained combustibles but have sit idle for awhile are likely candidates for explosion.

Thermal loading, in the form of sunshine, sets up a potentially explosive condition in partially empty tanks. Gases in the vapor space expand. At the same time, cooling by rain or cooler ambient temperature causes the gases to contract and condense. These daily cycles cause the tank to breathe through pressure relief valves or leaks in the wall joints and fittings. Repeated thermal cycling builds an explosive mixture of gases in the vapor space.

Drawing from the tank by pump also causes air infiltration. Air enters through vacuum relief valves, open hatches, missing hatches, and faulty seals. Wind blowing across the top of a flare stack also creates a suction inside the tank that draws air in. The intake can be significant on windy days. A wind blowing over an open hatch can also contribute to infiltration.

Flame arresters are needed in piping networks to prevent the spread of fire and explosion. When ignition occurs in a pipe, a flame heats and oxidizes vapor ahead of it in the pipe. A flame moving in a pipe creates a certain amount of restriction to venting of rapidly expanding products of combustion. Consequently, the flame pressure increases and velocity accelerates.

The pressure and velocity increase spirals until the detonation point is reached. At this point, a flame front is propagated by the shock wave compression-induced ignition of combustible gas and air mixture. Pressures exceeding one hundred times the original operating pressure are created. Detonation pressure may briefly exceed 1,000 psi. Turbulence-generating devices increase the acceleration between ignition sources and flame arresters.

Detonations travel at supersonic speed, typically greater than 5,900 fps, or at about Mach 5. Overpressures generally run around 350 psi. No wonder structural damage and equipment failure occur.

End-of-line flame arresters are typically installed at the top of a tank. Detonation flame arresters are more usually installed within piping systems. Explosions caused by misapplied flame arresters are, unfortunately, not uncommon. For protection between tanks and protection from a flare, use a detonation flame arrester if:

- The distance between the ignition source and the arrester (called the run-up distance) is greater than the tested limitations of the end-of-line arrester (typically 15 feet).

- The piping system has bends, elbows, or tees.
- The piping system is subject to lightning strikes.

Ensure the flame arrester has proven capability by obtaining from the manufacturer the U.S. Coast Guard letter of acceptance for the device. The Coast Guard letter will specify:

- Dates of letters from the manufacturer
- Names of agencies that provided information on the device
- A statement that the device meets the standard in Appendix A to 33 CFR 154
- The model number of the device
- The diameter of the device
- The maximum experimental safe gap measurements of the vapor control lines in which the device may be used
- The absolute operating pressure under which it may be used
- The length of time the arrester will resist endurance burn. This parameter is designated either Type I (two hours) or Type II (fifteen minutes)
- The laboratory certification report number upon which the acceptance is based
- Other models with different capabilities that should be distinguished from the model described

The Lower Flammability Distance (LFD) is the distance from a release through which the concentration of the released gas is in the flammable range. This is an important parameter in emergency planning, as it delineates the extremely dangerous zone as the result of a spill. Kumar and Luo, under the tutelage of Bennett, found that a simple gas model developed by the Ontario Ministry of the Environment (OME) performed best in predicting LFD. This is not an endorsement and you are referred to their article for further information about the OME model and others they tested.

Dangerous Materials

Maintain flammable liquids in approved containers—that is, in containers that have been tested for such use by one of the nationally recognized testing laboratories. When use of the material makes such containment impractical, Figure 20-8 lists the generally accepted quantities of flammables inside a building.

Figure 20-8
Acceptable Quantities inside a Plant Building

Twenty portable 660-gallon tanks:	Class III-B Liquids
and Two portable 660-gallon tanks:	Class I-B, I-C, II, or III-A Liquids
and 120 gallons in containers:	Class I-B, I-C, II, or III-A Liquids
and 25 gallons in containers:	Class I-A Liquids

An alternative to Figure 20-8 is a one-day supply, no matter what the quantity. Either Figure 20-8 or a one day supply, though, not both. Rekus points out the propensity of inexperienced OSHA compliance officers to cite facilities for violating flammable quantity limitations if any two categories of Table 20.4 are utilized. You may, under the fire code, have all those categories present, up to the sum of quantities allowable. With OSHA you may have to go to an informal hearing and argue your case.

You may build special inside storage areas and store more than the amounts in Figure 20-8 by adhering to the parameters of Table 20-3.

Inside rooms are fully enclosed inside a building and have no exterior structural walls of their own. *Cut-off rooms* have one or two walls in common with the building exterior walls. *Attached buildings* have three exterior walls and share only one wall in common with the main building.

Table 20-4 gives the requirements for fire doors for the rooms and structures of Table 20-3.

You may also place storage cabinets in your plant for flammable liquids. These cabinets are designed to protect the contents for up to ten minutes during a fire as summarized in Figure 20-9.

Single storage cabinets are limited to 60 gallons of Class I or II liquids or 120 gallons of a Class III liquid. A good practice is to limit one fire area to one storage cabinet. A fire area is protected from other areas by a 4-hour

Table 20-3
Inside Storage Areas
Flammable Liquids and Liquid Warehouses

Fire Resistance, hrs	Interior Walls[A], Ceilings, Intermediate Floors	Roofs	Exterior Walls
Inside Storage Rooms			
Floor Area <150 ft²	1		
Floor Area <500 ft²	2		
Cut-Off Rooms			
Floor Area <300 ft²	1	1[B]	
Floor Area >300 ft²	2	2[B]	2[C]
Liquid Warehouses	4[D]		2[E] or 4[F]

A. Between liquid storage area and any adjacent area not dedicated to storage.
B. Roofs of attached one-story buildings may be lightweight, noncombustible construction if the separating interior walls have a minimum 3-ft parapet.
C. Where other portions of buildings or other properties are potentially exposed to fire.
D. Standard fire wall.
E. When exposing walls are located more than 10 ft but less than 50 ft from an mportant building or adjoining property line that can be built upon.
F. When exposing walls are located 10 ft or less from an important building or adjoining property line that can be built upon.

firewall with no openings such as unprotected doors and windows. For most of us, this is the entire plant. If for some reason you need more flammable storage you may have up to three storage cabinets in the same fire area if they are separated by a 100-foot distance.

Contain Spills

To prevent fires from migrating to other areas, always ensure that flammable liquids can be contained if spilled anywhere within the facility. Especially prevent flammable liquids from entering areas where ignition sources

Table 20-4
Fire Door Ratings

Fire Resistance Rating of Wall[A], hr	Fire Resistance Rating of Door, hr
1	¾
2	1½
4	3[B]

A. As required by Table 20-3.
B. One fire door is required on each side of interior openings for attached liquid warehouses.

Figure 20-9
Storage Cabinet Construction

Steel Cabinets

- 18 gauge steel minimum thickness of walls
- Riveted or welded joints
- Double-wall
- Insulating wall gap ≥ 1½ inches
- Door sill raised two inches
- Door lock catches at top, bottom, and middle

Wooden Cabinets

- Exterior grade plywood 1 inch minimal thickness (a type that does not delaminate under fire conditions)
- Rabbet joints fastened in two directions with wood screws
- At least a 1 inch rabbet overlap if more than one door
- Raised sill or pan capable of holding 2 inches or more of spilled liquid

are present. The fire code exempts certain materials (small quantities, water miscible liquids, liquids that are more dense than water, and highly viscous liquids) from spill containment, but keep in mind that environmental rules may also apply before deciding not to install spill controls.

Static Electricity

Static electricity is generated when electrons build up in unequal amounts on two different surfaces. When the buildup becomes large enough, electricity flows in the form of a static discharge or spark. The static discharge transfers by means of either *conduction* or *induction.* To transfer by conduction the object with the static charge and the object to which the charge is discharged must in contact with each other. Static discharge by induction occurs across an air gap.

Static charges build up in gases, on dust particles, in liquids, on pipe walls, on surfaces of machinery, and even on people, among other things. A charged object transfers electrons to a non-charged, or lesser-charged, object until an equilibrium of charge exists on both objects.

When low conductivity fluids, such as hydrocarbons, flow in a pipe, static charge is generated. The streaming current, I amps, is determined as

$$I = k \times 10^{-6} V d^{n} \qquad [20.15]$$

where

k = a constant
V = average fluid velocity, m/s
d = inside pipe diameter, m
n = an exponent

Values of 3.75 for k and 2 for n are commonly used in practice.

The Expert Commission for Safety in the Swiss Chemical Industry reports these resistivities for poorly conductive and conductive liquids (Table 20-5).

Material transfer of powders, which are fine particulates, also generates static electricity. Table 20-6 lists some typical charges of organic powders based on the process unit operation involved.

Electrical capacitance in farads is a measure of any object's ability to store a charge. Table 20-7 lists the capacitance of some common objects.

Bonding and Grounding

To avoid exploding aerosols, potential static charges must be either dissipated or prevented by *bonding or grounding.* The difference in electrical charge between the non-charged state of any object and its charge in any

Table 20-5
Resistivity of Conductive Liquids

Conductive Liquids	Resistivity, Ωm
1,2-Dichloroethane	10^7
Benzoic acid ethyl ester	10^7
Methanol, ethanol	10^6
n-Propanol, n-butanol	10^6
Acetic acid ethyl ester	10^6
cis-1,2-dichloroethylene	10^6
Acetic acid	10^5
Pyridine	10^5
Acetonitrile, propionitrile, Benzonitrile	10^5
Acetone, butanone, Cyclohexanone	10^5
Isobutanol	10^5
Isopropanol, t-butanol	10^4
Formic acid ethyl ester	10^4
Anhydrous acetic acid	10^4
Propionaldehyde	10^4
Nitrobenzene	10^4
Glycol, glycolmonoethyl ether	10^3
Dimethyl formamide	10^3
Acetaldehyde	10^3
Formic acid	10^2
Poorly Conductive Liquids	**Resistivity, Ωm**
Carbon disulfide	10^{16}
Carbon tetrachloride	10^{15}
Diesel oils	10^{13}
Gasoline and white spirit	10^{13}
Cyclohexane	10^{13}
Benzene, toluene, xylene	10^{13}
Mesitylene	10^{13}
Diethyl ether	10^{13}
1,4-Dioxane	10^{12}
Anisole	10^{11}
Stearic acid dibutyl ester	10^{10}
Sebacic acid dibutyl ester	10^8
Bromobenzene, chlorobenzene	10^8
Dichloromethane	10^8
Chloroform	10^8
Propionic acid	10^8

Table 20-6
Static Charge in Powder Operations

Process Unit Operation	Charge, coulomb/kg
Sieving	10^{-9}-10^{-11}
Pouring	10^{-7}-10^{-9}
Scroll feed transfer	10^{-6}-10^{-8}
Grinding	10^{-6}-10^{-7}
Micronizing	10^{-4}-10^{-7}
Sliding down an incline	10^{-5}-10^{-7}

charged state is called *potential* and is measured in volts dc. In bonding, two objects are electrically connected so that the potentials on their surfaces remain in equilibrium. This is accomplished by attaching a bonding wire or strap as a connection between the two objects. Another way to bond two objects is to maintain them in contact with each other. An example of this practice is in the bonding between containers that is required when a flammable liquid is transferred from one container to another.

Bonding, by itself, is typically inadequate protection against static discharge. Contact can not always be maintained. Paint and undercoatings can interfere with the bonding contact. Bonding straps can work loose. The equilibrium potential between the bonded objects can be greater than the static charge of a third object. Great care must be taken, therefore, when making

Table 20-7
Capacitance of Common Objects

Object	Capacitance, farad
Small scoop, can, tools	5
Buckets, small drums	20
50-100 gallon containers	100
Person	200
Automobile	500
Tank Truck	1000

the bond connection, to ensure that a good contact is made, check the surfaces at the connection points for metal to metal contact. In the case of bonding plastic containers, make sure dirt or grease does not prevent a direct contact between the strap attachment and the plastic container. Always keep bonded containers closed, except to make transfers. Close the container first before disconnecting its bonding wire.

Poorly conductive materials also accumulate static charges. Capacitance, as a term, has no meaning when dealing with materials of very low conductivity. Therefore, the proper measure of static potential with these materials is the measure called charge density.

Ignition Energies

Ignition energies for gases and vapors present in explosive concentration range from about 0.02 millijoule (mJ) for hydrogen and acetylene to 0.1-0.3 mJ for most hydrocarbons. Almost all dust clouds require 1 mJ or more energy. The majority of materials require 5 mJ or more energy.

The energy, E Joules, available in a static spark is

$$E = \frac{1}{2} CV^2 \qquad [20.16]$$

where

C = capacitance, farads
V = voltage

Another way to estimate available energy is

$$E = \frac{1}{2} QV \qquad [20.17]$$

where

Q = charge transfer, coulombs

Mancini reports the voltage requirements summarized in Table 20-8 to produce the energy levels to ignite materials.

Grounding

In grounding, an object is electrically connected to electrical ground potential, or the earth. This is the surest method of controlling static charges.

Table 20-8
Voltage Required to Produce Ignition Energy

Igniting Material	Energy Required, mJ	Volts
Many dusts, most gases and vapors	10	10,000
Most gases and vapors, very sensitive dusts	0.40	2,000
Hydrogen, acetylene	0.15	1,300

Therefore, when bonding two objects, ground one of them. Check the effectiveness of bonding and grounding by measuring electrical continuity.

Controlling Static Electricity Explosions

Figure 20-10 summarizes the conditions necessary for an explosion to be ignited by static electricity.

Figure 20-10
Conditions for Explosion by Static Discharge

1. A flammable gas or vapor or an aerosol in the presence of sufficient oxygen
2. A static charge generated by mass transfer
3. Accumulation of potential
4. Discharge to achieve equilibrium:
5. Starting with a sufficient voltage difference
6. A suitable gap across which a spark can develop
7. Spark energy exceeds the minimum ignition energy requirement of the fuel/air mixture

An aerosol is a finely divided substance mixed with air. The substance may be composed of solids, such as metal fines or fumes or grain dust, or liquids, such as a fine mist. Theoretically, some combination of fine particle size, the amount of oxygen present, and the amount of energy available in a static discharge is explosive for any material.

To prevent an explosion caused by static discharge, simply reverse the items in Figure 20-10. Eliminate the possibility of a flammable atmosphere by either removing the presence of the flammable material or adjusting its concentration out of the explosive range. This can be accomplished by reducing the oxygen content so that the material concentration is greater than the upper explosive limit (UEL). Such an atmosphere is probably imminently dangerous to unprotected humans, so a more practical method would be to dilute the concentration of the material (to less than the lower explosive level—LEL) by blowing in more air. Either procedure can be dangerous in itself, so unless the situation is an emergency, other procedures are recommended.

The safest way to control this type of explosion hazard is to use bonding and grounding as described above to control the amount of static electricity that accumulates and to relax the charges already accumulated. Impurities, such as rust, dirt, and metal fines, contribute to electrostatic charges in liquids. As much as practical, keep liquids clean. Also, the speed of transfer of liquids through pipes, pumps, filters, or funnels is directly proportional to the generation of static electricity. Slow the process down as much as feasible. Transfer liquids slowly from one container to another and reduce the amount of aerosol production. In other words, prevent splashing, spraying, or misting of the liquid. When connecting bonding straps for a transfer, allow time for *charge relaxation* before proceeding.

Two heuristics are in order:

The more material moved, the more charge generated.

The longer a charged material is left at rest, the closer it gets to equilibrium with its surroundings.

Figure 20-11 summarizes some precautions and preventive measures against explosions to be taken in specific industries.

Figure 20-11
Precautions and Prevention in Specific Industries

Dust or Powder Handling:

- Ground conductive parts of equipment, such as pipes, conduits, containers, and funnels.
- Ground personnel, especially around sensitive metal powder, with clothing and footwear.
- Conductive and grounded equipment is essential if dust or powder is handled in the presence of a flammable gas or vapor.
- Handle large quantities of powders in closed systems if flammable gases or vapors are also present.

Electric Component Manufacturing:

- Ground personnel with wrist straps and specially designed clothes, including undergarments that suppress electrostatic accumulation and discharge.
- Insulate work areas with floor mats, table mats, and air-ionization equipment.
- Apply conductive paints and coatings to the floor.
- Treat bare floors regularly with topical anti-static floor treatments and conductive or anti-static floor finishes.
- Floor coverings should be vinyl or anti-static carpeting.
- Require conductive footwear made of leather with conductive foot and heel straps.

Explosives Manufacturing:

- Employees should never rub any two objects together.
- Ground personnel with specially designed clothes, including undergarments that suppress electrostatic accumulation and discharge. This clothing is 50% polyester, 48% cotton, and 2% steel, which is mercerized, super pre-cured, and preshrunk.
- Use special equipment to collect dust.
- Bond and ground everything.
- Locate each process in a separate explosion-absorbing bunker to isolate it from other processes.

Rail Cars, Tank Trucks, and Marine Vessels:

- Attach bonding cables to bright, shiny metal before opening the dome top. Score the metal surface to ensure good electrical connection.
- Ground all parts of the equipment.
- Ground all nearby personnel with proper clothing and footwear.
- Transfer liquids or finely divided solids in cloced systems.
- If necessary, apply an inert gas (such as nitrogen) blanket during loading and unloading.
- Minimize splashing, spraying, swirling, and misting.

Figure 20-11 *(continued)*

- Limit flow rate until the nozzle or inlet valve is well-covered with liquid.
- Upon completion of filling, close the dome top and remove the bonding cable, in that order.
- Bonding is the first and last step in the operation.
- If bonding is not first and last, then whatever you do may be the last thing you do.

Life Safety

One thing about life safety codes, as pointed out by Dessoff: these codes concern themselves principally with safe and quick building evacuations and controlling the spread of fire. Not only must you not split hairs with these codes but also you must go well beyond their requirements.

> You need at least one evacuation warden for every twenty employees.

The evacuation plan required at 29 CFR 1910.35-38 requires not only written evacuation procedures but escape route assignments. This may be a general assignment based on the area of the plant the evacuating personnel find themselves in at the time of the emergency. In very small facilities, you may be able to make individual assignments.

Some employees may be required to shut down critical operations. For these employees, include special instructions that explain when to initiate the shutdown and when to abandon the process, if at all. On the other hand, some chemical processing plants have operations that must be kept running for the safety of all. The employees who operate these critical processes also need special instructions.

Once evacuation is complete, how are you going to account for all hands? How will you know if someone may be trapped inside or otherwise need to be rescued? Provide these instructions in your procedure. Who will perform the rescue? How will they be trained? When? How often? Who will perform first aid and CPR (cardiopulmonary resuscitation)? How will they be trained? When? How often?

> You need at least one first aid/CPR volunteer
> for each twenty employees.

What means have you provided for employees to report fires and other emergencies? Reporting the emergency promptly is the vital initial step of any emergency response plan.

Disaster Planning and Management

Disasters include fires and explosions but cover far more. Natural disasters are part of disaster management. Moving quickly, effectively, and efficiently when such events happen can mitigate, if not prevent, further disaster from happening. Planning must involve internal and external emergency responders.

Encourage the management leadership team to appoint an authorized spokesperson for the facility. This person deals with media while everyone else deals with the emergency and its aftermath.

Appoint and train several teams of people to assist with disasters. A fire brigade, emergency response team, spill control team, hazardous materials incident team, emergency medical response team, evacuation wardens, rescue team, search team, and salvage team are among those to consider. These suggested teams are not necessarily exclusive and some of their duties may be shared, so you may not need one of each. Figure 20-12 lists typical duties of an emergency response team.

Training

Training is the key to prompt and effective emergency operations. The time to impress duties on people is not when the place is coming down around your ears. Perform drills as a key to effective and safe operations. The purpose of drills is instill discipline through training and practice. Discipline is the difference between action and panic. Train and drill each mem-

Figure 20-12
Typical Duties of an Emergency Response Team

- Select and use various types of fire extinguishers effectively.
- Provide first aid and CPR.
- Protect themselves and others from potential consequences of bloodborne pathogens.
- Conduct shutdown procedures.
- Control and mitigate chemical spills.
- Select and use respirators and self-contained breathing apparatus (SCBA) as needed.
- Conduct search procedures for potential victims.
- Conduct emergency rescue operations.
- Respond to hazardous material emergencies in accordance with 29 CFR 1910.120.

ber of the team to react appropriately without confusion. If you can do this, then the chances of that member of the team reacting effectively to the emergency, is greatly increased. Hold drills at random intervals but announce the event as a drill. Otherwise, you will have real panic and possible injuries may occur.

Each employee must receive certain emergency training from their first day at your plant. Figure 20-13 summarizes this training. Figure 20-14 lists when the training needs to be conducted.

Tornado Response

One of the more devastating and costly natural disasters is a tornado. The weather conditions that spawn tornadoes are well understood but not always predicted by scientists. Recent advances allow weather scientists to predict tornado activity with greater accuracy than in the past and for the first time, communities are getting some advance warning to prepare. For industry, the best way to prepare for a tornado is to be ready all the time and to increase readiness as the probability of a twister increases.

Figure 20-13
Emergency Action Training

- Evacuation procedure
- Alarm system recognition
- How to report emergencies
- Shutdown procedures
- Types of potential emergencies

Figure 20-14
When to Conduct Emergency Action Training

- Upon hiring, as close to day one as practical
- Before using emergency equipment for the first time
- Upon installation of new processes that may create an emergency or become critical during an emergency
- Upon implementing new, updated, or revised procedures
- After an exercise, drill, or actual emergency uncovers inadequate employee or management emergency performance

Damage from tornadoes can come in the form of the cyclonic winds, strong straight-line winds, large hail, severe lightning, or driving rain—all which can accompany tornadoes.

The tornado plan should include the assignment of four tornado spotters to take posts at good vantage points for watching for funnel clouds. A spotter may see a funnel cloud before it appears on radar. Sometimes the weather information during a tornado watch is several tens of minutes behind real time. Spotters give you an advantage in situations where weather radios are limited. However, do not post spotters on the roof, as they could be struck by lightning.

If you have a one story building with no windows, as a lot of industrial facilities do, forget the spotters. You have nowhere to place them without their lives being in danger. A security officer in a guardhouse, where he is protected, can serve as a single spotter.

Evacuation should be a part of your tornado plan. Assign evacuation wardens if you have not done so already. However, do not evacuate during the threat of a tornado. Instead, take shelter inside the building. Evacuate immediately after the tornado passes until a damage assessment on the building can be made. Do not evacuate if the tornado passed by without affecting your building. In that case, it is still safer inside.

An early warning system, such as weather band radios in the guard house, the maintenance superintendent's office, and the plant manager's office, will give you a chance to make last minute preparations and alert key persons. Telephone systems and public address systems may fail in bad weather, so outfit key persons with walkie-talkies.

Training is recommended at least annually. The best time to train is just prior to the start of the active storm season when tornadoes might be spawned.

Disaster Recovery

Planning for the recovery from disaster has several purposes. For one thing, it helps get your facility back in productive operation as quickly as possible. The greatest benefit of that is that it helps prevent further damage and loss, especially in human terms. Once the fire department or other public authority turns the plant back over to you, have a plan in place for promptly and efficiency restoring the plant to as near normal operation as feasible as quickly as possible.

References

Beiti, Wesley H. "Make Fire Safety High Priority before, Not after, Fire Occurs." *Occupational Health & Safety.* August 1990, 19-22.

Boehm, Richard. "Smoke—A Major Killer in Building Fires, and What to Do about It." *Safety and Health by Design.* ASSE Engineering Division Newsletter. Date unknown, pp. 8-10.

Booher, Lindsay E. and Brian Janke. "Air Emissions from Petroleum Hydrocarbon Fires during Controlled Burning." *AIHA Journal.* May 1997, pp. 359-365.

Brubaker, Art. "Tornado Safety." *Occupational Health & Safety.* October 1997, pp. 150-154.

Daugherty, Jack E. *Industrial Environmental Management: A Practical Handbook.* Rockville, MD: Government Institutes, Inc., 1996.

Dessoff, Alan I. "Plan Now for Fire Safety . . . or Pay Later." *Safety + Health.* September 1996, pp. 52-55.

Expert Commission for Safety in the Swiss Chemical Industry. "Static Electricity: Rules for Plant Safety." *Plant/Operations Progress.* January 1988, pp. 1-22.

Fagel, Michael J. "Creating a Disaster Plan." *Occupational Hazards.* October 1997, pp. 141-144.

Hot Work Policy, Precautions and Permission. Norwood, MA: Factory Mutual Engineering Corporation, 1993.

Kane, Steven F. "An Equation for the Safe Venting of Hydrogen." *Safety and Health by Design.* ASSE Engineering Division Newsletter. 1993, pp. 18-23.

Kumar, Ashok, Jie Luo, and Gary F. Bennett. "Statistical Evaluation of Lower Flammability Distance (LFD) Using Four Hazardous Release Models." *Process Safety Progress.* January 1993, pp. 1-11.

Lenoir, Eric M. and John A. Davenport. "A Survey of Vapor Cloud Explosions: Second Update." *Process Safety Progress.* January 1993, pp. 12-33.

Mancini, Robert A. "The Use (and Misuse) of Bonding for Control of Static Ignition Hazards." *Plant/Operations Progress.* January 1988, pp. 23-31.

Mercx, W.P.M., C.J.M. van Wingerden, and H.J. Pasman. "Venting of Gaseous Explosions." *Process Safety Progress.* January 1993, pp. 40-6.

Rekus, John F. "Burning Issues about Flammable Liquid Storage." *Occupational Hazards.* November 1997, pp. 27-32.

Solving the Mystery: Static Electricity. Virginia Beach, VA: Coastal Video Communication Corporation, 1990.

Vickers, Kingston, and Trevor Knittel. "Vapor Control System Designs." *Occupational Health & Safety.* June 1993, pp. 53-58.

West, A.S. "Chemical Reactivity Evaluation: The CCPS Program." *Process Safety Progress.* January 1993, pp. 55-60.

21

CONTRACTOR SAFETY MANAGEMENT

CONTRACTUAL RELATIONSHIPS

Although, according to the Occupational Safety and Health Act, each employer is responsible for furnishing a safe and healthy work place for his or her employees, a catch exists when more than one employer occupies a workplace, however temporarily the situation may be. Interpretation of the Act, case law, experience of compliance officers, and enforcement precedence puts a burden on the prime or general contractor to ensure the safety and health of all contract employees. Similarly, host employers are also at least partially responsible for all hands on board.

Therefore, you should establish a construction safety policy. Require all contractors and subcontractors on your property to abide by your safety rules and procedures. Give them an option to use their own rules and procedures under one inflexible condition: they certify in writing that they have rules and procedures in place that are at least as stringent as yours. Further, make them certify that they have implemented these rules and procedures by having conducted the required training. Rietze points out that four general types of contractual relationships exist.

In the first type, the owner has a contract with the general contractor who is the construction manager. Subcontracts are strictly between the general contractor and the subcontractors. The general contractor, or construction manager, is responsible for safety oversight of the project.

In the second case, the general contractor is exempt from overall safety responsibilities but the subcontracts are between him and the subcontractors. The owner provides safety oversight of the project.

In the third situation, the owner assigns the safety oversight to the general contractor but subcontracts are between the owner and the subcontractors.

The final type of agreement is where the owner and general contractor have a contractual relationship and the owner has a separate agreement with subcontractors. However, the general contractor is responsible for safety oversight.

Regulatory Relationships

To reiterate the Occupational Safety and Health Act: each employer shall furnish to each of his or her employees employment and a safe place of employment, which are free from recognized hazards. The *OSHA Field Inspection Manual* asks the compliance officer whether the general contractor or prime contractor has implemented procedures to ensure that all employers working under him or her provide adequate protection for their employees. Also, the regulations state that the prime contractor is never relieved of his or her overall responsibility for compliance with safety standards for all work performed under the contract. The prime contractor and subcontractors are jointly held responsible for subcontracted work. Of course, general contractors are not happy with this situation.

A controlling employer, or host employer, is often held responsible with the general contractor at work sites where an injury or fatality occurs. It is not uncommon for host employers, the Owners as contractors and subcontractors refer to them, to be cited for the same violation of OSHAct or an OSHA standard as the contractor and subcontractor.

Risk Identification and Assessment Prior to Bid

Alert contractors up front that they will be held responsible for safety performance by including information in the request for quotation or request for proposal. This means that you must identify those hazards in your facility that will be a risk to contractors and subcontractors in order to disclose them in the bid request documentation.

Bid Invitation Preparation

The bid invitation needs to disclose those risks that were uncovered as discussed in Section 21.3 above. Some simple questions you may ask in the invitation to bid are:

- What is your incident rate (IR) as determined from your OSHA log for the previous three years?

- Do you have a written safety program?
- Is the content of required training documented?

Let bidders know that the successful bidder will have to produce documentation to support his or her claims.

Contractor Selection

In addition to the financial aspect, evaluate bidders by their safety performance. Some simple questions are in order. For one thing, what is the contractor's experience modification rating (EMR)? The Workers' Compensation insurance carrier can provide this information. Contractors with poor safety performance in the past tend to repeat their poor performance. Beware. The low bid that a contractor supplies you could turn out to be one large liability. Bell and Bell give us a good yardstick for selecting contractors:

- EMR ≤ 1.0
- OSHA Incident Rate ≤ 12.0
- Lost Word Day Rate ≤ 6.0

Contractor Performance Measurement

The very minimum performance of a contractor or subcontractor is that he or she religiously observes your plant safety rules and follows written safety procedures. Your procedures must always be the default. If the contractor is missing a relevant procedure or you think his or her procedure is inadequate, make yours the one to be followed. Ensure the proper implementation training is conducted, even if you have to do it yourself. A competent contractor will have all relevant safety procedures in place and training will have already been conducted before beginning work at your facility.

If you, as the owner, are not happy about the safety performance of the contractor or one or more of the subcontractors, do not correct or discipline their employees directly. Hold the contractor or subcontractor responsible for disciplining his or her own employees. Many owners and contractors hold daily meetings to discuss the project and this is where the matter belongs. You may also write a note, a formal memo, a letter, or even a report detailing the observed behavior or situation. Do not, however, provide specific solutions. Allow the contractor or appropriate subcontractor the freedom to take care of the matter in his or her own way. You may, however, require a written program addressing your safety concerns. You may also require that he or she designate by name a safety representative to attend construction briefings.

Develop an Action Plan

One thing that can help when managing contractor safety on your property is to provide the contractor a layout and road map of the job site. This can be provided at an initial safety briefing where you

- Trade copies of each other's safety manuals.
- Review each other's OSHA log for the current and previous three years.
- Discuss federal or state citations. (Failure to disclose can be grounds for terminating the contract.)
- Exchange new hire orientation information.
- Discuss training needs.
- Exchange substance abuse policies.

Figure 21-1 summarizes the elements of a good contractor safety program.

Figure 21-1
Contractor Safety Program

1. Contractor safety policy
2. Construction safety review
3. Contractor screening
4. Prebid conference relative to safety performance requirements
5. Preconstruction safety review of contractor
6. Contractor safety policy
7. Contractor safety program
8. Contractor's safety representative

Figure 21-1 *(continued)*

9. Contractor and joint training
10. Employee orientations
11. Weekly toolbox meetings
12. Manager safety meetings
13. Inspections
14. Joint safety walkthroughs with contractor
15. Site safety inspections
16. Contract enforcement
17. Accident reporting and investigations
18. Housekeeping

That does not look too different from what you must do internal to your own operation, does it? It should not because it's the same, except you have a contractual relationship with an outside firm.

References

Breiland, Donald J. and Lola Fraser. "Excavation Safety and the Competent Person." *Professional Safety.* September 1991, pp. 28-31.

Ellis, Teri. "That First Day." *Occupational Health & Safety.* April 1998, pp. 26-29.

Fagan, John L., Tyrone Monte, Darlene A. Powell, and Charles J. Crocini. "Contractor Review Committee: One Hospital's Approach to Facilities Development Quality & Safety." *Professional Safety.* May 1998, pp. 33-35.

Hislop, Richard D. "A Construction Safety Program." *Professional Safety.* September 1991, pp. 14-20.

Holt, Robert G. "Preconstruction Surveys." *Professional Safety.* September 1991, pp. 21-24.

Rietze, R. Benjamin. "Proactive Construction Management: Dealing with the Problem of Subcontractor Safety." *Professional Safety.* January 1990, pp. 14-16.

Smith, S.L. "Under Construction: Safety at Multiemployer sites." *Occupational Hazards.* May 1998, pp. 56-61.

22

Off-the-Job and Family Loss Control

Why Would a Company Be Interested in Off-the-Job/Family Issues?

For every job-related injury, eighteen or more injuries occur off the job. Thirty-five fatalities occur off the job for every on-the-job fatality. An employee who is killed off the job is just as unavailable for work as one who is killed on the job. I do not mean to be callous, but the point is that what happens to our employees off the job is always a concern of ours. We are concerned with off-the-job behavior to the extent that it impacts on work performance. Productivity-Quality-Safety: the big three.

These days, if an employee is injured or handicapped off the job we must find a way to bring them back, in most cases, under the Americans with Disabilities Act (ADA). However, for every such person who is productively employed, many more are at home unable or unwilling to work.

So, *wellness programs* are popping up in factories all over the country. It is becoming a fad to have a wellness effort in your plant. Obviously, in the case of a fatality off site, you lose a friend and coworker but also a valuable resource your company no longer has. Let us take a look at some other issues and how they relate to work in hopes you will see the benefits of implementing a wellness program.

This is not a requirement of OSHA. Wellness efforts are totally grassroots initiatives and mostly involve medical screening and education. An attractive benefit is that wellness programs are typically not expensive and help trim medical and dental insurance costs by keeping employees healthier, improving their morale, and reducing absenteeism and turnover.

A good program includes fitness screening as well as group education and awareness training for several health and wellness topics. These topics include nutrition, weight control, smoking cessation, and substance abuse control.

WORK-RELATED DIRECT ISSUES

The most direct wellness issue is getting to work in one piece day after day for a working lifetime. Don't hold me to this statistic but I think the current cost of our love affair with cars and trucks is 57,000 deaths per year. With several hundred people driving an average of thirty miles one way each workday for forty years, chances are that more than one employee in your plant will be killed on the highways at some time. Therefore, in my opinion, a wellness program does not make sense unless you couple it with a safe driving program.

With the phenomena of road rage being observed in all parts of America in the late twentieth century, teach your employees to be patient and cool in all sorts of traffic situations. Enough people are being killed already through vehicular accidents. As a society, we should not have to add homicide statistics to that.

Today more and more employees are working from their homes. This is especially attractive to young parents who want to spend the formative years of their children's lives at home and close to them. The number of people working at home is increasing. However, I think that situation only works for clerical and business workers. Industrial employees are still needed at their machines and do not have the luxury of making widgets at home.

To the extent that you have employees working out of their homes, you have a vested interest in the safety and well being of that person at home.

Health and well being are being more often recognized as directly job-related than in the past. Besides offering a benefits package that includes family medical and dental, many companies are offering other, less costly, and preventive wellness benefits, such as blood pressure screenings. Some companies have even installed automatic blood pressure machines so that employees may take their own blood pressure. Cholesterol screening is another wellness benefit offered at some proactive factories. Weight control and fitness programs are another easy-to-offer, relatively cheap benefit for employees. None of these things are very costly and all of them improve health and therefore increase productivity and availability of the employees who participate.

Vision health management is directly related to work performance in many cases. Machinists certainly need to maintain their vision in order to work safely and effectively.

Work-Related Indirect Issues

What a person eats and drinks and what he or she does with free time affects productivity and availability. While it is true that we really do not want to know what a person does with his or her free time, much less attempt to manipulate his or her choices, we can and perhaps should offer incentives that encourage wise choices. These choices should include balanced diets, plenty of rest and exercise, and other healthy lifestyle decisions. Today employees are more concerned than ever about the possibility of occupational cancer, but the odds are 1,000:1 that any given person will develop a lifestyle cancer instead. Cancer and other deadly diseases rob us of valuable human resources. No one is irreplaceable but replacement is painful and comes at a price. Then there is the loss of friends and coworkers, a price that cannot be measured. For very few dollars, then, we can offer wellness programs that encourage our friends and coworkers to live healthier and longer.

Diet

A balanced diet is said to be 40 percent complex carbohydrates, 30 percent protein, and 30 percent fat. Extremely low fat diets can be dangerous and one should consult a physician before undertaking one. Also see a physician before beginning a zero cholesterol diet. Some cholesterol is necessary for proper metabolism. Proteins build muscle and carbohydrates provide energy that the muscles burn. Our modern diet has become mostly simple carbohydrates so more complex carbohydrates are better than too low a content.

How many calories should you eat? The daily calories required to just maintain your weight is equal to fifteen times your body weight in pounds. To lose one pound of weight per week, eat five hundred calories less than the maintenance requirement per day. To lose two pounds per week, eat one thousand calories less than maintenance calories per day. Keep in mind that many factors affect the loss of weight and loss due to reduced caloric intake will only be an average. Some weeks you may not lose any weight.

One thing that affects weight loss is whether the loss is coming from protein tissue or fat. Fat is lighter than protein. Sometimes when we think we are making progress at weight loss we are really burning too much protein. At other times when we think we are not doing well on a diet we may actually be replacing fat with protein, which burns more calories. Weight loss is so much more than watching a scale.

Exercise

Aerobic or cardiovascular exercise is highly beneficial but some anaerobic exercise is needed to build muscle, which burns fat. An all aerobic exercise routine is not good. An all anaerobic exercise routine is not good either. We need anaerobic to burn fat more effectively and aerobics to keep our hearts strong and healthy.

For aerobic exercise, achieve your target heart rate (THR). A resting heart rate (RHR) is determined by taking your pulse the first thing when you get out of bed for three or four mornings in a row and taking the average. Take the pulse lying down in bed right after you wake up. Take the pulse for one full minute instead of the fifteen or twenty second pulse often taken by nurses and doctors. By taking a full minute sample you do not multiply the pulse you get by any number. The THR is then calculated as

$$\begin{aligned} THR_{Low} &= [(220 - Age - RHR) \times 0.55] + RHR \\ THR_{High} &= [(220 - Age - RHR) \times 0.75] + RHR \end{aligned} \qquad [22.1]$$

During aerobic exercise your heart rate should fall between these two calculated numbers.

Drug and Alcohol Abuse

The cost of alcohol and substance abuse in the workplace is staggering. Alcohol and illegal substances are so prevalent in our society that they are a mainstay of our culture. People abuse alcohol and/or drugs who need a high in order to relax or believe they are smarter, stronger, or happier under the influence of these substances. Abusers put on personality masks to hide their condition and other insecurities. However, masks only give the illusion of hiding our problems: family relationships, low self-esteem, financial difficulties, and feelings of inadequacy all will show through after awhile.

Addicts have given themselves over to a habit. They are not in control of their own lives and are thereby not free persons. Addicts spend a great deal of their time planning on getting high or drunk instead of concentrating on the task at hand. They also spend a lot of time worrying about how to get money to buy alcohol or illegal substances. A certain amount of time is spent making the arrangements to buy the pleasure substances, but an equal amount of time is spent worrying about how to procure them without getting caught and arrested. Therefore, addicts are slaves to something that consumes their time and seeks to destroy them in other ways, too.

Alcohol

Alcohol is legal, socially acceptable, and relatively inexpensive. People who drink regularly may have a drinking problem if they answer yes to any of the questions in Figure 22-1.

Figure 22-1
Self-Test for Alcohol Abuse

Lose time from work?
Want a drink in the morning?
Have trouble sleeping?
Drink to feel more confident or outgoing?
Feel easily frustrated?
Find you are anxious or oversensitive?
Blame others for your problems?
Drink alone?
Let family or job responsibilities slide?
Forget what happened when you were drinking?
Find you have lost weight?
Find your mind is not working quickly?
Have violent mood swings?

Even if the employee abuses alcohol on his or her own time, the health cost alone is staggering. Figure 22-2 lists some of the negative health effects of alcohol abuse.

The employee profile for workers abusing drugs or alcohol is summarized in Figure 22-3.

Figure 22-2
Health Effects of Alcohol Abuse

- Cirrhosis of the liver
- Alcoholic hepatitis
- Cancer of the liver
- Other liver damage
- Enlarged heart
- Congestive heart failure
- Other heart disease
- Ulcers
- Gastritis
- Malnutrition
- Delirium tremens
- Disorientation
- Memory impairment
- Hallucinations
- Cancer of the mouth
- Cancer of the esophagus
- Cancer of the stomach
- Brain damage
- Psychosis
- Fetal damage (if pregnant)

Figure 22-3
Substance Abuser's Profile

- Has 16x more absences than nonusers
- Uses 3x more sickness benefits
- Uses 8x more hospital days
- Is 5x more likely to file worker's compensation claim
- Involved in accidents 3.6x more often
- Perform at less than 70% of normal ability
- More likely to be involved in theft of company property

Figure 22-4 is a synthesis of Wilkinson's substance abuse behavior checklist and Bell and Bell's signs of job deterioration.

Drugs

Cocaine is an illegal substance that gives a temporary feeling of being superhuman. However, cocaine overconfidence ruins the judgment and reduces job performance. Unfortunately, the craving for cocaine keeps growing and the desire for anything else lessens. The use of cocaine invariably

Figure 22-4
Substance Abuse Behavior and Warning Signs

Absenteeism

- Excessive sick leave
- Monday and Friday absences
- Repeated absences of more than two days
- Excessive tardiness—morning or after lunch
- Excessive early departure from work
- Increasingly peculiar and improbable excuses for absences
- Higher than average rate of absenteeism
- Inventive/creative reasons for extended lunches
- Frequent sick days for minor illnesses
- Vague symptoms of illness: nausea, headaches
- Illness such as gastritis, sinus, stomach conditions

On-the-Job Absenteeism

- Continued absences from the work station
- Very frequent trips to the restroom/breakroom
- Long coffee breaks
- Vague, constant physical illness on the job
- Lethargy and inattention

Difficulty in Concentration

- Great effort to do any tasks
- More time than normal to complete a task
- Difficulty in recalling past assignments
- Poor memory in general
- Accidents, near misses

Inconsistent Work Pattern

- Extreme swings in productivity (high + low)
- Poor safety record
- Does not accept new responsibilities easily
- Misses deadlines (always blames others)
- Poor attention span
- Neglectful of details
- Poor work quality
- High mistakes ratio/poor judgment

Figure 22-4 *(continued)*

- Wasted material or equipment breakage
- Continued excuses for poor job performance
- Puts things off
- Avoiding supervisor or associates

Interpersonal Relations

- Unfounded resentment toward tasks, other people
- Constant berating of others
- Wild mood swings
- Oversensitivity
- Inappropriate indignation
- Irritability
- Edginess
- Intolerance of coworkers
- Patterns of verbal or physical assault on others
- Suspicion of coworkers

leads to crime, such as theft. Cocaine has health risks, too. It replaces food and sleep, leading to a fatigued body and mind. Nasal membranes break down with continued cocaine use, leading to risk of infections. Cocaine also lowers resistance to illness.

Another illegal substance causing problems in the workplace is marijuana. Marijuana users smoke it to be relaxed and happy, thinking it harmless. Certainly smoke in the lungs is not harmless. Marijuana slows down physical reflexes. Mental powers diminish. The drug makes its users forgetful. The details of life seem unimportant to the marijuana smoker, even sex is not interesting. Users lose track of time and their space and distance judgment is impaired. The chances of causing a serious accident increase while high on marijuana. Because of the smoke, the chances of getting lung cancer increase.

Amphetamines, called speed or uppers, are used to stay awake for extended periods of time. The feeling of stamina and drive are illusory as the amphetamines cause users to rush about wildly, exert themselves beyond their physical capacity, and/or make careless mistakes. Amphetamines are addictive. The more taken, the more that are needed to give that up feeling.

Sedatives or downers do the opposite of amphetamines. Tranquilizers, barbiturates, sleeping pills, and various painkillers are sedatives that give legitimate short-term relief from stress, depression, sleeplessness, or severe pain under the management of a physician. However, many people casually take these drugs until an abusive pattern is established. Downers cause the mind and reflexes of the abuser to operate slowly. Emotions are numbed. These drugs are addictive and eventually more have to be consumed to have the same effect on the mind and body.

Heroin users coined the term "junk" and became known as "junkies." Some people think heroin is not addictive if used only occasionally or taken by some means other than needle. Don't believe it! Heroin is *always* addictive. The addict is indifferent to his or her surroundings and becomes unable to do anything; the craving is so strong. The time between peak craving gets shorter and shorter and the abuser gets sicker. Besides being illegal, heroin can easily be overdosed with fatal consequences. Sharing needles has the deadly side effect of spreading hepatitis and AIDS. Jobs, families, friends, and health are neglected until the addict is alone and sick. Typically, the addict becomes a thief to support his or her habit.

Substance Abuse Programs

Many companies now provide employee assistance programs (EAPs) where alcohol and drug abusing employees can seek help. Intervention and control of substance abuse is never too late or too soon. EAPs refer employees to professionals who can diagnose and treat or refer to treatment. These services are confidential and voluntary. If the employee is worried about confidentiality, the answer to that dilemma is as close as the telephone book where human services, health, social services, and mental health organizations abound. Help can also be found with hospitalization, the employee's regular physician, or a clergyman.

Sometimes it is wise to find substance abusers because they typically do not report on themselves and can be a hazard in the workplace. Some agencies require reporting abusers anyway. The DOT and Federal Highway Authority, for instance. Also, FAA requires drug testing in work facilities. Other legal requirements may exist for your industry. Many industries, though, have no external requirements or guidance for drug testing.

Drug Screening

As you may imagine, drug testing can be controversial with employees. Some see it as an invasion of their privacy. However, when a person on a buzz can injure himself or others or, worse, cause a fatality, drug testing makes sense to protect the majority. You just need to ensure that confidentiality is preserved and the program is conducted ethically. Establish and preserve due process in enforcing your drug screening policies.

First of all, write a drug screening policy and post it for all employees to see. Have it approved by the senior operating executive on site. Have him or her present the policy in an "all hands" type meeting. Emphasize confidentiality, consistency, and firmness. Generally, "if we find drugs or alcohol in your urine once, we will offer you an assistance program. If we find a substance a second time or you quit the program before its conclusion, your employment here will be terminated.

Elements of a drug-screening program include prevention, education, enforcement, and treatment. Pre-employment screening is required by some federal agencies, but most industries are not affected. An attorney should help you decide whether pre-employment screening is right for your facility, as some legal issues about individual rights are involved.

Define and post the specifics of your screening program. What specifically will be tested for? Who will be screened? Generally, truly random testing of the entire workforce including executives is best. Your attorney can help you with establishing this. Employees should not be held to a standard that executives are exempt from. Keep the testing objective and fair. Do not allow drug screening to become a tool for vindictive managers to harass certain employees or to clean out those they perceive as undesirable from their staff.

Another legal issue is that drug screening has the potential for uncovering criminal activity. Therefore, make sure your attorney has briefed your management on how to respond if this occurs. You are obligated as an employer to report any criminal activity. Mere possession of certain types of drugs is criminal activity, for example. Also, if legal drugs are obtained illegally, possession of them becomes a criminal offense. This is why you need to get an attorney involved in your drug screening program, if only to supervise the writing of the policy and details of your screening procedure.

Another potential liability, and hence a need for assistance by an attorney, is the potential for damaging the name and reputation of an individual.

When positive results are reported from the testing laboratory, you need to have a procedure in place to control access to the information and describe how, where, and when the employee will be confronted. One way to prevent mistakes is to have each employee who undergoes screening fill out a form beforehand that allows them to disclose the use of legal drugs and medications. Believe it or not, consumption of food containing poppy seeds within 24 hours of the test can lead to positive results. Many over-the-counter medicines can also give positive results. Therefore, be very careful about what you do with drug screening information.

Ergonomic Issues

The number one issue of home safety and health is back problems. Poor mattresses, bad posture, strenuous hobbies or sports, gardening—many activities contribute to bad backs. Coupled with specific back straining activities at work, these away from work situations can lead to problems. Who is to say whether a compensable back injury is more due to home-related injuries or job-related injuries? The best preventives are lifting technique training and back care awareness. Home computer users should also be given carpal tunnel syndrome awareness training.

Driver Safety

The Federal Highway Administration regulates the testing of commercial drivers for alcohol and drug abuse but no one tests commuters and pleasure drivers, at least before the fact. When a highway safety officer or traffic policeman pulls a driver over for any reason, a sobriety test may be ordered. Unfortunately, no one requires a driver to be tested before getting into the vehicle. More unfortunately, the worst may have already happened by the time a sobriety test is required: an accident.

Wellness

Doesn't it make sense to assist our employees in protecting their health and well being so they can continue to come to work everyday? A health fair is an easy way to protect your bottom line and not too expensive, either. Invite local hospitals, clinics, and private practitioners to come in and help screen employees' weight, blood pressure, prostrate cancer, and other health complaints. Follow-up can be done as part of routine medical reimbursable by company medical insurance. The health fair at least presents some of the basics for employees who might otherwise not have done it on their own.

References

Bell, Jerry and Pat Bell. "Alcohol in the Workplace." *Professional Safety.* February 1989, pp. 11-15.

Kuhlman, Raymond L. "The Need for Off-the-Job Safety Programs." *Professional Safety.* March 1986, pp. 13-16.

Polakoff, Phillip L. "Proper Nutrition Gaining Importance in Workplace Health-Promotion Effort." *Occupational Health & Safety.* December 1991, p. 36.

Sofian, Neal S. "Health Promotion Can Be a Valuable Strategy to Assist in Cost Containment." *Occupational Health & Safety.* December 1991, pp. 26-27.

Vincoli, Jeffrey W. "Drug Testing Programs in Industry." *Professional Safety.* October 1988, pp. 12-17.

What Everyone Should Know about Substance Abuse. Madison, CT: Business & Legal Reports, Inc., 1996.

Wilkinson, Bruce S. "Signs and Symptoms for Supervisors." Mississippi Manufacturers' Association. 3rd Annual Safety Conference and Exposition, May 7, 1997.

INDEX

D

E

F

G

H

I

J

K

L

M

N

O

P

W

GOVERNMENT INSTITUTES ORDER FORM

4 Research Place, Suite 200 • Rockville, MD 20850-3226 • Tel (301) 921-2323 • Fax (301) 921-0264
Internet: *http://www.govinst.com* • E-mail: *giinfo@govinst.com*

3 EASY WAYS TO ORDER

1. Phone: **(301) 921-2323**
Have your credit card ready when you call.

2. Fax: **(301) 921-0264**
Fax this completed order form with your company purchase order or credit card information.

3. Mail: **Government Institutes**
4 Research Place, Suite 200
Rockville, MD 20850-3226
USA
Mail this completed order form with a check, company purchase order, or credit card information.

PAYMENT OPTIONS

❑ **Check** (*payable to Government Institutes in US dollars)*

❑ **Purchase Order** (this order form must be attached to your company P.O. Note: All International orders must be pre-paid.)

❑ **Credit Card** ☐ VISA ☐ MasterCard ☐ AMERICAN EXPRESS

Exp. ___/___

Credit Card No. ______________________

Signature ______________________

Government Institutes' Federal I.D.# is 52-0994196

CUSTOMER INFORMATION

Ship To: (Please attach your Purchase Order)

Name: ______________________

GI Account# (*7 digits on mailing label*): ______________________

Company/Institution: ______________________

Address: ______________________
(please supply street address for UPS shipping)

City: __________ State/Province: __________

Zip/Postal Code: __________ Country: __________

Tel: ( )

Fax: ( )

E-mail Address: ______________________

Bill To: (if different than ship to address)

Name: ______________________

Title/Position: ______________________

Company/Institution: ______________________

Address: ______________________
(please supply street address for UPS shipping)

City: __________ State/Province: __________

Zip/Postal Code: __________ Country: __________

Tel: ( )

Fax: ( )

E-mail Address: ______________________

Qty.	Product Code	Title	Price

Subtotal______
MD Residents add 5% Sales Tax______
Shipping and Handling (see box below)______
Total Payment Enclosed______

❑ **New Edition No Obligation Standing Order Program**

Please enroll me in this program for the products I have ordered. Government Institutes will notify me of new editions by sending me an invoice. I understand that there is no obligation to purchase the product. This invoice is simply my reminder that a new edition has been released.

Within U.S:	**Outside U.S:**
1-4 products: $6/product	Add $15 for each item (Airmail)
5 or more: $3/product	Add $10 for each item (Surface)

15 DAY MONEY-BACK GUARANTEE

If you're not completely satisfied with any product, return it undamaged within 15 days for a full and immediate refund on the price of the product.

SOURCE CODE: BP01